Physlets

Teaching Physics with Interactive Curricular Material

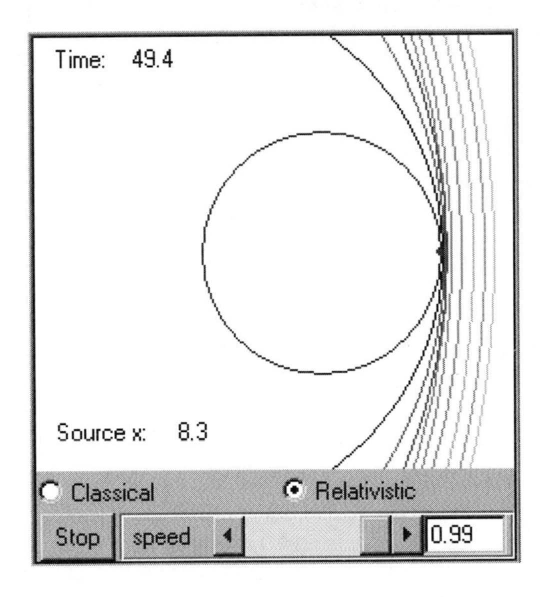

Wolfgang Christian
Davidson College

Mario Belloni
Davidson College

Published by Prentice Hall, Upper Saddle River, New Jersey 07458

Library of Congress Cataloging-in-Publishing Data

Christian, Wolfgang.
Physlets : Teaching Physics with Interactive Curricular Material / Wolfgang Christian, Mario Belloni.
p. cm.
Includes bibliographical references and index.
ISBN 0-13-029341-5 (pbk.)
1. Physics-Computer-assisted instruction. 2. Internet in education. I. Belloni, Mario. II Title.
QC30 .C48 2001
530'.078'5-dc21

Editorial Director: *Paul F. Corey*
Executive Editor: *Alison M. Reeves*
Editorial Assistant: *Christian Botting*
Executive Managing Editor: *Kathleen Schiaparelli*
Production Editor: *Kim Dellas*
Marketing Manager: *Eric Fahlgren*
Art Director: *Jayne Conte*
Cover Designer: *Defranco Design, Inc.*
Manufacturing Manager: *Trudy Pisciotti*
Manufacturing Buyer: *Michael Bell*
Assistant Vice President of Production and Manufacturing: *David W. Riccardi*
Assistant Managing Editor, Science: *Alison Lorber*
Media Editor: *Mike Banino*
Formatter: *Vicki L. Croghan/Envision, Inc.*

Published by Prentice Hall, Inc.
Upper Saddle River, New Jersey 07458

ISBN 0-13-029341-5

10 9 8 7 6 5 4 3 2

Prentice-Hall International (UK) Limited, *London*
Prentice-Hall of Australia Pty. Limited, *Sydney*
Prentice-Hall Canada Inc., *Toronto*
Prentice-Hall Hispanoamericana, S.A., *Mexico City*
Prentice-Hall of India Private Limited, *New Delhi*
Prentice-Hall of Japan Inc., *Tokyo*
Pearson Education Asia Pte. Ltd.
Editora Prentice-Hall do Brasil, Ltda., *Rio de Janeiro*

ei Prentice Hall Series in Educational Innovation

Physlets: Teaching Physics with Interactive Curricular Material by Wolfgang Christian and Mario Belloni, both of Davidson College

Just-in-Time Teaching: Blending Active Learning with Web Technology by Gregor M. Novak, Indiana University-Purdue University, Indianapolis; Evelyn T. Patterson, The United States Air Force Academy; Andrew D. Gavrin, Indiana University-Purdue University, Indianapolis; and Wolfgang Christian, Davidson College.

Tutorials in Introductory Physics by Lillian McDermott, Peter S. Shaffer, and the Physics Education Group at at the Department of Physics, University of Washington

Peer Instruction by Eric Mazur, Harvard University

Peer Instruction for Astronomy by Paul Green, Harvard University

Ranking Task Exercises in Physics by Thomas L. O'Kuma, Lee College; David P. Maloney, Indiana University-Purdue University, Fort Wayne; and Curtis J. Hieggelke, Joliet Junior College

Contents

Preface

The World Wide Web makes it possible to transmit multimedia-enhanced documents interactively in a platform-independent fashion using Hypertext Markup Language, html. These documents are prepared and transmitted as text documents and can, therefore, be prepared with any text editor. Yet the html browser displays full multimedia information, including animated text, graphics, video, and sound. The recent introduction of the Java programming language by Sun Microsystems makes it possible to add platform-independent programs to this multimedia stew. Java accomplishes this trick by specifying a relatively simple Virtual Machine (VM), which can be implemented on any computer architecture (i.e., UNIX, Macintosh, or Windows [Meyer 1997]). Although this VM does not provide as rich a set of tools as the native operating system, the virtual machine can have a user interface with buttons, a drawing canvas, and other graphical elements. There may be virtue in simplicity. Small, platform-independent programs are well suited for instructional purposes such as homework problems. These applets can be embedded directly into html documents and can interact with the user. This is accomplished with a scripting language such as JavaScript. We refer to the Java applets written at Davidson College for physics pedagogy as Physlets. This book demonstrates the use of Physlets in conjunction with JavaScript to deliver a wide variety of Web-based interactive physics activities.

The goal of this book is to enable you to incorporate Physlets in your instruction, whether you are a relative Web novice or are ready to write pages of JavaScript. Clearly, this is more than just a technical how-to book; we hope to give you some ideas about the new possibilities that Physlets offer. It often happens that the most valuable applications of new technologies are new teaching paradigms. But it takes considerable time and effort for these paradigms to become apparent. The examples presented in this book and on the accompanying CD are designed to make the transition to using Physlets quick and easy. This text provides examples of classroom demonstrations, traditional and Just-in-Time Teaching homework problems, pre- and postlaboratory exercises, and interactive engagement activities. Of course, if you already know how you want to use Physlets, you may turn to Part Three of this book, a reference to Physlet methods, and start scripting. But even hard-core programmers will appreciate the ease with which a preexisting Physlet problem, described in Part Two and available on the CD, can be modified for use in a new context.

CONTENTS

Part One gives an overview of the pedagogy and the technology. After a brief introduction ("What Is a Physlet?"), we will argue that new methods are needed in the teaching of physics. If you already believe this to be true, you may feel free to skim Chapter 2 ("JiTT and Physlets," by Evelyn Patterson and Gregor Novak) and Chapter 3 ("PER and Physlets," by Aaron Titus and Melissa Dancy). In subsequent chapters, we will describe the underlying technology and how to install Physlets locally on your desktop or Web server. The core technology of Java and JavaScript is discussed in Chapter 5. Since Physlets are flexible and users can write their own problems, Chapter 6 gives a tutorial on how to script three of the most used Physlets, *Animator*, *Efield*, and *DataGraph*.

In Part Two, we give examples of curricular material that can be used as in-class exercises and homework problems in introductory and advance physics courses. There are over 100 of these examples in Part Two. These examples and an additional 80 problems are available on individual html pages on the CD that accompanies this book. The Additional Resources portion of the CD contains even more examples of curricular material from other institutions.

Part Three provides resources for instructors who are interested in scripting beyond the tutorial in Part One. These resources include a detailed description of the methods for version 4 Physlets: *Animator*, *Bar*, *BField*, *Circuits*, *DataGraph*, *DataTable*, *Efield*, *EnergyEigenvalue*, *Faraday*, *Hydrogenic*, *Molecular*, *Optics*, *Poisson*, and *SurfacePlotter*.

COMPANION WEBSITES

Many of the Physlet problems provided in Part Two are included on Prentice Hall's Companion Website for Douglas Giancoli's two physics texts, *Physics: Principles and Applications* (5th edition) and *Physics for Scientists and Engineers* (3rd edition). The site is located at http://www.prenhall.com/giancoli. These resources are also available on Prentice Hall's Companion Website for *College Physics* (4th edition), by Jerry Wilson and Tony Buffa, at http:www.prenhall.com/wilson.

ACKNOWLEDGMENTS

There are a great many people and institutions that have contributed to our efforts, and we take great pleasure in acknowledging their support and their interest.

We thank our colleagues Dan Boye, Larry Cain, Tim Gfroerer, and John Yukich at Davidson College for their use of Physlets in the classroom and the productive discussions that developed from this use. Larry Martin at North Park University was an early adopter of Physlets and has provided numerous suggestions for improvements to both the programs and the text. Andy Gavrin at Indiana University–Purdue University, Indianapolis, has helped us to more closely tie Physlets to the Just-in-Time Teaching technique.

Mur Muchane has provided invaluable computer and network support throughout this project and Laura Cupples helped design and organized the Physlets CD.

W.C. would like to thank the numerous students who have worked with him over the years developing programs for use in undergraduate physics education. Some of our best Physlets are the result of collaborative efforts with student coworkers. In particular, we would like to single out Mike Lee, Cabel Fisher, and Jim Nolen.

M.B. would like to thank Anne Cox, Edward Deveney, Harry Ellis, Bill Junkin, and Steve Weppner for many useful and stimulating discussions about the incorporation of Physlets with existing curricular material.

Special thanks to Evelyn Patterson at the United States Air Force Academy; Taha Mzoughi at Mississippi State; Aaron Titus of North Carolina A&T; Loren Winters, Taylor Brockman and Jeremy Portzer at the North Carolina School of Science and Math; Robert Beichner, John Risley, Margaret Gjertsen, Jeff Saul, Scott Bonham, Duane Deardorff, David Abbott, Rhett Allain, Melissa Dancy at North Carolina State University; Larry Martin, Tait Swenson, and Robin Trautman at North Park University; and Morten Brydensholt at Orbit. All of these people contributed Physlet problems that appear on the Additional Resources part of the CD and and on the Davidson Web site.

We also thank Melissa Dancy and David Hestenes for the inclusion of the Physlet-based Force Concept Inventory (FCI) on the CD.

Workshops have been an especially fruitful arena for the give-and-take of ideas with fellow faculty. The Physlet strategy could not have grown and matured without these opportunities and the exchange of ideas that they afforded.

Some people have been such frequent contributors of time and ideas that we have brought them in as the authors of Chapters 2 and 3 of this book. However, we would like to thank Evelyn Patterson, Gregor Novak, Aaron Titus, and Melissa Dancy again, both for their writing and for the many valuable ideas we have gained during our associations with each of them.

We would like to thank Larry Cain for the many hours he spent reading the manuscript and making suggestions. Any mistakes that remain are likely the result of changes made since his last inspection.

Both of us express our thanks to Alison Reeves and her coworkers at Prentice Hall for supporting the development of Physlets while Java was still an untested technology, for encouraging us to write this book, and for securing permission to include problems from the Prentice Hall Companion Website in this text. Numerous others at Prentice Hall have helped in the production process, but we would especially like to thank Kim Dellas, Mike Banino, and Alison Lorber.

We also wish to express our sincerest thanks and apologies to those who have encouraged us the most: our spouses, Barbara and Nancy.

This work was partially supported by the National Science Foundation under contract DUE-9752365.

Foreword

David Hestenes
Arizona State University

The explosive growth of the Internet is driven by commercial forces. Wolfgang Christian, Mario Belloni, and their collaborators have taken the lead in taming the beast for the purposes of physics education. This book is an essential primer for anyone who wants to join in. It is informed by the genuine expertise that comes from having made all the big mistakes, so it can guide you quickly to a productive path. It brings you the balanced perspective on pedagogical and technical aspects of software design and use that you need for effective teaching and research. It gives you a rich store of examples and practical tools that enable you to get to instructional applications quickly. An implicit goal of this book is to build a community of users and researchers to harness the Internet for physics education. Its *Physlet philosophy* provides a practical plan of action.

The *Physlet strategy* for exploiting the Internet is based on a sober evaluation of our opportunities and our limitations. First, its adoption of Java as the programming language saddles us with the limitations of Java for programming physics but gives us free access to many desirable features afforded by the intense development of Java for commercial purposes. This book shows how much can be accomplished within those constraints. Second, the Physlet strategy directs us to "think small and simple." This is justified by a frank assessment of our limited knowledge about designing educational software as well as practical considerations about how to manage incremental innovation. As Physlet problems are fairly easy to create and deploy, they provide a means for rapid prototyping and testing of ideas about educational software. In particular, they can be used as sharp tools for investigating precisely what it takes to get a physics concept across to students. In this way, the "virtue of simplicity" becomes an issue in learning research on the design and use of Physlets. The upshot is that Physlets are equally valuable as tools for instruction and for research. Both uses are strongly coupled. To use Physlets effectively in instruction, one must be informed about the relevant educational research.

Federal funding today is stimulating a proliferation of Internet resources for science education. Many Web sites are glitzy and enticing but pedagogically flawed. How can teachers and students separate the good stuff from the bad? Teachers are encouraged to incorporate Web stuff into their courses, but will this contribute to greater fragmentation of an already fragmented curriculum? The problems are many and difficult. Webmasters and teachers alike can profit from this book, whose purpose is to sharpen their judgment of Internet resources. Funding agencies need to recognize the critical role of educational research in developing effective science education software. The future is up for grabs!

Foreword to Physlets for the Physics Instructor

Edward F. Redish
University of Maryland

You hold in your hand more than just another book about computational physics. Physlets are a tool environment—a method that will allow you to integrate the computer into your classroom in a way that is easy and powerful for both you and your students. Even better, it's a tool that can help us together build a community of physics instructors using computer tools and working together to improve physics teaching throughout the world.

DO WE REALLY NEED COMPUTERS IN A PHYSICS CLASSROOM?

The computer changes everything. Sure, physicists have been using computers longer than almost anyone. Back in the 1960s we learned to program in FORTRAN, punching cards and waiting—sometimes hours or even days—for each incarnation of our program to come back from the mainframe and reveal typing mistakes or compilation errors. In the 1980s we pushed the envelope, being among the first to use parallel processing supercomputers and doing calculations our thesis advisors had never imagined were possible. But since the appearance of the IBM PC in 1981 and the Apple Macintosh in 1985, powerful information processing has become available to more than just the mathematical elite.

Personal computers have had a strong attraction for children ever since they first appeared. Today it's not only the children of scientists who get their hands on computers early. Thousands of young men and women under the age of 30 have become millionaires as a result of their computer skills. The *Digest of Education Statistics* reports that in 1998, 89% of U.S. schools had access to the Internet, and that the average number of computers per school was 75. As of this writing in April 2000, nearly 60% of homes have a computer. Within a few years, essentially every one of our students will have experienced using the computer from infancy.

The development of the Internet in the 1990s added a level of enabling and facilitation that drives both the capability of the computer and the general public's interest in using it. Where do you think I got the statistics in the last paragraph? I called up my favorite search engine and asked it to find the information on the Web. More than 1000 references turned up, many of them containing exactly the information I was looking for. When I want to find a citation for a book to reference in a paper I'm writing, where do I go? I no longer walk the 300 meters to the library's card catalog; I connect to amazon.com.

The information environment is changing; not just for professional scientists, but for everyone. We know very well that changing the way children are raised changes the way that they think. Reading to children from an early age strengthens their skills with words and makes them into readers; this, in turn, opens their horizons and broadens their thinking. Having two parents who speak to a child in two different languages produces bilingual children who are truly fluent in multiple languages in a way that is almost impossible to reproduce for those of us who learn second languages later in life. What changes will having access to information produce in the current generation of children? What will these children be like?

Physicists are consistently among the most creative and inventive of all scientists. Physicists discovered and invented quantum mechanics, relativity, the transistor, black holes, string theory, and dark matter. But as teachers, we tend to be among the most conservative, repeating the content and methods that we had received as students from our teachers and that they had from theirs. For the coming generations of students, that's not going to be good enough. We are going to have to understand their ways of thinking and their learning styles. We are going to have to find new ways to hold their interest and keep them excited about learning physics. Increasingly, the tools we use to achieve those goals are going to involve computers.

HOW CAN WE USE COMPUTERS IN THE CLASSROOM?

Using computers in the physics classroom is not a trivial or obvious exercise. When I got my first PC in 1981, I was excited about the prospect of using it in my teaching. In 1983, when I was chair of Physics and Astronomy, my colleague John Osborn, then chair of Mathematics, and I petitioned to create the first student microcomputer laboratory at the University of Maryland. But when I looked at the software then available for instruction, I was appalled. The simulation software was filled with glitz and graphics, but essentially assumed the students understood the basic ideas and were just looking for neat new ways of reexpressing what they knew. This completely missed the point of instruction and treated students as if they were just smaller (and less gray) physics professors. The data acquisition software seemed designed to replace the student in the lab and to turn the lab into a data collection facility rather than an instructional environment. The programming environments were hard to learn and deeply sophisticated, like FORTRAN and C, or were easy to learn but built bad programming habits, like BASIC.

In a very real sense, the Physlet method and WebPhysics are direct descendents of the ideas my colleagues and I worked on in the mid-1980s but adapted and extended to modern academic situations and modern computing environments. In the years 1985–1990, my colleagues at Maryland (Jordan Goodman, Bill MacDonald, Charles Misner, and Jack Wilson) and I got together to explore what we could do with the computer in our classes and laboratories. Our Maryland University Project in Physics and Educational Technology (affectionately M.U.P.P.E.T., for short) explored building modular programming environments for students, using spreadsheets to do physics calculations, using simulations, and using computing to facilitate student research projects at an early stage.[1] In subsequent years, Bill MacDonald and his collaborators extended these ideas to upper division physics in the Consortium for Upper-level Physics Software (CUPS) project,[2]and Jack Wilson and I extended it to building full modular learning environments in the Comprehensive Unified Physics Learning Environment (CUPLE) project (today, it's used in the environment called Studio Physics).[3]

WHAT MAKES SOFTWARE USABLE IN A PHYSICS CLASSROOM?

Physlets make use of the immense opportunities made available today by modern computing environments and the strong interactions enabled by the Internet. What makes it especially powerful is that the Physlet environment is

- modular
- tool-based
- platform independent
- immediately distributable

In their construction of specific examples, the authors of Physlets focus on what has been learned about student learning through research. This focus on the student guarantees that you will find a large number of valuable examples to use as starting points for building new lessons.

In order for a particular computer simulation to be effective and receive widespread use, three conditions have to be met:

1. The simulation has to be *authentic*—that is, it must address real educational issues. It has to allow something to be taught in a way that the students who are going to use it can understand.
2. The simulation has to be *adoptable*—that is, it must be easy for an instructor to put into his or her class and easy for the students to learn to use. A heavy learning curve—for teachers or students—severely impedes its adoption.
3. The simulation has to be *adaptable*—that is, it must be easy to modify it to fit a particular instructional setting. No two classrooms are interchangeable. Each instructor creates his or her own coherent setting, choosing methods, approaches, and notations that are to some extent idiosyncratic. The software must have sufficient flexibility to deal with different environments.

The Physlet environment meets all three of these criteria.

---

[1]W. M. MacDonald, E. F. Redish, and J. M. Wilson, "The M.U.P.P.E.T. Manifesto," *Computers in Physics* **2**, No. 4, 23 (July/Aug. 1988); Edward F. Redish and Jack M. Wilson, "Student Programming in the Introductory Physics Course: M.U.P.P.E.T.," *Am. J. Phys.* **61** (1993) 222–232; Charles Misner and Pat Cooney, *Spreadsheet Physics* (Addison Wesley, 1991); E. Redish, J. Wilson, and I. Johnston, *The M.U.P.P.E.T. Utilities: Programming Tools for Turbo Pascal with Physics Examples*, (Physics Academic Software, Raleigh, NC, 1994).

[2]R. Ehrlich, M. Dworzecka, and W. MacDonald, "Text Materials to Accompany Simulations for the CUPS Project," *Computers in Physics* **7**, 508 (1993).

[3]J. Wilson and E. F. Redish, *The Comprehensive Unified Physics Learning Environment* (Physics Academic Software, Raleigh NC, 1994); J. M. Wilson and E. F. Redish, "The Comprehensive Unified Physics Learning Environment: Part i. Background and system operation," *Computers in Physics* **6** (March/April, 1992) 202–209; "Part ii. The basis for integrated studies," ibid., 282–286.

WHAT ARE SIMULATIONS GOOD FOR?

Pedagogically, simulations can be immensely valuable when used in a manner based on a good understanding of student thinking. Here are some general learning activities that simulations can provide usefully and effectively.

1. *Simulations can help students make sense of translation among representations.* In physics, we represent information about a physical system in many different ways: using words, equations, graphs, diagrams, tables of numbers, contour maps, vector plots, and so on. Strong evidence exists that many students have considerable difficulty, not only with creating these representations but in seeing how they express information about the system and how they are related to each other. Showing animations of ta dynamical system and tying it to a coordinated graph, diagram, or plot can, when used in conjunction with an appropriate lesson, significantly help students develop skills in using different representations to help them make sense of the physics.
2. *Simulations can help students understand equations as physical relationships among measurements.* Many students in introductory physics classes treat equations as if they were only a way to calculate a variable or determine a number as a solution of a set of equations. But physical equations represent relationships between various observations and measurements. By setting up a simulation in which students can vary parameters and see the effect of these variations, the students' view of am equation's role is powerfully enriched.
3. *Simulations can help students build mental models of physical systems.* In some cases, students don't have the experience or imagination to put together what they are reading in their texts and hearing from the lectures into a coherent, sensible picture. They memorize bits and pieces, but because these pieces are not linked into a consistent, self-supporting structure, they forget or confuse the parts. In physics, many of our coherent pictures are in the form of mental models—visions of interacting objects having qualities and measurable properties. Producing visualizations that display these characteristics can help students create these mental models.
4. *Simulations can give students engaging, hands-on, active learning experiences.* Educational research has demonstrated repeatedly that students learn much more effectively when they themselves are in control. Having simulations that students can use to explore a phenomenon on their own can produce more effective learning experiences.
5. *Simulations can serve as a sketchpad on which students can explain and describe their understandings to each other.* Educational research shows the value of having students explain what they are thinking, both to themselves and to each other. Two or three students working together to answer questions with a simulation can produce a powerful learning environment.

These are only a start. You will find examples in this book and on the Physlet Website, http://webphysics.davidson.edu/Applets/Applets.html, that can be used in each of these ways and in many others. They can be a valuable part of a rich and effective physics-learning environment.

WHERE ARE WE GOING?

The use of the Java/JavaScript programming language, as used here in the Physlet environment, makes it easy to adapt existing Physlets and to create new ones. Applications can be put up on the Web for the students to access in a computer lab or on their own time. But, in my view, the most important implication of choosing a Web-based technology is the way it facilitates sharing.

The idea of the Web is changing how we handle and deliver information. Powerful single sites (like the NASA Web site) provide vast sources of data that everyone can use. Also, work done by small groups and even individuals can be shaed effectively. Problems, lessons, and simulations created for a class in Colorado or Ohio can be used immediately in Florida and Maryland. As the community of Physlet builders grows, the value of learning how to use them and modify them will grow factorially. With Physlets, you can choose how deeply you want to get involved. You can just stick your toe in the water, adopting existing Physlets and adapting them to your own classroom. Or you can dive in—the water's warm!—and write your own Physlets. Each one can be small enough that the task of making them usable by others is not prohibitive. As you use Physlets more and more and participate in their exchange, you will be helping to build the activity of teaching physics as a community activity—one in which we learn from each other and, by sharing our work, help improve both our own teaching and the teaching throughout the physics community. Welcome to Physlets!

List of Figures

List of Scripts

List of Tables

PART ONE

PEDAGOGY AND TECHNOLOGY

CHAPTER 1

PEDAGOGY AND PHYSLETS

1.1 WHAT IS A PHYSLET?

Physlets—"**Phys**ics app**lets**"—are small, flexible Java applets (such as the Doppler Physlet shown in Figure 1) that can be used in a wide variety of World Wide Web (WWW) applications. Many other Physics-related Java applets are being produced around the world—some of them very useful for education. However, the class of applets we call "Physlets" has some attributes that make it especially valuable for the educational enterprise.

Physlets are simple. The graphics are simple; each Physlet is designed to deal with only one facet of a phenomenon and does not incorporate very much in the way of data analysis. This keeps Physlets relatively small—easing downloading problems over

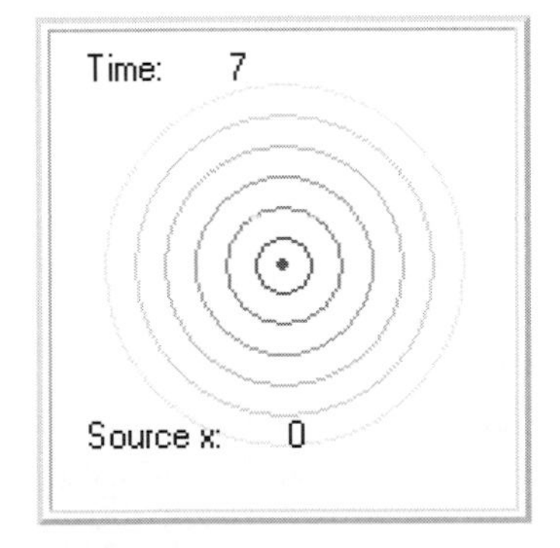

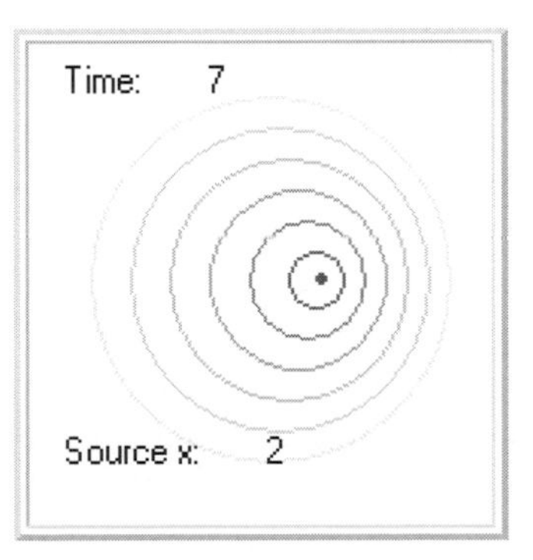

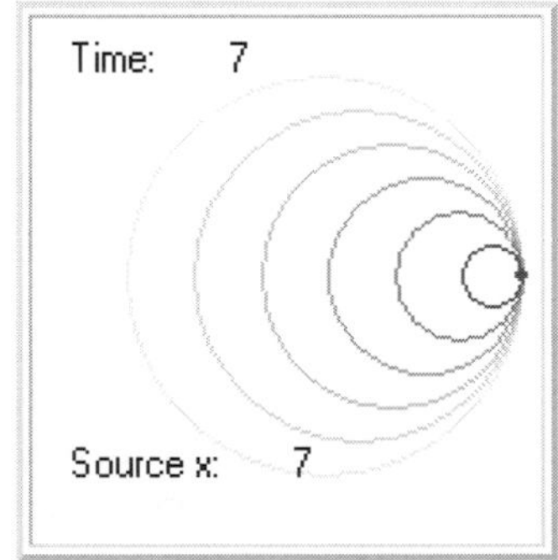

FIGURE 1: Doppler Physlet scripted with three different velocities.

slow network connections—and removes details that could be more distracting than helpful.

Physlets are flexible. All Physlets can be controlled with JavaScript. For example, *Animator* can be used for almost any problem involving forces or particle trajectories. Only the JavaScript—and not the Java—associated with the problem needs to be changed. Data taking and data analysis can be added when needed using interapplet communication.

Physlets are written for the Web. They can run on (almost) any platform and be embedded in almost any type of html document, whether it be a homework assignment, a personal Web site, or an extensive science outreach site. Physlets can be used as an element of almost any curriculum.

Physlets are freely distributable for noncommercial use. Physlet archives, that is, compressed archives containing compiled Java programs, can be downloaded from the Davidson College WebPhysics server: http://webphysics.davidson.edu/applets/applets.html. These files can be mirrored on a local hard drive or on a noncommercial Web server in order to provide students easy access to customized curricular material.

1.2 A NEW TEACHING PARADIGM

> Good educational software and teacher-support tools, developed with full understanding of principles of learning, have not yet become the norm.
>
> *[Bransford 1999]*

In 1991 Davidson College hosted the second National Science Foundation (NSF)-sponsored Conference on Computational Undergraduate Physics. Over 125 physicists from the United States and abroad attended the four-day event. A unique feature of this conference was that each participant was given a computer account and was asked to bring software and curriculum material to share. Within hours of the conference opening, 350 megabytes of programs (but almost no images or text) appeared on the server. Nearly every participant contributed a small homegrown DOS or Apple II program. During the conference, it was not unusual to see participants with stacks of floppy disks downloading files to take home in preparation for the anticipated computer-enabled educational revolution. It never came, or at least not in the form envisioned by the conference participants. The mainstream teaching community adopted little of the conference software, and almost none of it is in use today. In contrast, printed material from the earlier post-Sputnik curricular reform movement (the Berkeley Physics series, for instance) is still widely used by physics educators although the audience has changed and the pedagogy upon which it is based has gone out of fashion. Will this scenario be repeated and are we doomed, like the Greek hero Sisyphus, to forever push computational physics up the hill of curriculum reform? Can we expect widespread adoption of computation in the current curricular reform initiative? And, if so, what strategies should we adopt to ensure that computation-rich curricula being developed today will be adopted and be in widespread use a decade from now?

The rapid pace of hardware and operating system development made it difficult for text and software authors to produce computer-rich curricular material that was not obsolete shortly after publication. The half-life of a typical desktop computer was shorter than the textbook publication cycle, and this all too often led to minimal documentation, poorly tested nonstandard user interfaces, and idiosyncratic behavior. Computational physicists accustomed to programming in FORTRAN had little interest in page layout and user interface design. The tools used by scientists in day-to-day office work such as correspondence, class management, and professional publication interoperated poorly with programming tools and with educational software. Publishers were, therefore, reluctant to integrate computer usage into primary educational texts. An effective mechanism for creating and distributing media-rich documents from the desktop simply did not exist.

It is not surprising that many teachers were unwilling to invest the time and energy to overcome these obstacles. Since little research had been done on the effectiveness of computer-based instruction, this wait-and-see approach may have been wise. However, we believe that this throwaway cycle for educational software need not repeat itself and that key technologies are now available that enable authoring and distribution of curricular material that will withstand the test of time. The most promising technologies are based on virtual machines, meta-languages, and open Internet standards. These technologies are platform independent. The marriage of word processing and desktop publishing with Internet technologies is already taking place. The current generation of desktop applications is Internet aware (that is, they can embed html anchors and export static Internet content) and the next generation will be Internet centric. Authors already download documents from servers, edit the documents in a what you see is what you get (WYSIWYG) environment, and upload them back onto the server. Extensible markup language (XML) and cascading style sheets (CSS) will soon provide rich formatting and enable context-sensitive searching and indexing of these electronic documents [Goldfarb 1997]. Manufacturing, inventory control, and advertising, in effect, have provided the education community with a rich and flexible set of standards to enable electronic curriculum distribution.

1.3 INTERACTIVE ENGAGEMENT

Although the Internet lowers the barriers to authoring and distribution of educational software, its ability to deliver active content may, in the long run, be more significant. Much of the current curricular reform effort in physics is based on the idea of interactive engagement (abbreviated IE). A widely respected study, published by Richard Hake in the *American Journal of Physics*, compared the cognitive gains of IE classes to traditional lecture-based classes. Hake defines interactive engagement methods as

> [methods] designed at least in part to promote conceptual understanding through interactive engagement of students in heads-on (always) and hands-on (usually) activities which yield immediate feedback through discussion with peers and/or instructors.
>
> *[Hake 1998]*

Interactive engagement teaching methods take many different formats. Some efforts include the use of technology to make the lecture more of a two-way conversation [Mazur 1997, Novak 1999]. Others use recitation sessions as important supplements to instruction [Heller 1992, McDermott 1998]. Still other curricula focus on discovery learning in the laboratory, either as part of a larger course [Sokoloff 1995] or, in some cases, as the main component of the course [Laws 1997]. All of them, however, force the student to play a much more active role in the learning process, increase the amount of interaction with fellow students and instructors, and emphasize conceptual understanding as well as quantitative problem solving. Hake's study compared IE methods with traditional lecture methods at a variety of institutions and showed a significant, across-the-board improvement in students' conceptual understanding in IE classes. While the most dramatic differences are seen in the area of conceptual understanding, this does not require reduced ability in quantitative problem solving—the conceptual understanding lays the foundation, usually resulting in improved problem solving [Mazur 1997, Heller 1992, Thacker 1994].

1.4 MEDIA-FOCUSED PROBLEMS

In designing interactive material, it is useful to distinguish between media-enhanced problems, where multimedia is used to present what is described in the text, and media-focused problems, where the student uses the multimedia elements in the course of solving the problem. Multimedia-focused problems are fundamentally different from traditional physics problems, and Physlets are ideally suited for these types of problems. Consider an example. A traditional projectile problem states the initial velocity and launch angle and asks the student to find the speed at some point in the trajectory. This problem can be media enhanced by embedding an animation in the text, but this adds little to the value of the problem. Alternatively, this same problem could be a media-focused Physlet problem as shown in Figure 2. In this case, no numbers are given in the text. Instead, the student is asked to find the minimum speed

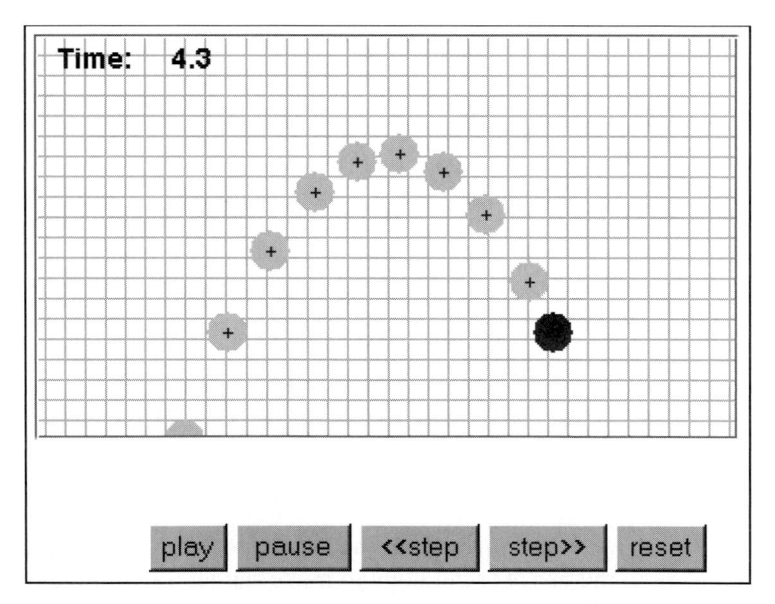

FIGURE 2: The *Animator* Physlet scripted to show a projectile problem.

along the trajectory. The student must observe the motion, apply appropriate physics concepts, and make measurements of the parameters he or she deems important within the Physlet. (A mouse-down enables the student to read coordinates.) Only then can the student "solve the problem." Such an approach is remarkably different from typical novice strategies, where students attempt to analyze a problem mathematically before qualitatively describing it (an approach often called "plug-and-chug" and characterized by the lack of conceptual thought during the problem-solving process) [Titus 1998]. Requiring students to consider problems qualitatively has been shown to have a positive influence on students' problem-solving skills and conceptual understanding [Larkin 1980, Leonard 1996].

1.5 APPROPRIATE TECHNOLOGY

Although the media-rich content and interactivity provided by technology is pedagogically useful, it lacks a human dimension that is important to effective teaching. Computer-assisted instruction (CAI) has, after all, already been tried on very elaborate proprietary systems. It is unlikely to be improved significantly by being ported to the Internet. To be truly effective, the communication capabilities of the computer must be used to create a feedback loop between instructor and student. A new and particularly promising approach, known as Just-in-Time Teaching (JiTT), has been pioneered at Indiana University and the United States Air Force Academy and further developed at Davidson College [Novak 1999]. It employs a fusion of high-tech and low-tech elements. On the high-tech side, it uses the World Wide Web to deliver multimedia curricular materials and manage electronic communications between faculty and students. On the low-tech side, the approach requires a classroom environment that emphasizes personal teacher–student interactions. These disparate elements are combined in several ways and the interplay produces an educational setting that students find engaging and instructive. The underlying method creates a synergy between the Web and the classroom to increase interactivity and allow rapid response to students' problems.

Although it would be foolish to predict the future direction of computer science and the computer industry, physics education research (PER) informs us that technological advances do not necessarily lead to improved learning. For example, streaming video is currently a hot technology, and both traditional broadcasters and software companies are competing to establish themselves in this market. PER has shown that merely watching video has little effect on student learning, and it is unlikely that streaming large video files will change this result [Beichner 1997]. Small cognitive effects have been shown to occur using video clips if the showing of the clip is accompanied with in-class discussion or if the clip is used for data taking and data analysis. Similarly, database technology has become ubiquitous in our society. It is used to store consumer-shopping profiles for corporate marketing departments and to manage Christmas card mailing lists at home. But little has been gained in attempts to use database techniques to track student learning and to tailor curriculum to individual learners. Other high-end technologies, such as virtual reality, three-dimensional modeling, and voice recognition, will almost certainly come online in the coming decade. However, their most enduring effect on education may be to drive the price/performance ratio of consumer, and hence educational, hardware even lower. These technologies are unlikely to have a significant impact on undergraduate education without a corresponding curricular development effort. In fact, current commercial technologies may already be good enough to implement the most effective teaching strategies. Unlike previously written educational software, software written using current Internet standards should be accessible for years to come. For computation to have a long-lasting impact on science education, it will need to be based more on successful pedagogy than on the latest compilers, hardware, or algorithms.

1.6 VIDEO, INTERACTIVE PHYSICS, AND PHYSLETS

When one first sees Physlets, one's first reaction is probably that it is like Interactive Physics™(IP), except that it is written in Java and runs in a Web page. When one first observes a simultaneous presentation of an animation and a graph, one's first reaction is probably that it is similar to video analysis (VA), where students simultaneously view motion and an object's position, velocity, or acceleration versus time graphs [Zollman 1994]. These observations are certainly valid in that Physlets share similar characteristics with both IP and VA; however, there are also significant differences, both technical and pedagogical, that are worth noting. Each program has its strength as shown in Table 1.

VA is a good choice when one wants to analyze a "real" phenomenon. VA is essentially a method for collecting and analyzing experimental data [Derby 1999]. Especially appealing to students is that they can analyze the physics of something in which they are interested or involved, such as sports. Analyzing "real-world" problems helps them realize that we often use simple theoretical models and that the real world is complex; after all, objects in the real world are not point particles and usually do not move with constant acceleration. Comparing theoretical predictions and experimental results helps students acquire laboratory analytical skills. In addition, VA allows them to analyze situations that may be otherwise difficult to model theoretically, such as bungee jumping. Perhaps students can even develop a better theoretical model based on the experimental data obtained by VA.

IP is a powerful tool for modeling mechanics and some electrostatics phenomena. The closest matching Physlet, *Animator*, does not model the moment of inertia of an object and therefore does not allow objects to rotate as does IP. IP is also an excellent choice when one wants to develop quickly a simulation using a graphical user interface (GUI). In a graduate physics education research course at North Carolina State University, participating graduate students used MBL (Microcomputer-Based Laboratory) tools to measure and plot the position, velocity, and acceleration versus time of a mass oscillating on a vertically-oriented spring. The acceleration versus time graph departed from the expected sinusoidal curve at both the maxima and minima of the acceleration curve. While most graduate students regarded the behavior of the graph to be a result of uncertainty in the calculation of the acceleration, one bright student modeled an oscillating massive spring using IP by connecting masses and springs in series such that the total mass and effective spring constant were the same as that in the experiment. The acceleration versus time graph of the massive spring modeled in IP was similar to the experimentally determined graph, showing the departure from sinusoidal behavior at the maxima and minima of the curve. Quickly building a simulation using IP's GUI is very easy and useful in instances such as this.

Physlets' strength is their size, scriptability, and modularity. A Physlet problem author can use scripting to add features or disable features of the animation in order to focus students on a particular task. The animation may be coupled to a graph, a bar chart, or a data table depending on what data the author wants the student to have access to and in what form. Likewise, the author may choose to show footprints in order to help the student read the animation like a motion diagram. In another case, vectors may be shown in order to focus students on a certain kinematic variable and associate it with the motion. In another instance, the author may allow the student to adjust initial conditions perhaps by dragging objects in the simulation or adjusting numerical values shown in standard hypertext markup language (HTML) text input boxes. The result of this flexibility is that the author can eliminate certain distractions and focus students on a certain task by requiring that they use the information and tools provided.

The scriptability of Physlets is ideally suited for instructors who wish to prepare media-focused physics problems. (It is usually not a good idea to have students script Physlets.) Students then collect data, either numerically or visually, from the animation to solve the problem. These problems can include the following features: adjusting initial conditions or constants; choosing the appropriate graph that corresponds to a given animation or description of motion; collecting data from measurements, graphs, tables, or bar charts; and critically observing a given phenomenon. Physlet problems are often "simple" problems, meaning that the situations shown in the Physlets involve relatively simple processes, much like textbook problems. Physlets may show idealized phenomena or they may be scripted to show unphysical phenomena so that the students must critically observe the situation and identify what is unphysical about it. Naturally, the fact that Physlets are Web based is a big advantage. Physlet simulations shown in class can be posted on the Web for students to view anytime and anywhere.

Physlets, IP, and VA can all be used in contexts other than traditional homework; however, it seems that the typical use of Physlets will be different from the typical uses of IP and VA. The delineation appears to be that Physlets are better used for physics questions where students must view a simulation and collect data, either numerically or visually, and IP and VA are useful for more involved projects where students must first master the technology to either build a simulation (IP) or obtain experiment data (VA). The breadth of topics addressed by Physlets makes them a better tool for topics such as magnetism, special relativity, thermodynamics, waves, or quantum mechanics.

TABLE 1: Comparison of Physlets, Interactive Physics, and video analysis.

	Breadth of topics	Analytic models	Unphysical phenomena	Data collection & visualization	Control of conditions or parameters	GUI interface	Scriptable
Physlets	X	X	X	X	X		X
IP		X		X	X	X	
VA				X		X	

CHAPTER 2

JiTT AND PHYSLETS

Evelyn T. Patterson and Gregor Novak

2.1 WHAT IS JiTT?

Just-in-Time Teaching (JiTT) is a pedagogical strategy aimed at many of the challenges confronting instructors in today's classrooms. Student populations are diversifying. In addition to the traditional 18-year-old recent high school graduates, we now have a kaleidoscope of "nontraditional" students: older students, students working part time, commuting students, and, at the service academies, military cadets. At a minimum, these students face time management challenges. They come to our courses with a broad spectrum of educational backgrounds, interests, perspectives, and capabilities that compel individualized, tailored instruction. They also need motivation and encouragement to persevere in what for many is a bewildering, unfamiliar task. Consistent, friendly support often makes the difference between a successful course experience and a fruitless effort, and often it means the difference between graduating and dropping out.

Education research has made us more aware of learning style differences and of the importance of passing some control of the learning process over to the students. Active learner environments yield better results [Hake 1998], but they are harder to manage than the lecture-oriented approaches.

To confront these challenges, the Just-in-Time Teaching strategy pursues three major goals:

1. To maximize the efficacy of the classroom session, where human instructors are present.
2. To structure the out-of-class time for maximum learning benefit.
3. To create and sustain team spirit. Students and instructors work as a team toward the same objective: to help all students pass the course with the maximum amount of retainable knowledge.

The JiTT pedagogy exploits an interaction between Web-based study and an active-learner classroom. Essentially, students respond electronically to carefully constructed Web-based assignments, and the instructor reads the student submissions just-in-time to adjust the lesson content and activities to suit the students' needs. Thus, the heart of JiTT is the "feedback loop" formed by the students' outside-of-class preparation, which fundamentally affects what happens during the subsequent in-class time. The students come to class prepared and already engaged with the material, and the faculty member already knows exactly where the students are and where classroom time together can be best spent.

Although JiTT can certainly be implemented fully using technically simple Web-based assignments, incorporating some Physlet-based questions can heighten the extent to which student understanding can be probed and encouraged. Responding to questions that involve watching or analyzing a Physlet animation often requires different skills and a different level of understanding than responding to static questions, and JiTT is ideally suited to help students improve their analysis skills and deepen their understanding. The JiTT strategy as applied in physics education is richer for the incorporation of Physlets.

To date, JiTT has been adopted (and adapted) by faculty at more than two dozen institutions across the country, from research universities such as Harvard to small liberal arts colleges such as Doane College. JiTT has been featured as an effective use of technology at National Science Foundation (NSF)-sponsored national workshops for new faculty, Project Kaleidoscope national workshops, and numerous other presentations, conferences, and workshops.

Although Just-in-Time Teaching makes heavy use of the Web, it is not to be confused with either distance learning (DL) or with computer-aided instruction (CAI). Virtually all JiTT instruction occurs in a classroom with human instructors. The Web materials, added as a pedagogical resource, act primarily as a communication tool and secondarily as content provider and organizer.

The JiTT Web pages fall into three major categories:

1. Student assignments used in preparation for the classroom activities. WarmUps and Puzzles, described in the next section, fall into this category.
2. Enrichment pages. We title these pages "What Is Physics Good For?"—"GoodFors" for short. These are short essays on practical, everyday applications of the physics at hand, peppered with URL links to interesting material on the Web. These essays have proven themselves to be an important motivating factor in introductory physics service courses, where students often doubt the current relevance of classical physics developed hundreds of years ago.
3. Stand-alone instructional materials, such as simulations and spreadsheet exercises.

2.2 WARMUPS AND PUZZLES

WarmUps and Puzzles are at the core of JiTT. They are short, Web-based assignments that prompt the student to think about a physics-related topic and answer a few simple questions. As can be seen from the following examples, these questions, when fully discussed, often have complex answers. We expect the students to develop their answers as far as they can. We finish the job in the classroom. These assignments are due just a few hours before class time. The responses are collected electronically and scanned by the instructor in preparation for class. They become the framework for the classroom activities that follow. In a typical application, sample (anonymous) responses are duplicated on transparencies and taken to class. In an interactive session, built around these responses, the content lesson is developed.

Students complete the WarmUp assignments *before* they receive any formal instruction on a particular topic. They earn credit for offering answers substantiated by their prior knowledge and whatever they managed to glean from the textbook. To earn credit, the answers do not have to be complete, or even correct, but they do have to represent honest, deliberate attempts to answer the questions.

In contrast to the WarmUps, Puzzle exercises are assigned to students *after* they have received formal instruction on a particular topic. They serve as the framework for a wrap-up session on a particular topic. One can loosely consider the WarmUp and the Puzzle to serve as bookends for the treatment of a given topic.

2.2.1 Categorization and Examples of WarmUps and Puzzles

The WarmUp, and to some extent the Puzzle, can be categorized as follows:

1. Concepts and vocabulary
2. Modeling, connecting concepts and equations
3. Visualization in general and graphing in particular
4. Estimation, getting a feel for the magnitude of things
5. Relating physics statements to "common sense"
6. Understanding equations—the scope of applicability

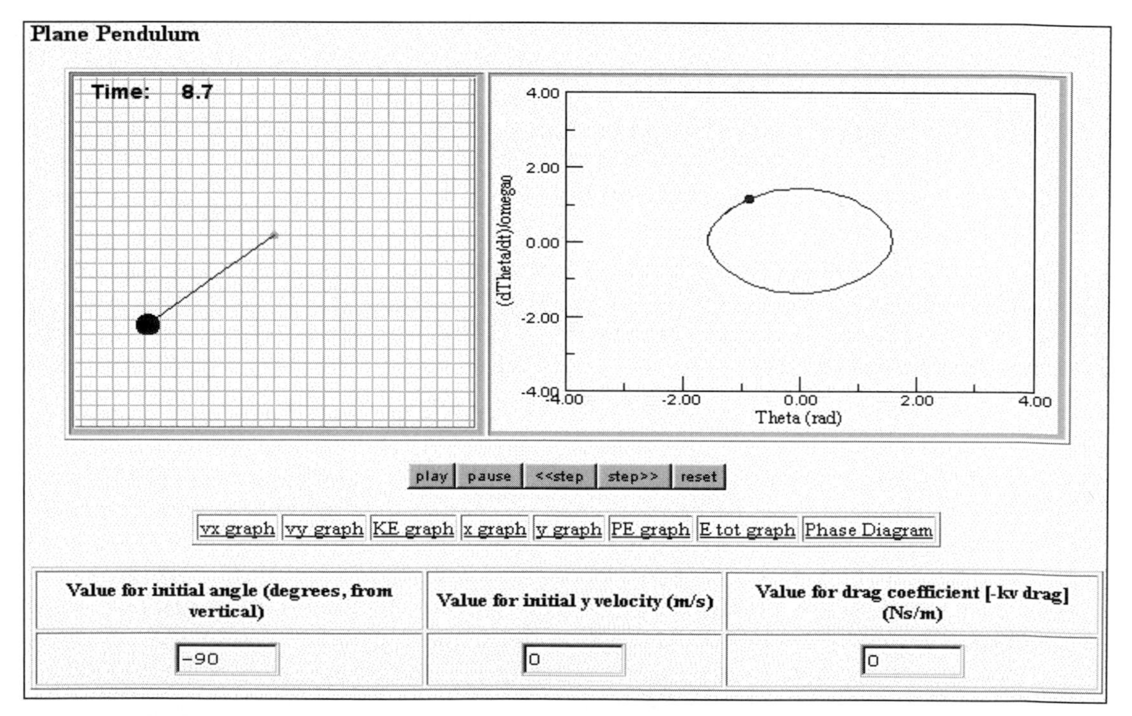

Figure 3: JiTT phase diagram question.

These are best illustrated by examples from each category. The actual WarmUp questions are italicized.

Category 1 (Concepts and Vocabulary):

Example A (from Introductory Mechanics): *Using aerobic exercising, people often suffer injuries to knees and other joints due to HIGH ACCELERATIONS. When do these high accelerations occur?*

Example B (from Junior Level Mechanics): *Consider the simple pendulum simulated in the panel on the left (See Figure 3.). By varying the initial angle and the initial* y *velocity of the pendulum bob, explore the phase diagrams traced out in the panel on the right. In your own words, try to briefly explain what a phase diagram is and what it represents. What kinds of motions of the plane pendulum correspond to the different curves/shapes on the phase diagram? How does the phase diagram change when the pendulum is damped?*

Category 2 (Modeling, Connecting Concepts and Equations):

Example A (from Introductory Mechanics): *In rewinding an audio or video tape, why does the tape wind up faster at the end than at the beginning?*

Example B (from Junior Level Mechanics): *The Physlet simulation below (see Figure 4.) provides an illustration of how tides are created. Use the animation to determine the time between high tides on Earth (tides due to the Moon effect only). What mathematical relationships or equations are you using? Briefly explain.*

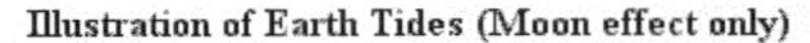

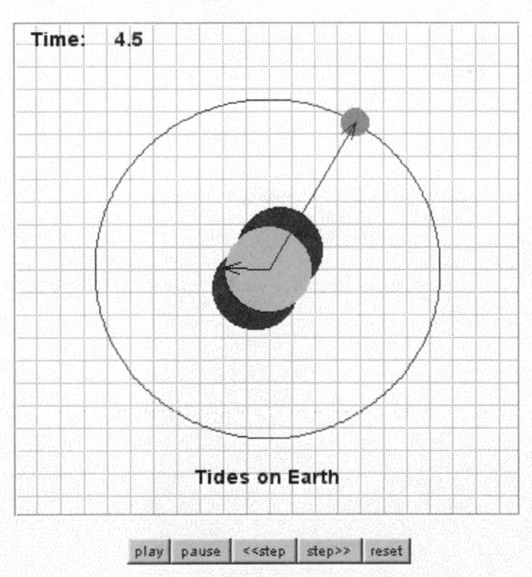

Physlet Problem Description

Use the animation to determine how long it should be between high tides on Earth. The black arrow originating at the Earth's center makes a revolution once every day. Consider the time to be in days.

FIGURE 4: JiTT tides question.

Category 3 (Visualization):

Example A (from Introductory Electricity and Magnetism): *The figure above shows two pictures of a setup including a business card, a converging lens, a mirror, and the camera. The setup is sketched in the upper right. The two pictures on the left differ only in how the camera lens was focused. All of the objects in the setup remained in place. Please describe the paths that light took from the business card to the camera. The areas labeled "Part A," etc., each correspond to a different path.*

Example B (from Junior Level Mechanics): *(See "Gravitation WarmUp Question 3," the "hole through the Earth" question, in Section 2.4 on Physlet-based JiTT questions.)*

Category 4 (Estimation, Magnitudes):

Example A (from Introductory Mechanics): *Estimate the magnitude of the tangential velocity of an object at your geographic location, due to the rotation of the Earth.*

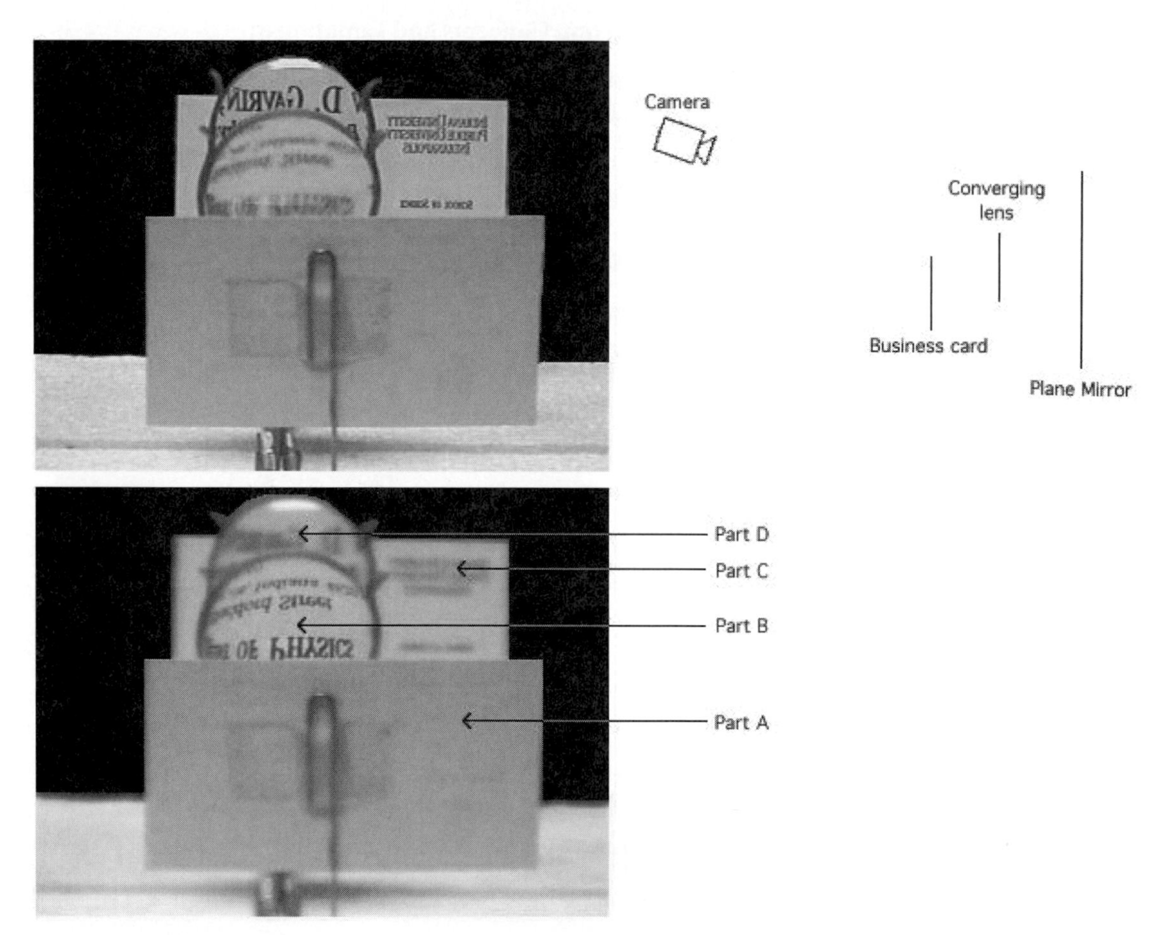

FIGURE 5: JiTT optics question.

Example B (from Junior Level Mechanics, about Central Force Motion): *Suppose you are, once again, an astronomer. This time you are observing a binary star system (two stars in orbit about one another as shown in Figure 6.). By carefully observing the motion of the pair of objects, you can infer the ratio of the masses. Give this a try by looking at the simulation on the Physlet page. By observing the motion of the red mass-blue mass system (feel free to "Pause" and "Step" the simulation!), estimate the ratio of the masses. Please be sure to state any assumptions you make and briefly explain the thought process you use to arrive at your estimate.*

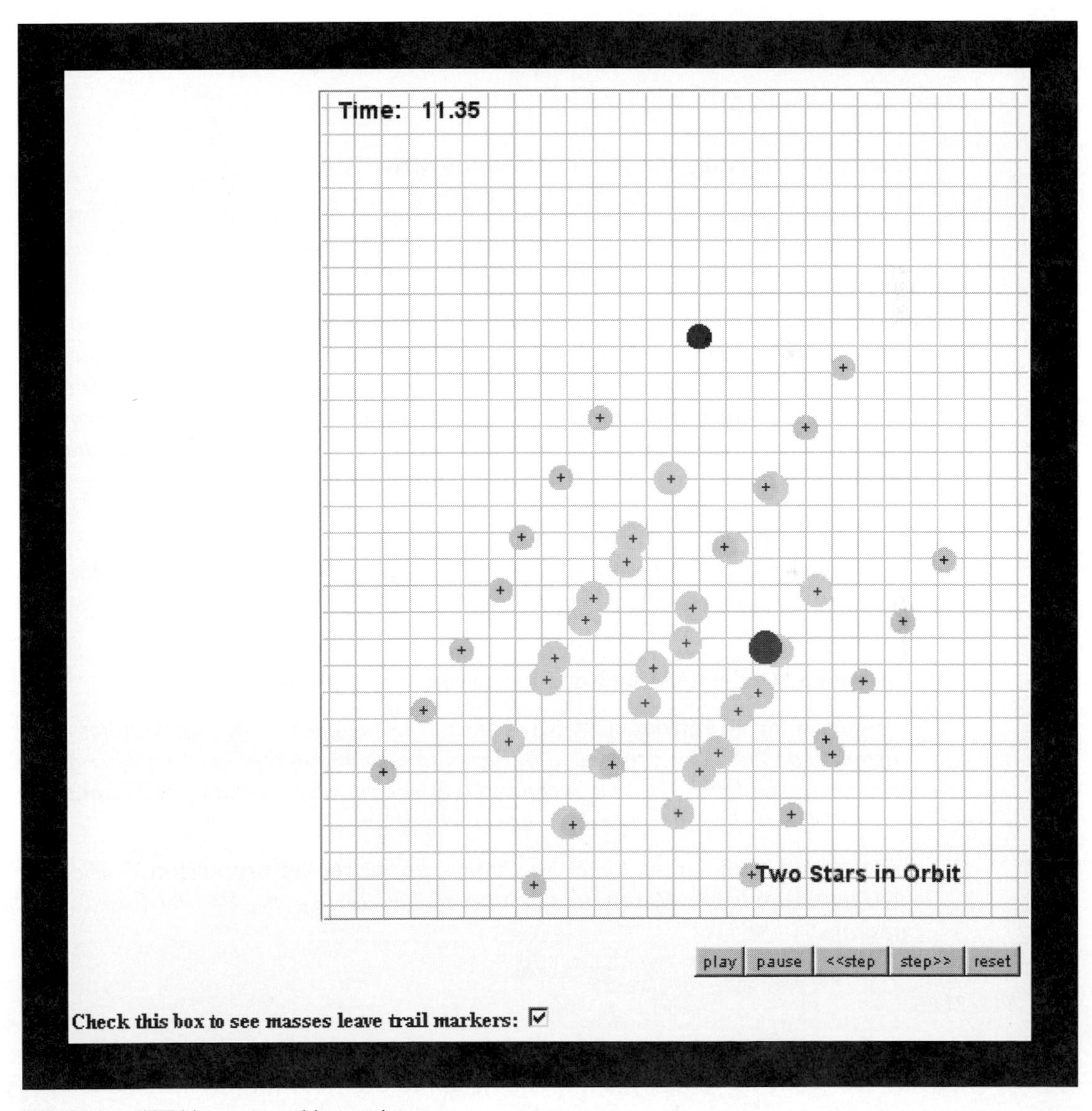

FIGURE 6: JiTT binary star orbit question.

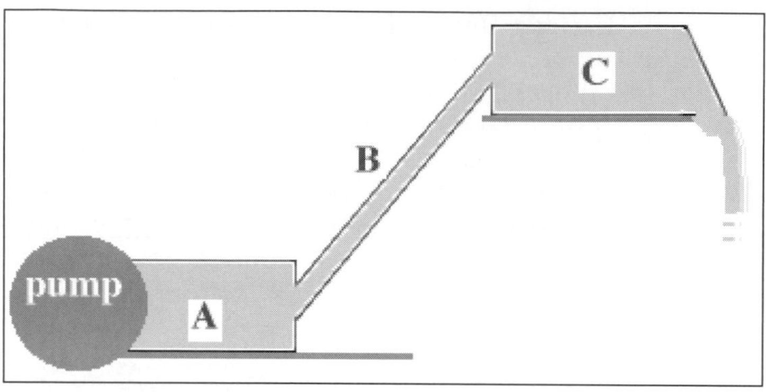

FIGURE 7: JiTT hydrodynamics question.

Category 5 (Relating Physics to "Common Sense"):

Example A (from Introductory Mechanics): *Water is pumped up a pipeline as shown in Figure 7. The water pours out at the top and to the ground. The pump is running at constant speed. Compare the water speed at the three points A, B, and C in the pipeline.*

Example B (from Junior Level Mechanics, about generalized coordinates and constructing Lagrangians): *In the Physlet simulation shown in Figure 8, a simple pendulum is attached to a mass that can slide freely on a horizontal frictionless surface. Before you run the simulation,* ***predict*** *what you think the motion of the system will look like, and describe your prediction as part of your answer. (Consider that the pendulum bob starts out at* $t = 0$ *hanging vertically but with an initial velocity to the left.) Also predict what you think the graph of* x *positions of the pendulum bob and the sliding mass will look like as a function of time, briefly explaining your answer.*

Now run the simulation. Comment on the correctness of your prediction. Do you think the simulation correctly obeys the laws of physics?

Category 6 (Understanding Equations, Scope of Applicability):

Example A (from Introductory Mechanics): *A skater is spinning with her arms outstretched. She has a 2-lb weight in each hand. In an attempt to change her angular velocity, she lets go of both weights. Does she succeed in changing her angular velocity? If yes, how does her angular velocity change?*

Example B (from Junior Level Mechanics, about rocket propulsion): *(See the "Rocket Propulsion" WarmUp question in Section 2.4 on Physlet-Based JiTT questions.)*

2.3 PUZZLES

Puzzle assignments are more complex than WarmUps but are given in the same spirit. Students are expected to analyze a situation, apply the relevant physics, and answer specific questions. The in-class activity, based on the student responses, is a summary review of the concepts learned.

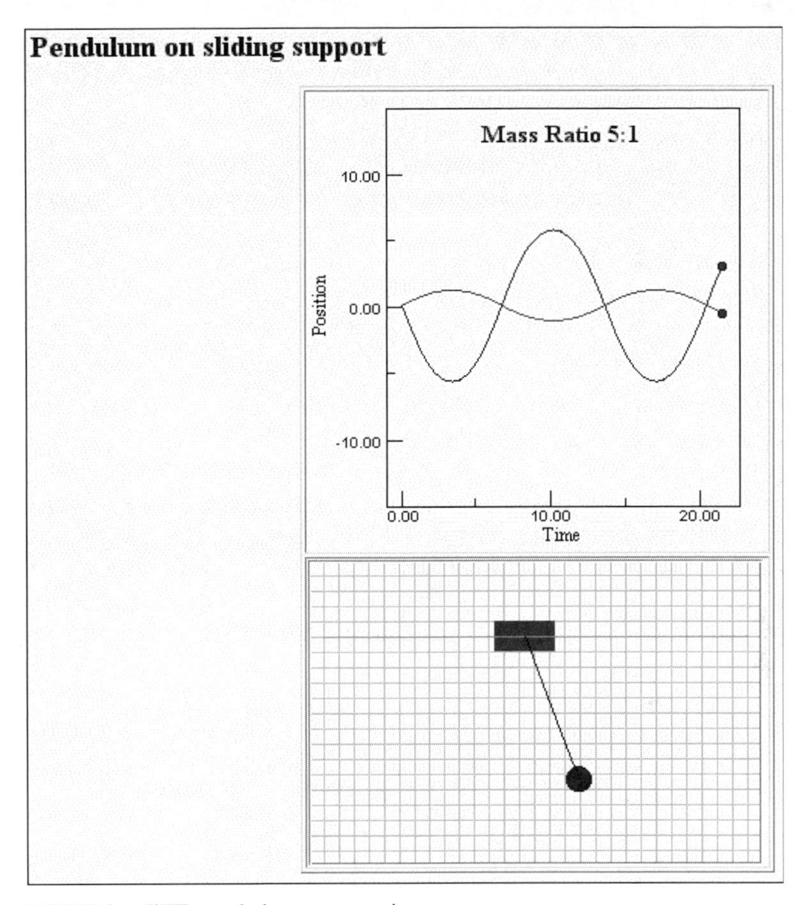

FIGURE 8: JiTT coupled masses question.

Puzzle Example 1: (*Note*: Compare this Puzzle to the Category 2, Example A, WarmUp about rewinding an audio or video tape.)

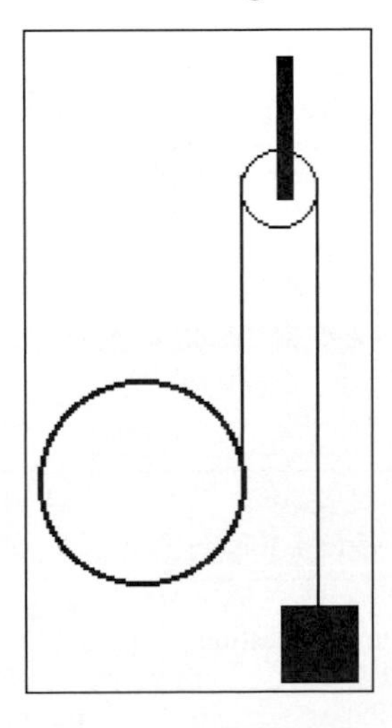

FIGURE 9: JiTT yo-yo Puzzle question.

You make yourself a yo-yo as shown in Figure 9 by wrapping a fine string around a thin hoop of mass M and radius R. You pass the string around a pulley and attach it to a weight, whose mass is exactly half the mass of the hoop.

You release the system from rest.

Describe the subsequent motions of the yo-yo and the weight. You may use equations to arrive at your answer, but you must state your result in plain sentences.

Puzzle Example 2: *Rank simulation 1 and simulation 2 as shown in Figure 10 from least to greatest in terms of the moment of inertia of the wheel, the tension in the string, and the total angular momentum about the wheel's axle after 4 seconds. The hanging weights have identical mass.*

It is interesting to compare Puzzle examples 1 and 2. The static Puzzle Example 1 involves nearly the same physics as the dynamic Physlet-based Puzzle Example 2, but what is required of the student in order to "solve" each Puzzle is quite different. In each case, the student must understand the concepts of moment of inertia, torque, angular acceleration, angular velocity, and the relationships among those quantities. In each case, it also behooves the student to draw free-body diagrams to consider the forces involved. The static Puzzle involves the concept of rolling without slipping (because of the pulley) and can be solved completely with equations and subsequent

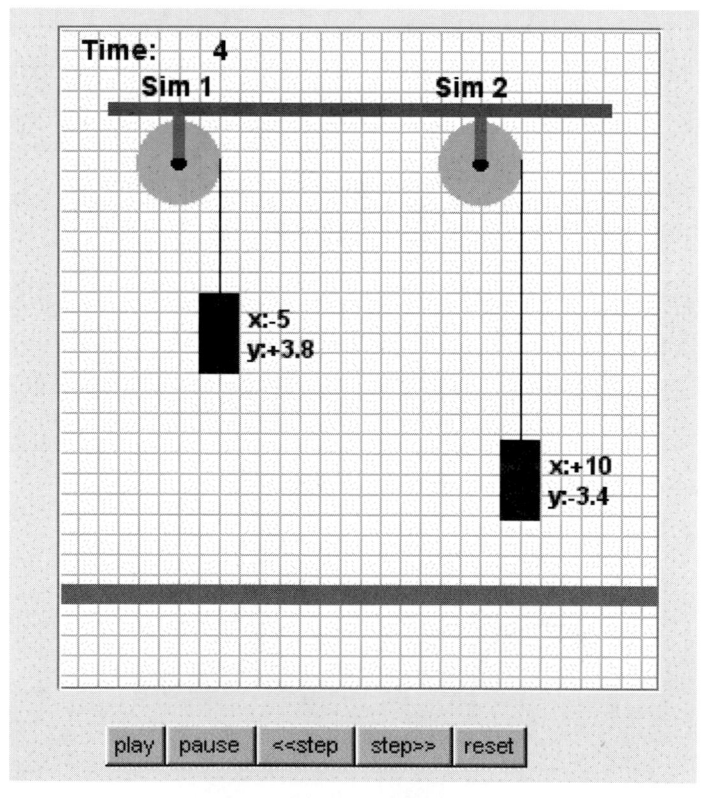

FIGURE 10: JiTT moment of inertia Puzzle question.

English sentences of explanation. The dynamic Puzzle, however, requires some visual analysis and understanding of how the speed with which the mass falls is related to the "physics quantities" like angular momentum and moment of inertia. It is clear from use of both static and Physlet-based questions that students who understand how to solve one of these sorts of questions do not necessarily know how to solve the other, so incorporating both types is an effective way to broaden and deepen all the students' understanding.

Puzzle Example 3: *Figure 11 shows a graph of the current in a series RLC circuit. There are five points marked on the graph, labeled A, B, C, D, E. Please state at which point (or points) each of the following quantities are a maximum, a minimum, or zero:*

- *The energy stored in the inductor*
- *The voltage across the capacitor*
- *The power input to the capacitor*
- *The power output by the inductor*
- *The charge on the capacitor*
- *The voltage across the resistor*

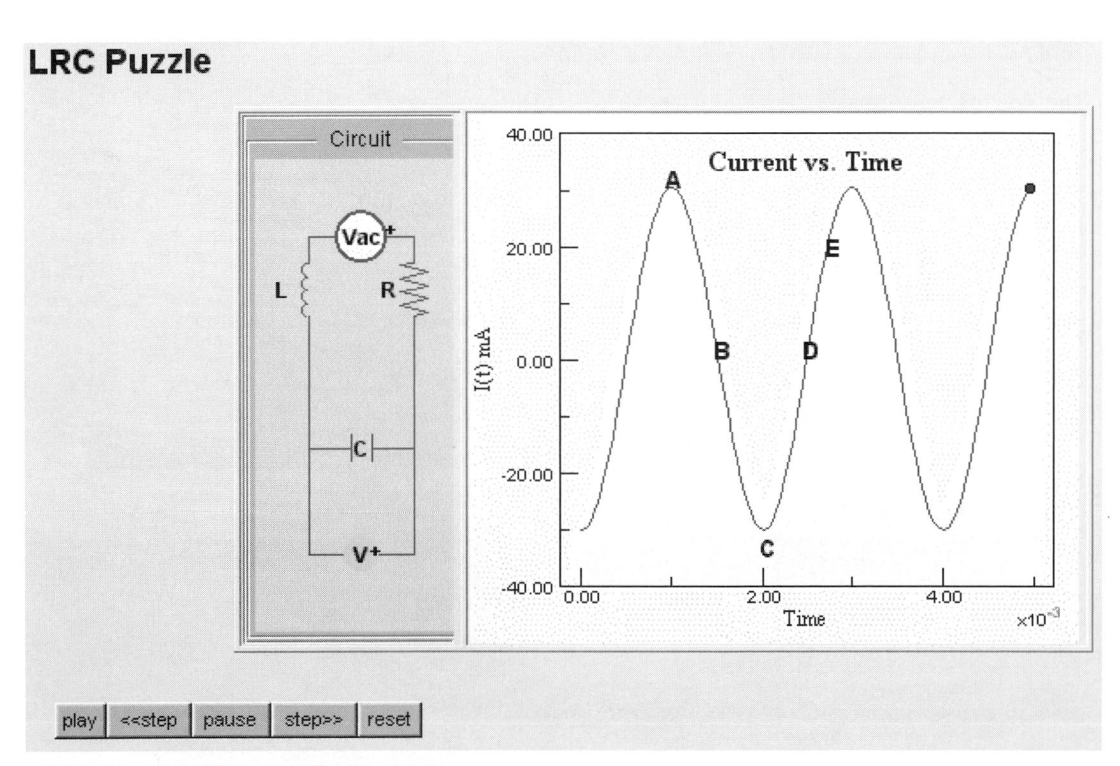

FIGURE 11: JiTT RLC circuit Puzzle question.

2.4 PHYSLET-BASED WARMUP QUESTIONS: A LOOK AT WHAT THEY OFFER

Physlet-based interactive animations can greatly enhance the usefulness of a WarmUp. As an example, consider the following WarmUp, intended to precede the second period devoted to projectile motion. Students have already worked with the equations of motion.

From the WarmUp Web page on the left, the student interacts with the simulation in the Physlet window on the right.

Projectile Motion WarmUp Question 1: *Without changing any of the settings, run the simulation (shown in Figure 12) a few times by clicking the "run again" button in the Physlet window and observe the graphs.*

In the box below this question answer the following questions:

A. *How do the components of the initial velocity* v_o *determine how long the projectile will be in the air?*

B. *How do they determine the range* $x - x_0$ *of the projectile?*

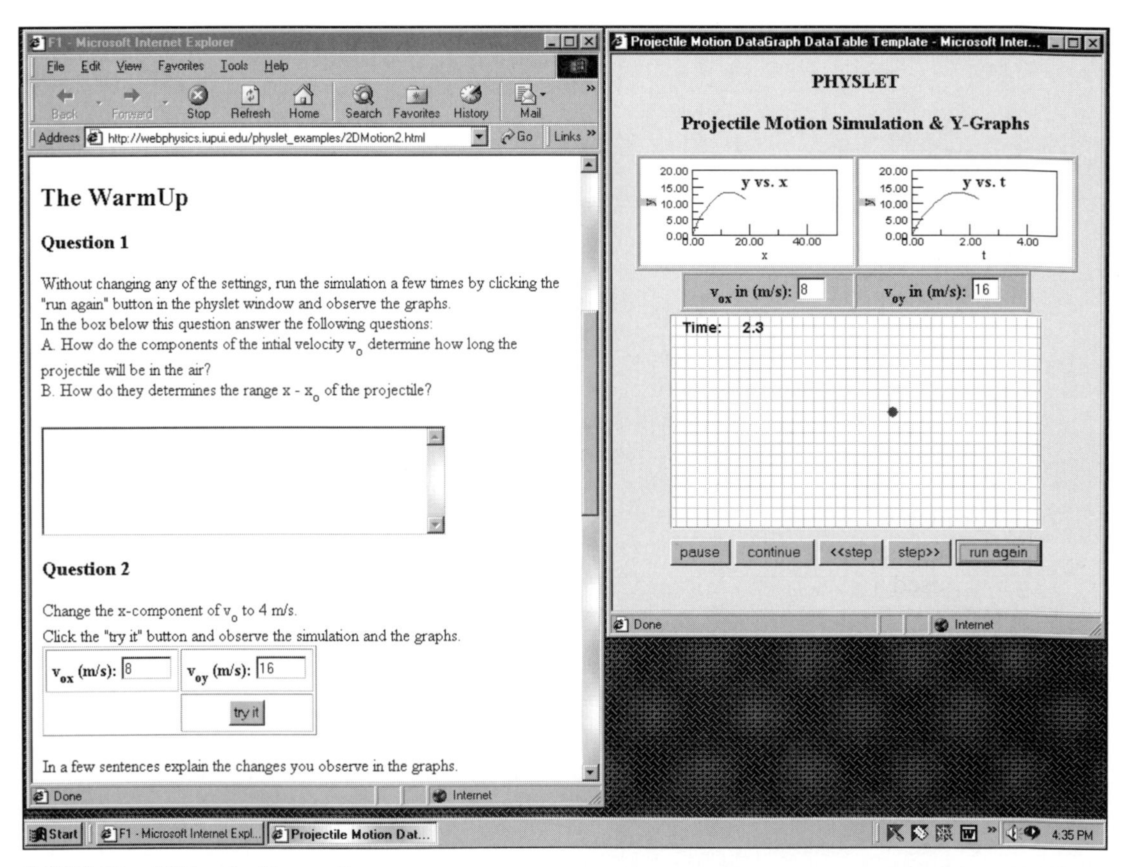

FIGURE 12: JiTT projectile question.

Projectile Motion WarmUp Question 2: *Change the* y-*component of* v_0 *to 8 m/s. Click the "try it" button and observe the simulation and the graphs. In a few sentences explain the changes you observe in the graphs.*

Projectile Motion WarmUp Question 3: *Write an expression for* y *as a function of* x.

This WarmUp exercise gives a student an opportunity to reflect on the kinematics learned thus far. In particular, the student is forced to think about graphs and their interpretations. Typical students in our experience have a difficult time distinguishing between trajectories and time graphs. Observing the two graphs in question 2, they are forced to explain why the y versus x graph changed, while the y versus t graph did not. The notions they develop are further reinforced in the third question, where both graphs change.

We believe it is essential that thorough classroom discussion and application activities follow this exercise while the notions that the students developed as they worked through the WarmUp are still fresh in their minds. The misinterpretations culled out of the student submissions can be addressed through the use of effective demonstrations. In fact, we have found that real-life demonstrations are needed to support the simulations. Students often accept the simulations, but they don't believe that they faithfully represent reality.

This type of WarmUp can be extended to a more elaborate self-study review tutorial. More detailed short questions can, for example, deal with the relationships between the initial speed v_0, the launch angle, and the components.

In general, Physlet-based WarmUp questions provide the ability to probe student understanding in ways not previously possible. Offered as part of a JiTT WarmUp, such questions "prime the pump" of the students' thinking processes, so that the students come to class with particular questions and issues in mind. The students' submissions to the questions provide valuable feedback and insights to the instructor, who can then tailor the in-class time to best meet the needs of the students, as evidenced by their submissions.

Entire WarmUps can be based on a single Physlet animation or a series of animations, as illustrated previously. Often, though, a topic or concept can be particularly well addressed via a WarmUp composed of multiple kinds of questions that represent a variety of approaches: explain/describe, calculate, visualize and interpret, etc. In this case, it is entirely possible that only one or two of the questions comprising the WarmUp will be Physlet based.

Consider, for example, the following junior-level mechanics course WarmUp composed of three questions dealing with universal gravitation and gravitational potential. (Question 3 of this WarmUp is an example of a Category 3, Visualization and Graphing, WarmUp question, as mentioned in Section 2.2.1.)

Gravitation WarmUp Question 1: *We can think of the gravitational field vector* **g** *as being the gradient of a scalar function* Φ *called the gravitational potential. The gravitational force is given by* F = m**g**. *We can also think about the gravitational potential energy of a body in the gravitational field and can express that potential energy in terms of the gravitational potential* Φ. *We can even think about equipotential surfaces, which are surfaces of constant gravitational potential.*

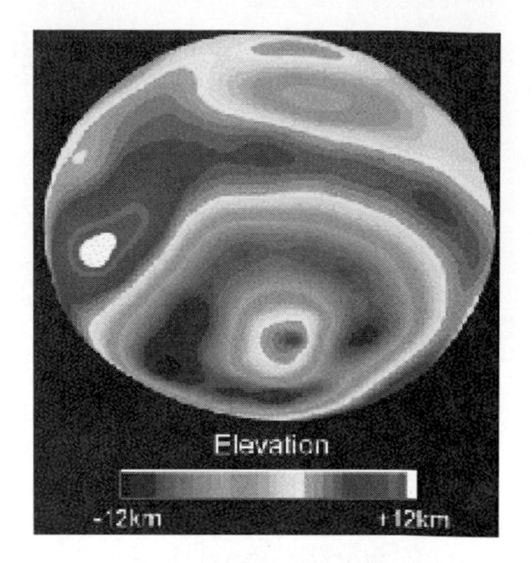

You've seen all of this before, but in the context of electric fields and electric potentials rather than gravity. What are all of the analogous quantities and equations from the electricity context? Is there anything whose analog you don't know or understand?

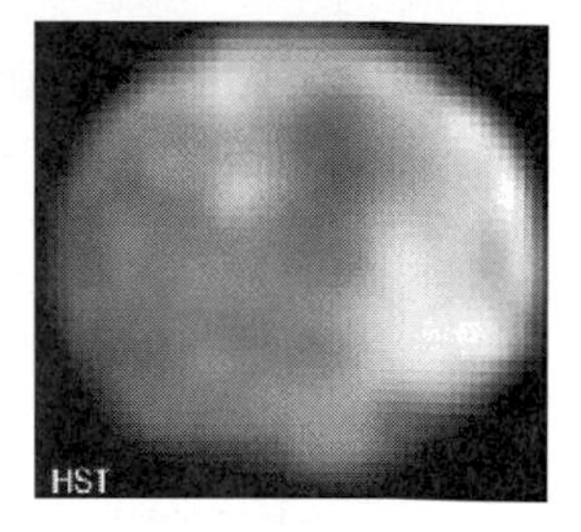

Gravitation WarmUp Question 2: *Suppose you are stranded on the surface of a small asteroid. (The image shown is an image of the asteroid Vesta3, taken by the Hubble Space Telescope!) Estimate how fast you'd have to jump off its surface in order to be able to escape from its gravitational field. (Your estimate should be based on parameters that characterize the asteroid, not parameters that describe your jumping ability.) Please be sure to explain your assumptions, as usual.*

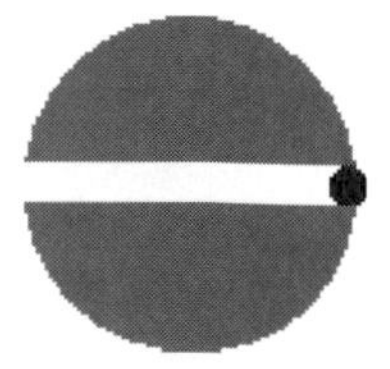

Gravitation WarmUp Question 3: *The Acme engineering company has been hired by a wealthy person to investigate the feasibility of a train that travels in a tunnel through the center of the Earth. The Acme engineers have programmed a simulation of this into a Physlet page so that you can better take a look. The Physlet shows the train (modeled by Acme as a spherical black ball) moving through the tunnel, and four possible graphs (one for each simulation) to represent the train's acceleration versus time.*[1]

Which of the four Physlet animations shown on the following page correctly shows the train's acceleration versus time?

These three questions approach the topic of gravitation from very different vantage points and probe different kinds of understanding. The first question requires a response composed of English sentences relating concepts and quantities in electricity and gravitation. Here, the important point is for the students to try to connect electricity and electrostatics quantities with which they are already familiar from their introductory physics courses to the new gravitation quantities (gravitational potential, etc.).

[1]See Section 8.4, "Gravity," for a related Physlet problem.

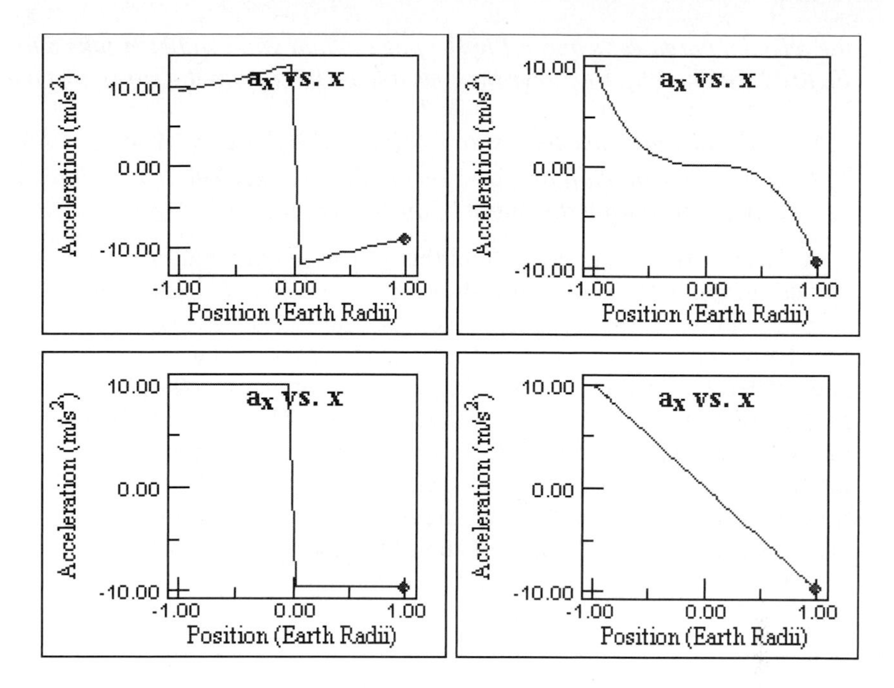

The second question requires both an understanding of the concept of escape velocity and the ability to estimate a reasonable mass and radius for a small asteroid. To estimate the jumping-off speed, a student needs to be able to apply the escape velocity formula, using reasonable values for the mass and radius of the asteroid. The third question, the Physlet-based question, probes the students' understanding of the universal gravitational force and Newton's second law. The object undergoes simple harmonic motion, and the students must look at the four acceleration versus time graphs to determine which one correctly characterizes such motion. Because the third question is Physlet based, the students watch the proposed acceleration versus time graphs evolve in time as they observe the object undergoing simple harmonic motion. They can readily identify points on the graphs with actual locations of the object in its motion, because they see both the graph and the object evolving simultaneously. This provides different and additional information from that given by a static paper representation of the four graphs.

Because a Physlet-based question inherently involves visualizing or watching something, it probes the students' understanding in a different way than a static, text-based question does.

Consider the following WarmUp example about rocket propulsion. (This question is an example of a Category 6, Understanding Equations and Scope of Applicability, WarmUp question, as mentioned in Section 2.2.1.)

Rocket Propulsion WarmUp Question: *An aero major friend of yours has designed a rocket and has asked you to take a look at the design. Your friend has programmed* *the relevant parameters into a Physlet page so that you can better take a look (See Figure 13.). The Physlet shows the rocket launching from the surface of the Earth.*

A. *What do you notice that is strange about this design? (What seems wrong with it?)*
B. *Use the information provided on the Physlet page (hint: including the elapsed time* t*) to estimate what the* dm/dt *of the rocket must be.*

What is the pedagogy behind this WarmUp question? As can be seen from the y versus t graph in the screenshot shown in Figure 13, the rocket burns fuel for about 10 seconds before it begins to lift off of the launch pad. (This is also evident in the left Physlet panel; the rocket begins to move upward at about $t = 10$ seconds.) The student is provided with values for the empty rocket mass (10,000 kg), the initial mass of the fuel (56,000 kg), the total mass of the rocket initially (66,000 kg, which is the sum of the previous two masses), the value for the speed of the exhaust speed of the rocket (2500 m/s), and the value for g, the acceleration due to gravity (9.8 m/s^2). To answer the question, the student must realize that the rocket thrust is insufficient to allow the rocket to accelerate upward until some of the initial mass of the rocket is lost due to burning off of "extra" fuel. To estimate the dm/dt fuel flow rate, the student must set the thrust [which is equal to the product of the exhaust speed and the fuel flow rate:

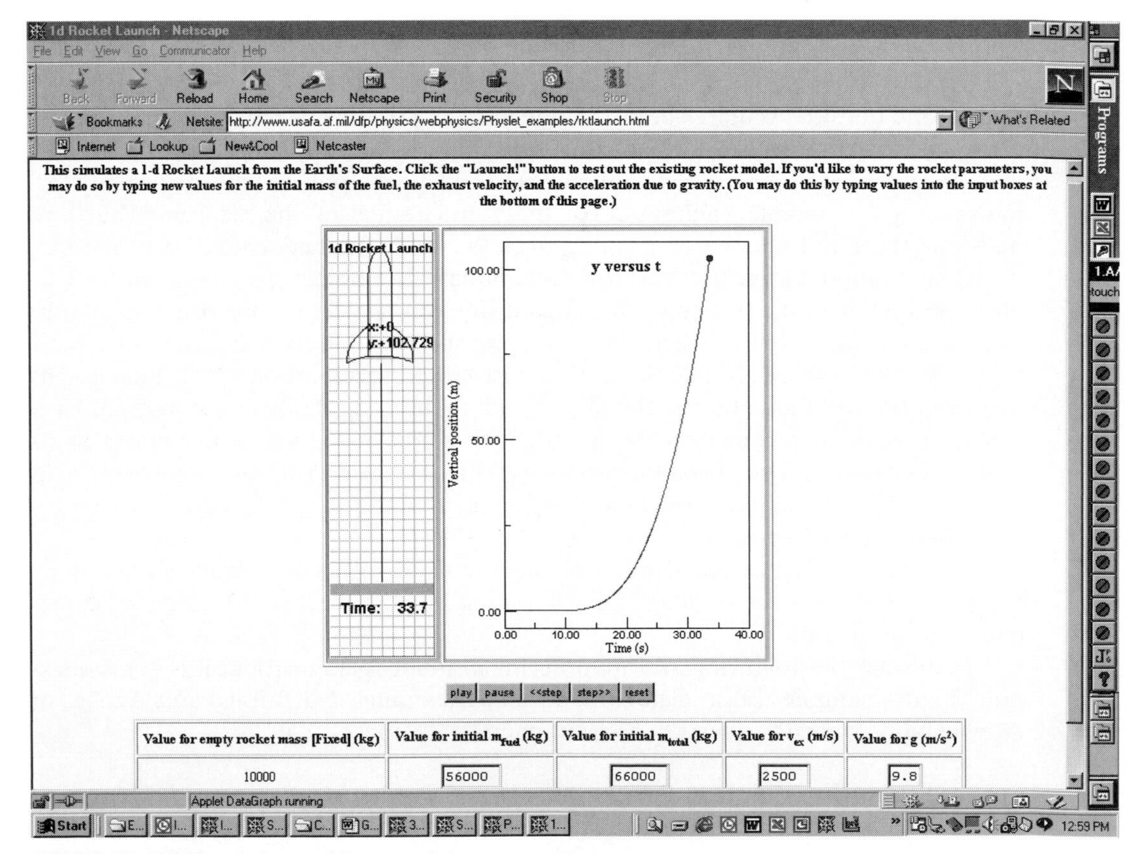

FIGURE 13: JiTT rocket propulsion question.

$T = v_{ex}*(dm/dt)$] equal to the force of gravity on the rocket at the time it actually begins to accelerate upward, and then solve for the unknown dm/dt.

2.4.1 Sample Student Responses

Student submissions in response to carefully crafted WarmUp questions provide a rich store of information about their levels of understanding of the topics at hand. Because the students are asked to explain their assumptions and sketch out the process by which they arrive at any estimates they make, the submissions are highly useful centerpieces for the subsequent classroom time together.

For example, consider a selected subset of responses to the rocket propulsion WarmUp question discussed in the preceding section. Most student responses fell into one of the three following groupings or categories:

Category A: Students who either focus on superficial features or see nothing "wrong" and who assume that, because the simulation happens to run for 40 seconds, all of the fuel (56,000 kg) is expended during that 40-second elapsed time period.

> "I don't see anything that is wrong. dm/dt is 1400 kg/s."
>
> "It's only got one booster/fuel tank rolled into the same structure. Also, they usually don't launch from the ground, but up 100 ft or so, so that the thrust can spread out and build. dm/dt = 1400 kg/s 56,000kg/40s (to burn the fuel)."

Category B: Students who understand that the rocket doesn't have enough thrust and/or has too much fuel initially to take off, but who make faulty or unclear starting assumptions to estimate dm/dt.

> "The rocket takes entirely too long to get off the ground. After 40 secs the rocket has only reached 200 meters in the vertical direction (I can run this distance 12–15 seconds faster and I'm a distance runner). Since the rocket mass probably isn't the problem here because of the fuel requirements, my aero major friend probably hasn't designed a powerful enough engine for this rocket, or has miscalculated his exhaust velocity. A good thrust to weight ratio would fix this problem. The mass lost rate is approximately equal to thrust/ $-u = -1.48E4$ kg/s assuming a thrust of 37E6 N. To get the exact thrust, I'd have to fit a function to the rocket's displacement versus time graph and take the second derivative of the graph to find out its acceleration, then multiply that value by its instantaneous mass to find the thrust. This would involve Newton's 2nd law."
>
> "It seems that the rocket is very slow in taking off. It may not have enough thrust. 1000 kg/40 s = 25 kg/s = dm/dt."
>
> "It looks like it has to burn a lot of fuel before it is light enough to leave the ground, because the rocket does not leave the ground for 9 seconds. dm/dt must be close to 22 kg/s."
>
> "The rocket doesn't actually lift off until just before 9 seconds. Take the equation $v = -gt + u \ln(m\_0/m)$. Maybe u is not sufficient to make v positive, so the rocket will stay on the ground untill enough fuel is burnt off . . . it should be designed so that it lifts of as soon as it reaches a full burn rate (paraphrasing what the book says on pages 92 & 93). Assume the thrust is about 1,000,000 N, then dm/dt = thrust/$-u = -400$ kg/s."

Category C: Students completely grasp the question and can answer it correctly.

"Whoever designed the rocket didn't give it enough thrust . . . that is, it weighs too much initially to leave the ground. As the fuel burns off, it becomes light enough to fly but not until then. So, you just manipulate the equation from the book to solve for m, which turns out to be 63,712, so, 66,000 − 63,712/9 sec = 254.2 kg/s."

"It seems odd that the rocket sits on the pad for nearly 10 seconds (probably turning the launch pad into slag) while the motor burns off fuel until the upward force generated by the thrust is finally greater than the weight of the rocket. Why waste gas when one could simply load the rocket with less fuel in the first place? What with the high pump prices today. Referring to eq 2.132, I would estimate dm/dt to be approximately 216 kg/s. (This is eerily like problem 2-54.)"

2.5 JITT IN REVIEW

In this chapter, we have examined WarmUps and Puzzles, two kinds of Web assignments from the pedagogical arsenal of the Just-in-Time Teaching strategy. Both types of assignments are used to prime the students for the classroom activity that follows. The assignments ask the students to think deeply about the topic to be studied. This creates a need to know that is a strong motivating factor for taking the subsequent classroom activity seriously and approaching it with interest. In many instances, as we have shown, Physlet-based animations can make a substantial contribution to the value of the assignment. In some cases, the simulation is the central, essential component.

Other Web resources employed in the JiTT-based curriculum include enrichment pages with Web links, preparatory assignments that prepare students for the hands-on lab, and various communication pages that put students in touch with one another and with the instructors. For more detailed information, please examine our book, *Just-in-Time Teaching: Blending Active Learning with Web Technology* [Novak 1999], or visit our Web sites:

http://webphysics.iupui.edu/jitt.html
http://webphysics.iupui.edu/introphysics

For more examples of using Physlets in support of JiTT, please visit

http://www.usafa.af.mil/dfp/physics/webphysics/Physlet_examples/

CHAPTER 3

PER AND PHYSLETS

Aaron Titus and Melissa Dancy

Physlets turn traditional problems into interactive problems. They allow the student to see a dynamic phenomenon, control the simulation, and collect data. But what evidence does physics education research (PER) provide that this type of problem is different from traditional problems?

Physlets, that is, the applets themselves, are tools that allow the teacher to create innovative types of physics questions, problems, simulations, and interactive activities. As a tool, a Physlet may be used effectively or ineffectively within a given task. Effectiveness likely depends on many factors, such as how well the task targets known student difficulties, how students use visual cues given by the Physlet, how important visualization is to the given task, and the appropriateness of the Physlet to the given task. There are undoubtedly a large number of factors and tasks to study. This chapter presents the results of PER that utilized Physlet-based activities. The research focus is on students, not on the Physlets themselves.

Following a description of the purposes of instructional graphics and previous research on animation, we report on three primary studies that we have conducted using Physlets (all three of these studies concern students' responses to numerical Physlet problems or conceptual Physlet questions in mechanics that use the *Animator* Physlet): (1) comparison of students' answers on Physlet problems to their answers on traditional numerical problems; (2) think-aloud interviews with students as they solve Physlet problems; and (3) comparison of students' answers on the Force Concept Inventory [Hestenes 1992] to students' answers on similar conceptual Physlet questions.

3.1 EFFECTIVENESS OF ANIMATION

An essential feature of Physlet problems is that they include a dynamic visual; that is, students see the animated evolution of a phenomenon as opposed to the more typical static visual included with paper-based media. It is important then to review what other researchers have discovered about animation.

Research on animation can be neatly organized according to the purpose of the animation. In general, instructional graphics satisfy five purposes, also known as the five applications (or functions) of graphics [Rieber 1994]: cosmetic, motivation, attention getting, presentation, and practice. Cosmetic and motivation are in the affective domain, while the others are in the cognitive domain.

Graphics used for cosmetic purposes are used to "dress up" the text. An example is the use of a visually stimulating Physlet such as Doppler on a course home page. The

Physlet will undoubtedly make the page look neat. Unfortunately, learning does not take place directly as a result of cosmetic graphics. Rieber notes, "At their best, cosmetic graphics help maintain student interest. . . . At their worst, cosmetic graphics distract student attention from other important material." If Physlets are used predominantly for this purpose, they will likely have little impact on the teaching and learning of physics.

Graphics may also be used for motivational purposes, appealing to the viewer's attitudes. For example, an animation of a pendulum shown next to text describing the motion of a pendulum may motivate the student to read the text. It's important for learners to see material as exciting and relevant, but there is a danger in overusing graphics for motivational reasons. Although learners may be motivated by novel graphics, they may also become saturated as they are inundated with such visuals. As a result, motivating visuals can quickly lose their instructional impact. Educators must be cautious about falsely attributing learning to the design of multimedia-enhanced applications when it is mostly due to novelty [Rieber 1994].

Although sometimes difficult to distinguish from motivational graphics, visuals can also be used to gain attention. These are known as attention-gaining graphics. Their primary difference from motivational graphics is that they are not designed to influence the attitude of the viewer (motivation) but rather to focus the viewer's attention. Attention-gaining graphics may directly influence students' learning; thus they are classified according to the cognitive domain (or outcome) of learning. A good example of an attention-gaining graphic is an *Animator* simulation that shows the ghost images of an object as it transverses the screen. The ghost images may direct students' attention to the displacement of the object during equal time intervals, thus helping them observe how the average velocity during those intervals is changing. Likewise, velocity and acceleration vectors may be shown with the animated object in order to help students focus on how these variables are changing.

Probably the most common use of instructional graphics is for presentation. For instance, an animation collision between particles may be used to illustrate the collision described in the text of a problem. Physlets are not only used to present a dynamic visual, but also to present data. Data may be presented within a graph, within a table, or within the animation itself. Although calling data an "animation" is arguable, the interesting feature here is that the data are observed to change along with the animation. Therefore, the learner can make a connection between the "state" of the data and the "state" of the animation.

Finally, instructional graphics may be used for practice activities. This purpose is especially suited to interactive computer simulations, where the user receives feedback based on his or her input. Physlets can easily be used for practice activities, where students control the animation, turn on or off certain parameters, drag an object on the screen, or change initial conditions.

Not only can Physlets be used for all of the aforementioned purposes, but Physlets can be used in a variety of tasks. For instance, they can be used in the context of a problem or they can be presented alongside lesson content. Most research on animation in education has studied the effect of animation on students' reading comprehension and its effect on students' understanding as a result of practice activities [Rieber 1994]. There seems to be only one study investigating the influence of animation in testing [Hale 1985]. The purpose of Hale's study was to design a test that would minimize certain validity problems with printed tests. An ideal test should assess students based on their knowledge or problem-solving skills rather than their ability to visualize a problem. Unfortunately, their conclusions, which pertain mostly to the display of animations, do not seem to apply very well to our use of Physlets. As a result, there is great need for research into the effect of animation in problem-solving tasks.

Rieber's findings on animation used as presentation of lesson content can be useful in guiding the use of Physlets as presentation in a problem. According to Rieber, there are two prerequisites if animation is to have positive effect on learning outcomes: (1) There must be a need for external visualization, and (2) learning of the described phenomenon must require an understanding of how an object's properties change with time or position. Rieber recommends that "animation should be incorporated only when its attributes are congruent to the learning task." That is, the previous attributes should be satisfied. Yet even if these attributes are satisfied, it does not guarantee a positive effect of animation on learning, which is why Rieber calls them necessary conditions and not sufficient conditions.

In addition, Rieber found that "when learners are novices in the content area, they may not know how to attend to relevant cues or details provided by animation." That is, students may overlook important details in the animation unless specifically drawn to vital information using verbal cues or "chunking" strategies, where the animation is presented in stages, thus helping the student to focus on important details. For example, it is possible that in a Physlet animation the mere presence of a velocity vector may cause students to observe how the velocity varies with the motion in the animation. Or perhaps, showing a graph of some variable versus time will cause students to consider that variable as important in the problem, even though they may not have otherwise considered it.

3.2 COMPARISON OF STUDENTS' PROBLEM SOLVING

One of the conditions for animation to be effective in presentation of lesson content is that external visualization must be essential to learning the material. If we apply this same principle to the effectiveness of animation in the context of problem solving, then external visualization should be essential to solving the problem. In other words, the animation may not be an advantage over a static image in all cases. It is possible to use a Physlet animation to show a dynamic visual of a phenomenon; however, if the need for external visualization is not established, the animation will likely be ineffective.

It is therefore desirable to make viewing the Physlet absolutely required to solve the problem. This is accomplished by embedding data necessary to solve the problem within the Physlet. As a result, data required to solve the problem are not obtained from the text of the question but are obtained by direct observation, collection, or calculation. This has two advantages: (1) Viewing the Physlet, and thus visualization in a way, becomes necessary to solve the problem, and (2) students may be challenged to adopt a more expertlike approach that is evidenced by visualizing the problem before attempting a mathematical solution. Visualization is an important first step in a good problem-solving strategy [Heller 1992].

A first-order attempt to understand the difference between students' answers on Physlet problems and traditional numerical problems was to compare the proportion of correct responses [Titus 1998]. In Physlet problems, the data necessary to solve the problem are embedded in the animation and not in the text of the question. It was expected that students would perform more poorly on Physlet problems because characteristics of Physlet problems make them more difficult. According to Schoenfeld,

> Problem difficulty can be measured by applying a ranking of 0 (easier) or 1 (more difficult) to the following characteristics: problem context, problem cues, given information, explicitness of question, number of approaches, and memory load (Heller et al., 1992). Physlet problems in general rank as more difficult than traditional problems based on the criteria of given information and problem cues. In the case of Physlet problems, there are a large number of data points that can be collected. The student must decide what data points are necessary to solve the problem. The fact that there are no numbers given in the text of the question that prompt the method of solution also makes Physlets more difficult [Schoenfeld 1985, p. 13].

In most testing situations students are asked to work problems similar to those they have been trained to solve. As a result, the context keeps them in the right arena, even when they are unable to solve the problems. The problems asked of students were certainly within their capability and were often technically easier than problems they solved in other classes. They were not put in a context that oriented the students toward the "appropriate" solution methods. Time and time again the students working such nonstandard problems would go off on wild goose chases that, uncurtailed, guaranteed their failure. The issue for students is often not how efficiently they will use the relevant resources potentially at their disposal. It is whether they will allow themselves access to those resources at all. Therefore, students accustomed to solving traditional problems simply may not apply the same resources to Physlet problems.

To investigate students' success at solving Physlet problems compared to traditional numerical problems, the proportion of correct responses on each set of problems was calculated and compared for a large group of introductory calculus-based physics students [Titus 1998]. A randomized posttest-only control group design was utilized. Students were given separate assignments covering topics of work, energy, collisions, rotation, rolling, and oscillations throughout the semester as each topic was covered. A sample Physlet problem and its corresponding traditional problem are shown in Figure 14. The treatment group received Physlet problems; the control group received similar questions with no animation and with numbers given in the text of the question (i.e., "traditional" problems). Differences in proportions of correct responses for the two groups were tested for significance. It was determined if the null hypothesis—that the treatment group does as well as or better than the control group—could be rejected.

Students in five sections of introductory physics participated in the study as part of their homework requirement. The course represented a typical course that could be taught at any college or university that utilizes large-enrollment lectures. Students in the study were physical science and engineering students. Students from all sections were randomly assigned to one of two groups. Students did not see Physlet problems solved in class; therefore, solving a Physlet problem was never explicitly modeled.

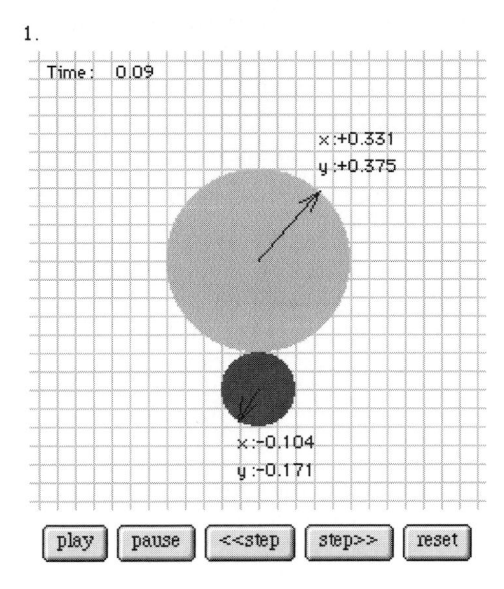

Start Animation(**Click here for help using the animation.**)

A large wheel (i.e. a flat cylinder) is used to rotate a small wheel of the same density and thickness as shown in the animation (position of a point on the edge of each wheel is shown in meters and time is shown in seconds).
(a) What is the ratio of the moment of inertia of the green wheel to the moment of inertia of the red wheel (I_{green}/I_{red})? 39.0625
(b) What is the ratio of the rotational kinetic energy of the green wheel to the rotational kinetic energy of the red wheel(K_{Rgreen}/K_{Rred})? 6.25
(c) What is the torque on the small wheel? 0 N·m

1. The edge of a large wheel (i.e. a flat cylinder) of radius 0.6 m rests against a small wheel of radius 0.3 m such that turning the large wheel causes the small wheel to turn (just like gears!). The two wheels have the same density and thickness. The large wheel turns with a constant rotational frequency of 4π rad/s and the small wheel turns with a constant frequency of 8π rad/s
(a) What is the ratio of the moment of inertia of the large wheel to the moment of inertia of the small wheel (I_{large}/I_{small})? 16
(b) What is the ratio of the rotational kinetic energy of the large wheel to the rotational kinetic energy of the small wheel(K_{Rlarge}/K_{Rsmall})? 4
(c) What is the torque on the small wheel? 0 N m

FIGURE 14: A Physlet-based problem and corresponding text problem.

On 12 of 15 problems, the proportion of correct responses on Physlet problems was less than the proportion of correct responses on traditional numerical problems. On the remaining three questions, there was no significant difference between the groups. To gain insight into whether students thought that the Physlet problems more difficult, students were asked, "Are homework questions with animations more difficult than other homework questions?" Ninety-four percent reported that the Physlet problems were always or usually more difficult. When asked why the Physlet problems were more difficult, most explanations fell into two broad categories: (1) determining what information was needed; and (2) being able to collect accurate data. The second category was largely due to the fact that answers were collected and graded online and that students were not shown in class how to solve the problems.

The first category supported our belief that Physlet problems are more difficult because students must determine what data were necessary to solve the problem. The typical cues that students had come to depend on were not available in the Physlet problems. A few of the students' comments are as follows. Spelling and grammatical errors have not been changed in order to accurately reflect students' responses.

> Because we are not just given numbers to play with in the calculatlor or put in a formula when you have an animation you got to figure out what numbers to use. I guess thats good for us, helping us to understand problems closer to real life situations.
>
> b/c they involve more hands on approach rather than the usual word problem that gives you the boundries and criteria
>
> It becomes tedious to gather all the required information necessary. The same effect could be given by just giving us a word problem that states the same information that can be obtained from the animation but still require us to filter out what is necessary and what is not.
>
> Well, you are not given all the information, so you have to decide which information to gather from the animation and then use what you find out to help solve the problem.
>
> They are more difficult because you do not have the numbers right in front of you. Instead they are in the form of the objects moving, and it is hard to get the real numbers.
>
> You have to gather the information on your own and then apply it. Where as the information in other homework has the info for you.
>
> Sometimes it is hard to interpret the animations because you can't tell what information is given and what information is necessary to solve the problem. Also, you can't check your answers like most of the other problems and you don't know how to go about solving them sometimes.
>
> Animation is more difficult because in addition to solving the problem you may have to determine time, displacement, velocity, etc. . . . by viewing, whereas in a standard problem that information has to be given. Also, since you are not given all the pieces of information needed to work the problem (you have to get information from viewing) it is difficult to know what you are looking for, i.e. what you need to be able to solve the problem.
>
> You have to obtain data, it is not merely given to you for "plug n' chug". This is sometimes difficult. Also, it is harder to define what method/formulas you will be utilizing in the problem.

The fact that students may consider Physlet problems to be more difficult than traditional problems should not be considered negative. On the contrary, it supports the notion that Physlet problems cannot be solved by so-called weak methods [Larkin 1980]. After all, the nature of Physlet problems requires a more expertlike solution strategy. Problem-solving strategies recommend that one should visualize the problem, create a physics description, and plan a solution before executing the plan mathematically [Heller 1992]. To solve a Physlet problem correctly, the student should (1) observe relevant information from the motion, (2) develop a physics description of the situation, (3) apply appropriate principles, (4) decide what quantities are needed to solve the problem, (5) decide what measurements are needed, and (6) take data. The acts of observing the motion, deciding what measurements are needed, and collecting data are rarely needed to solve traditional problems. This is what makes Physlets challenging but also what makes them valuable.

From our experience, we do not believe that Physlets in and of themselves help students develop expertise in problem solving. However, we do believe that combined with instruction in problem solving, Physlets can be a useful tool to encourage students to apply a more expertlike approach. We believe that Physlets should be used within an effective learning environment that includes modeling, guided inquiry, and cooperative groups in order to have maximum benefit.

3.3 THINK-ALOUD INTERVIEWS OF STUDENTS SOLVING PHYSLET PROBLEMS

Students indeed find Physlet problems more challenging than traditional problems. But where in the problem-solving process do students run into difficulty? What prevents them from succeeding at solving a Physlet problem? Is it that students don't know how to begin the problem; or do they know the appropriate concepts and equations but not which data to collect; or can they not determine the appropriate concepts and equations to apply to the situation? To shed light on these questions, we looked in depth at students' thought processes while solving Physlet problems [Titus 1998].

A procedure fairly standard in the study of human thought processes as well as problem solving in physics was employed in this study, namely, verbal protocols. Despite issues of internal validity that must be properly addressed, verbal protocols are believed to provide the best insight into students' mental processes [Russo 1989]. There are three methods of verbal protocols: talk-aloud, think-aloud, and retrospective. A review of thinking-aloud data gathering procedures is provided by Shapiro [Shapiro 1994].

The verbal protocol used in this study was a think-aloud protocol. Five students were asked to solve three Physlet problems and to speak aloud their thoughts as they solved the problems. Transcripts of their verbalizations were later analyzed. Based on the analysis of those transcripts, the following observations were made.

3.3.1 Measurement and Data Taking

Making measurements before having a plan for solving the problem may be a way of redescribing the problem to facilitate the development of a solution, a quality that is encouraged in many recommended problem-solving strategies [Heller 1992, Larkin 1980]. However, some students may focus on data and how to plug data into the "right" equation without considering the problem conceptually.

Consider one student's verbalization of the projectile motion problem shown in Figure 15.

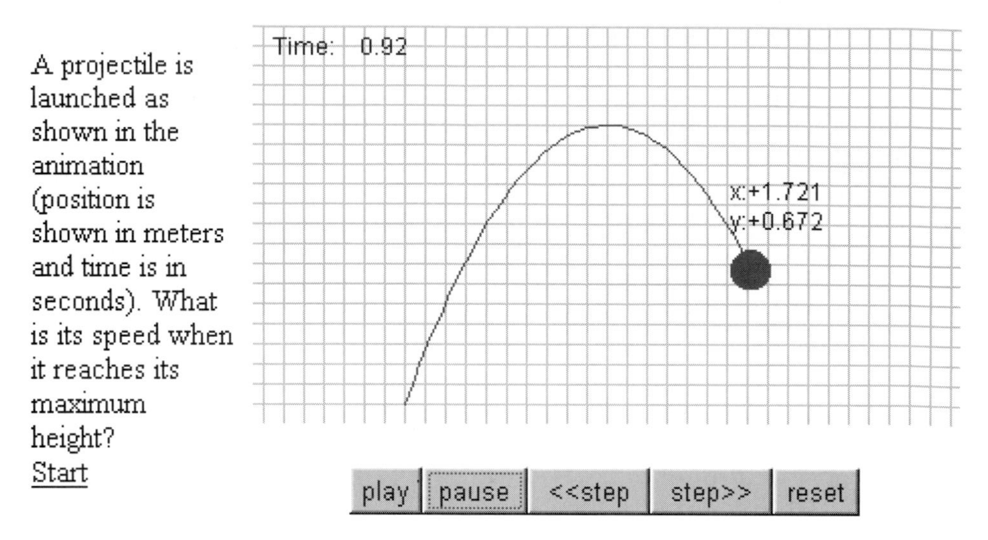
A projectile is launched as shown in the animation (position is shown in meters and time is in seconds). What is its speed when it reaches its maximum height?
Start
Time: 0.92
x:+1.721
y:+0.672
play
pause
<<step
step>>
reset

FIGURE 15: Think-aloud interview problem.

> Alright, I am looking for the speed as it reaches the maximum height as the projectile is launched. So we are just going to start it and watch the pretty ball bounce and step back to near the height to find out where the *y* component is at its greatest which seems to be at 1.4. So its *y*-velocity at that point is zero and the *x* velocity really hasn't changed. So you can actually just measure how long it goes for one second. Its *x* position is 1.908 meters so it moved 1.908 meters in one second so it would be, velocity would be 1.908 m/s.

Cueing off the phrase "maximum height" in the question, the student immediately steps the animation to where the projectile is at the maximum height and measures its *y*-position. The student realizes that the *y*-velocity of the projectile is zero and the *x*-velocity of the projectile is constant throughout its motion, and then proceeds to measure its *x*-velocity. Although the student initially measured the maximum height, the student returned to a conceptual approach and then recognized that the *x*-velocity was needed to solve the problem.

Measuring data before determining a clear method to solve the problem is not necessarily bad. As we saw in the preceding example, it may help the student formulate his or her ideas and eventually solve the problem correctly. However, this is not always the case. Another student first measured the position and time data at the beginning, at the end (when the projectile hit the ground), and at the maximum height. Subsequently, the student searched the textbook for an equation for which this data could be used to solve the problem. In the end, the student could not solve the problem.

In the first case, data collection was used as a way of discovering what principle was needed. Once the necessary principle was discovered, additional data were collected. In other words, the principle determined what data were truly necessary. In the second case, data were used as a guide to determine which equation to apply. Data were used to determine what principle was necessary. The first case is more "expertlike," and the second case is more "novicelike."

Viewing the animation and making measurements may cause students to alter loosely held conceptual beliefs. Students may state an important concept in solving a problem but not have the confidence to follow through with that concept. Making measurements or viewing the animation may result in confusion, where the student does not know which direction to take when solving a problem.

For instance, in solving the projectile motion problem, one student stated that

> there's no vertical component of the velocity. There is only the horizontal component which remains constant throughout the flight.

But later after viewing the animation, the student had difficulty analyzing the x- and y-motion independently, as evidenced by the following comment:

> It does have a velocity, so I'm not thinking that velocity is equal to 0 which sometimes happens when you throw it up and it comes down.

This student later solved for the horizontal velocity using, $v = v_0 - gt$ (obviously confusing x- and y-components), where v_0 was assumed to be zero (seemingly this was said just to make the calculation easier; no reason for setting it to zero was given), and t was the clock reading at the maximum height.

3.3.2 Conceptual Understanding

In some Physlet problems, conceptual understanding of "basic" principles besides those directly assessed by the problem is required. For instance, in Physlet problems that use *Animator*, such as those given in these interviews, a basic understanding of position, velocity, and acceleration is imperative to being able to use measurements to calculate important quantities.

Often more than one method can be used in solving a Physlet problem. The solution, however, always depends on the student's ability to measure position and calculate displacement, velocity, or acceleration. As a result, students should have a solid understanding of these concepts and how to calculate velocities and accelerations given position and time data. In all of the instances where students in our interviews obtained a correct answer, they showed a proper understanding of these concepts.

One student's solution to the projectile motion question is of particular interest. The student initially attempted to find the solution to the problem by measuring the maximum height and initial velocity of the projectile and using a kinematics equation. However, the student realized that one could simply find the velocity of the projectile at its maximum height by measuring the small x- and y-displacements around the maximum height, calculating the approximate instantaneous x- and y-velocities at the maximum height, and calculating the magnitude of its velocity using the Pythagorean theorem. Later the student realized that the y-velocity is zero at the top and thus only needs the x-velocity, and then showed that one obtained the same answer either way. Even though the student did not realize, or remember, this principle before taking data (which would have made the solution simpler), a solid understanding of velocity and how to measure it given its x- and y-components gave the same result.

In our experience using Physlet problems on homework, we have found that many students have difficulty understanding the difference between average velocity and instantaneous velocity and how to obtain an approximate instantaneous velocity by calculating an average velocity during a small time interval. For example, an object may be shown to accelerate in one dimension. To find the acceleration, students may measure the initial position and the position at some later clock reading, calculate the average velocity during that interval, and then calculate the acceleration by dividing the change in velocity by the time interval.

3.3.3 Incorrect Solutions

Like most physics problem-solving exercises, proper conceptual understanding is fundamental to solving a Physlet problem. As a result, if one has ideas about physical phenomena that are inconsistent with physical principles or if one does not understand the language of physics and how physicists use that language to describe physical occurrences, then one will have difficulty solving a Physlet problem. This was evident in most of the incorrect solutions submitted by students in our interviews.

When a student has a conceptual difficulty, he or she may make assumptions that are inconsistent with the problem and animation. Students often use novice approaches to problem solving, whether it is a traditional or a Physlet problem. Because all data are not given in the text of a Physlet problem, students may have difficulty obtaining the necessary data and determining the principle needed to solve a problem. To facilitate a solution using whatever equation he or she chose, a student might make assumptions that are unphysical and even inconsistent with what is shown in the animation. A good example is the solution cited earlier, where the student set the initial velocity to zero in the calculation even though it was inconsistent with what was shown.

The overlying theme of our observations of student solutions to Physlet problems is that conceptual understanding is key to solving Physlet problems. Without strong conceptual understanding, students in the interviews were prone to guess, search for the "right" equation, and lack direction. It seems, from these observations, that Physlet problems may be very useful for identifying whether students can apply conceptual understanding to solving numerical problems. Physlet problems generally cannot be correctly solved using "plug-and-chug" methods. The fact that data are not given in the text of the problem requires that students apply proper conceptual understanding to the solution before analyzing data. Therefore, it also seems that Physlet problems may be useful for encouraging a "concept-first" approach to solving problems, where students consider the concepts or principles to be applied to the problem before making calculations.

3.4 PHYSLET-BASED FCI

The studies discussed so far have focused on student solutions of numerical Physlet problems. However, Physlets can also be used to probe students' conceptual understanding. Physlets allow students to respond to what they see, rather than what they read, which should tap into their intuition. Physlets also eliminate the additional step of translating from words or graphs, which is difficult for many students.

We have written a Physlet version of the recently revised Force Concept Inventory (FCI) [Hestenes 1992, Mazur 1997]. An example Physlet question and its

The positions of two blocks at successive 0.20 second time intervals are represented by the numbered squares in the diagram below. The blocks are moving toward the right.

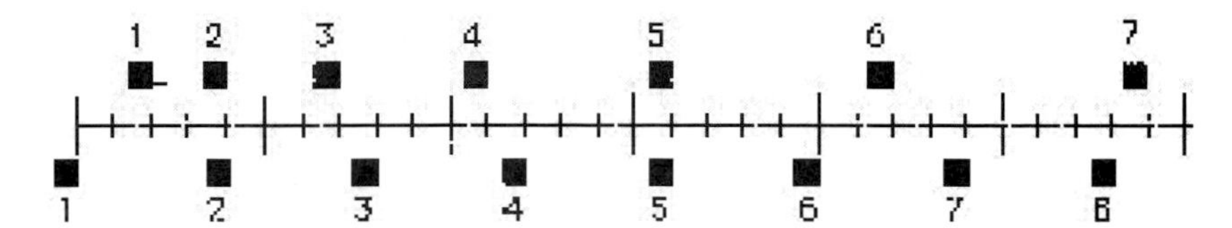
1 2 3 4 5 6 7
1 2 3 4 5 6 7 8

Do the blocks ever have the same speed?

- A) No
- B) Yes, at instant 2
- C) Yes, at instant 5
- D) Yes, at instant 2 and 5
- E) Yes, at some time during interval 3 to 4

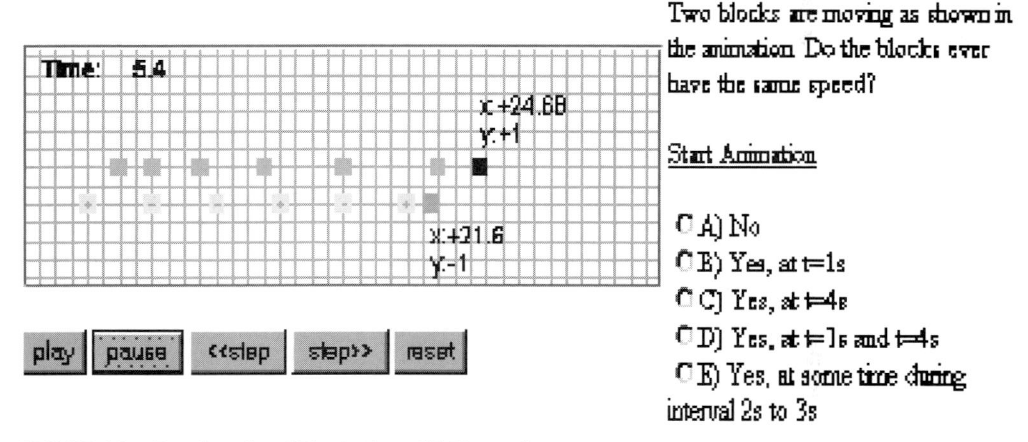
Time: 5.4
x:+24.68
y:+1
x:+21.6
y:-1
play
pause
<<step
step>>
reset
Two blocks are moving as shown in the animation. Do the blocks ever have the same speed?
Start Animation
A) No
B) Yes, at t=1s
C) Yes, at t=4s
D) Yes, at t=1s and t=4s
E) Yes, at some time during interval 2s to 3s

FIGURE 16: Text-based and Physlet-based FCI question.

corresponding FCI question is shown in Figure 16. The Physlet FCI was given to approximately 400 university and high school students as a pretest, thus eliminating the effect of specific instruction from the results [Dancy 2000]. Each student was randomly assigned to a group, which determined whether he or she saw the original or Physlet version of each FCI question. This gave us a control group (answered the question in its original form) and a treatment group (answered the Physlet version of the question) for each question on the FCI. We were then able to compare responses to the original and Physlet questions. Table 2 shows significant results from the Physlet-based FCI, taken as a pretest, when comparing responses given to animated and traditional versions of each question.

TABLE 2: Physlet FCI and text FCI comparison.

Question #	% Correct (N) Animation	Traditional	Z
1	57% (135)	71% (265)	2.70
7	80% (134)	64% (262)	3.22
14	76% (134)	51% (262)	4.86
19	34% (131)	51% (256)	3.14
20	24% (131)	47% (257)	4.40
26	21% (126)	5% (251)	5.01

$p < 0.01$

There were six questions for which the likelihood of a correct answer was statistically different at the 0.01 level on a two-tailed z-test. For three questions (7, 14, 26), the Physlet group performed better, and for the remaining three questions (1, 19, 20) the control group had a larger number of correct responses.

All six of these questions came from a group of 14 questions, which required the student to activate and view the animation over time. The remaining questions (none of which showed significant differences) could be answered without playing the animation since all necessary information was given in the written question statement or in the first frame of the animation. It should also be noted that all six of the significant questions asked about the motion (i.e., path, velocity, or acceleration) of an object. There are 12 FCI questions that deal with the forces on an object. None of these had significant differences between the traditional and animated groups. We also looked to see if the effect of adding animation to a question depended on whether or not the original question had a picture. There was no correlation.

Physlets should be an integral part of the question and not just a cosmetic addition. We found that the Physlet only had an effect on student responses when the question was designed so that students had to interact with the Physlet to answer the question and when the question concerned dynamical quantities. This result is consistent with the findings of Rieber [Rieber 1994], discussed earlier, that the animation will only be effective if external visualization is essential.

At first, this result may seem obvious and trivial, but there is a valuable lesson to be learned. Too often, multimedia has been added to instruction just because it was available. Our results indicate that a Physlet, used without a clear goal and purpose, will just be a more time-consuming and complex alternative to traditional methods. In order for a Physlet to be valuable for student learning and assessment, it must be designed around the needs of the student. It should contain information essential to the goal and not be just a flashy addition.

It is also important to note that, for the most part, adding animation to the FCI questions had no effect. When it did affect the answers students gave, it was just as likely to increase the likelihood of a correct response as to decrease it. More research is needed to determine why students were affected by the Physlet under certain circumstances. This issue is being taken up as part of Dancy's doctoral dissertation, which should be available by December 2000 [Dancy 2000].

Early results of think-aloud interviews with students answering FCI and Animated FCI questions indicates that the animated questions can reduce memorized responses (more reliance on intuition and less on memorized words) and can reduce difficulties students have in correctly interpreting the question statement. These interview findings are consistent with the result, from the large data sample, that the FCI is correlated with verbal ability while the Animated FCI is not significantly correlated. In other words, part of what the traditional FCI measures is students' ability to understand what they read. The Animated FCI reduces this dependency on verbal ability.

3.5 CONCLUSIONS FROM PER

Physlet problems are dynamic problems. Not only do they help students visualize a situation, they can encourage the student to solve a problem like a physicist solves a problem—to consider the problem conceptually, to decide what method is required and what data to collect, and finally to analyze the data. It is akin to an open-ended laboratory experiment, where students are not given instructions but merely a question. They must decide what data to collect and how to collect it most efficiently. This quality seems to make Physlets well suited for evaluating students' application of conceptual understanding to numerical problems and for helping students identify weaknesses in conceptual understanding (novice solution strategies leave the student in despair). This also makes Physlets challenging compared to traditional problems. Like other "nontraditional" problems, students must be equipped to solve such problems [Heller 1992]. Teaching and modeling a good problem-solving strategy is important, and using other teaching strategies, like modeling and cooperative grouping, is also recommended [Wells 1995]. We believe that Physlet problems are a good context for students to apply learned problem-solving strategies.

When developing a Physlet-based problem it is important to ascertain a clear purpose of the problem. Using a Physlet cosmetically merely to enhance visualization of a question is gratuitous. For maximum benefit, Physlet questions should require students to interact with the Physlet in order to answer the question. Students should be required to collect data, either numerically or visually, from the Physlet. If interaction is required, Physlets may influence how a student responds to a conceptual question. Therefore, in some cases, Physlets may be a more valid way to measure conceptual reasoning.

Based on our results, we do believe that Physlets can be a valuable tool for conceptual diagnostics as long as they are designed around the needs of the student. We have investigated using Physlets to alter existing conceptual questions. However, the greatest potential of Physlets will probably come from using Physlets to ask questions about things and in ways that cannot be done on paper.

CHAPTER 4

A TOUR OF PHYSLETS

This chapter gives a sampling of how Physlets can be used. Although the images give an impression as to how the animations evolve, you are encouraged to run them directly from the CD. Running the tour is, of course, a good check to ensure that the Physlet problems on the CD—and on the Davidson WebPhysics server—run on your computer before you start scripting. You may also want to try running the Physlet tour using different browsers since some browsers may have a speed advantage on various platforms.

Insert the CD and open the index.html file located in the root of the CD using a Java-enabled Web browser. Click on the hyperlink to the tour in the table of contents (TOC) in the left frame. The TOC will change to reflect this tour's contents and the first example will load into the right-hand frame. Later, you should check to see if the Physlet tour also loads from the Davidson College WebPhysics server if you have Internet access. The Web address is http://webphysics.davidson.edu/physletprob.

4.1 EXAMPLES

Example 1: *Animator* is probably our most versatile Physlet. Both geometric objects, such as circles and rectangles, and images can be scripted either to follow analytic trajectories or to obey Newton's laws of motion. In this example, written by Aaron Titus, a Ferris wheel is simulated by a collection of colored squares moving along a circular path with constant angular velocity as shown in Figure 17. Each square represents a chair on the Ferris wheel. A typical question might ask students to compare the net force on the rider at point (A) and at point (B). Although the answer, $F = mv^2/r$, is easy, students will

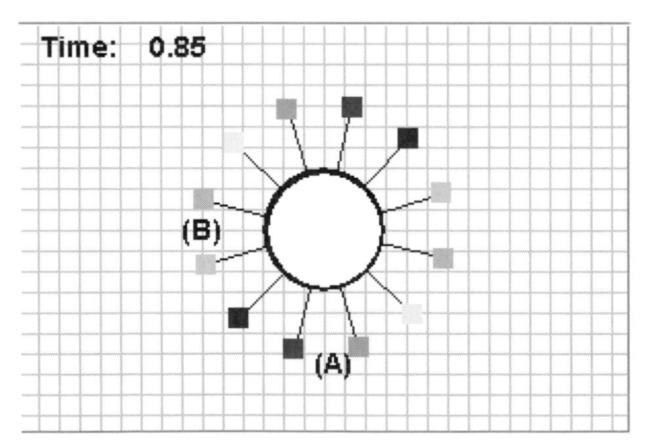

FIGURE 17: A Ferris wheel.

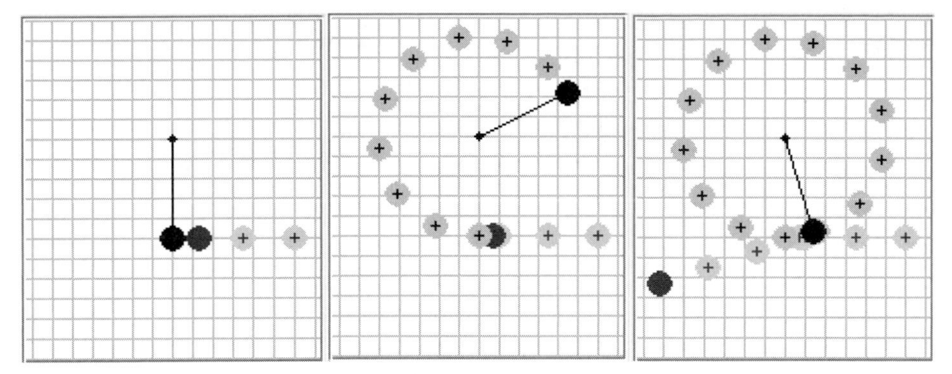

FIGURE 18: Angular momentum of a point mass.

often include frame-dependent forces since they have personal experiences riding Ferris wheels or get confused by the varying cause of the centripetal force.

Example 2: Figure 18 shows how *Animator* can be used to simulate a collision between a projectile and a tethered mass. The collision is dynamic. That is, the two particles interact via internal interactions rather than according to an analytic trajectory. The collision is constructed so that the interaction between the masses is perpendicular to the lever arm of the constrained mass, thereby conserving angular momentum about the pivot. This simulation can be used as an effective in-class Peer Instruction or an out-of-class Just-in-Time Teaching (JiTT) problem to determine if students understand that free particles carry angular momentum.

Example 3: Animation be used to model both physical and nonphysical behavior. This Physlet problem was written by Peter Sheldon to test students' understanding of fluids and buoyant forces. Figure 19 shows a wooden block floating between water and oil; the oil is siphoned off. Only one of the three situations shown is physical.

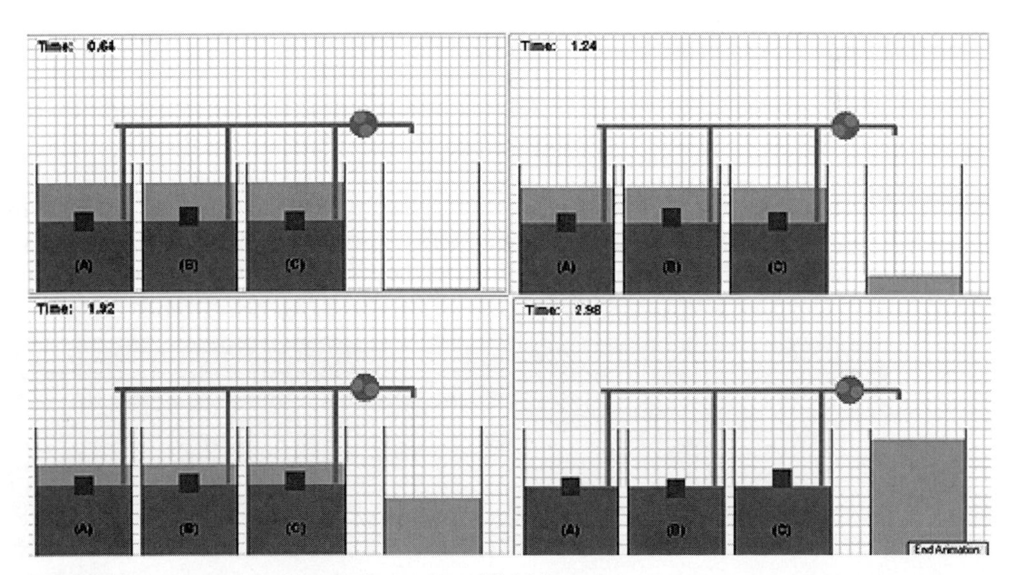

FIGURE 19: *Animator* problem in the context of fluid dynamics.

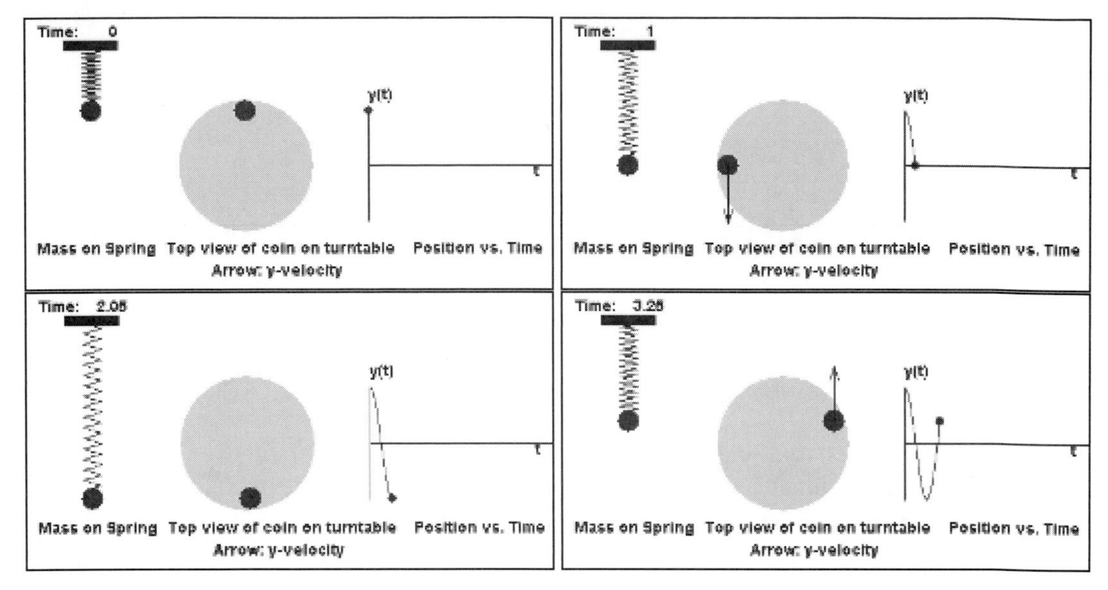

FIGURE 20: The equivalence of oscillation and rotation.

Example 4: *Animator* scripts can become very elaborate. We often use Physlets for lecture demonstrations. As explained in most textbooks, Figure 20 shows that both the oscillations of a mass on a spring and the vertical position of an object undergoing uniform circular motion can be described sinusoidally. These types of motion are in fact isometric. The features of the *Animator* applet are used to make this clearer to students than the static images that appear in textbooks.

Example 5: Figure 21 models the conversion from kinetic energy to internal energy in a system. A heavy ball with an initial kinetic energy of 4000 J is trapped inside a box with rigid walls containing a cylinder constructed of small, lightweight spheres. The vertical bar on the right of each frame shows the kinetic energy of the ball. To the right are 80 small objects representing a solid that will be crushed and vaporized through impact with the disk. The simulation uses the Molecular Dynamics package.

Example 6: Students often have difficulty with functions of two variables, such as the wave equation. Figure 22 demonstrates how two waves add using the superposition principle. The interactive version on the CD allows the user to enter an analytic

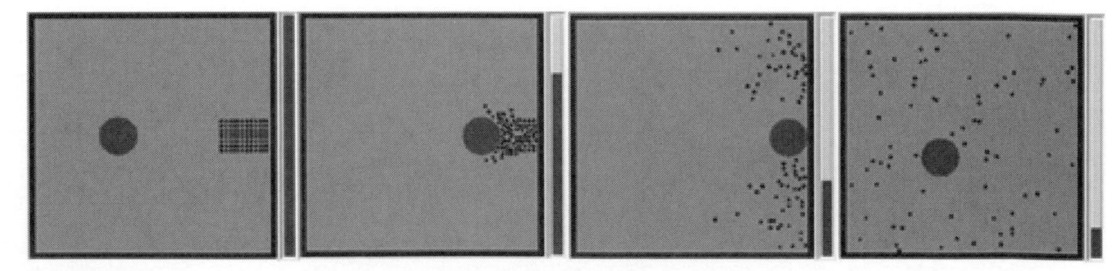

FIGURE 21: Conversion from kinetic energy to thermal energy.

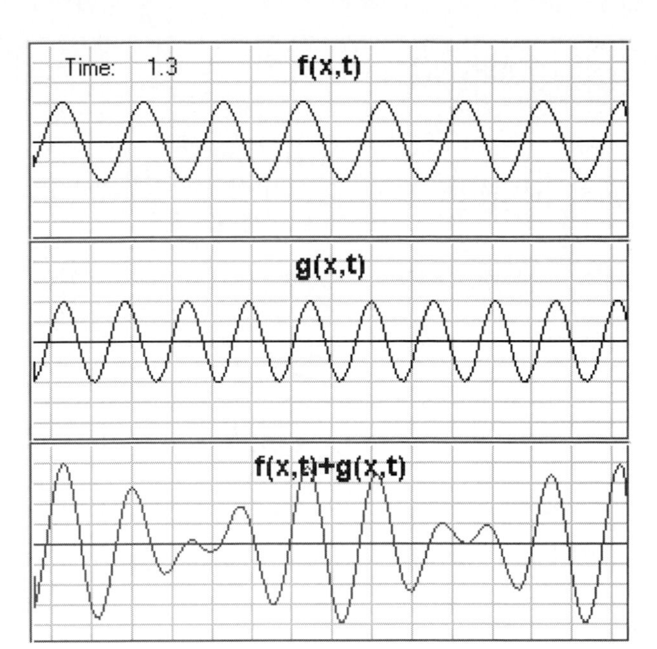

FIGURE 22: Superposition of two traveling waves.

expression, $f(x, t)$, for the second wave function. This simulation is used at Davidson College in conjunction with a laboratory exercise on waves and sound.

Example 7: *EField* was designed to model the Coulomb interaction between particles. Figure 23 shows how this Physlet is used to reinforce the operational definition of charge ("opposites attract, likes repel"). Figure 23(a) shows charges, with arrows representing the total force acting on each charge. Figure 23(b) shows the potential energy of each charge. As you move the charges, these visual clues change appropriately. The accompanying question asks you to determine the sign of each

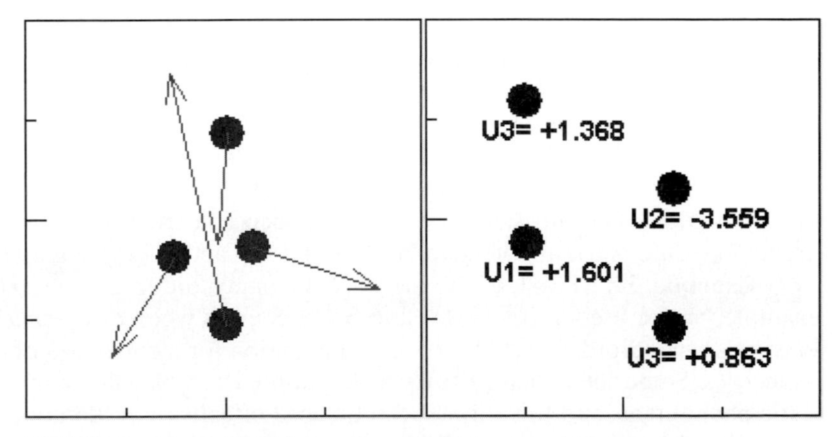

FIGURE 23: The Coulomb interaction with (a) forces and (b) potentials.

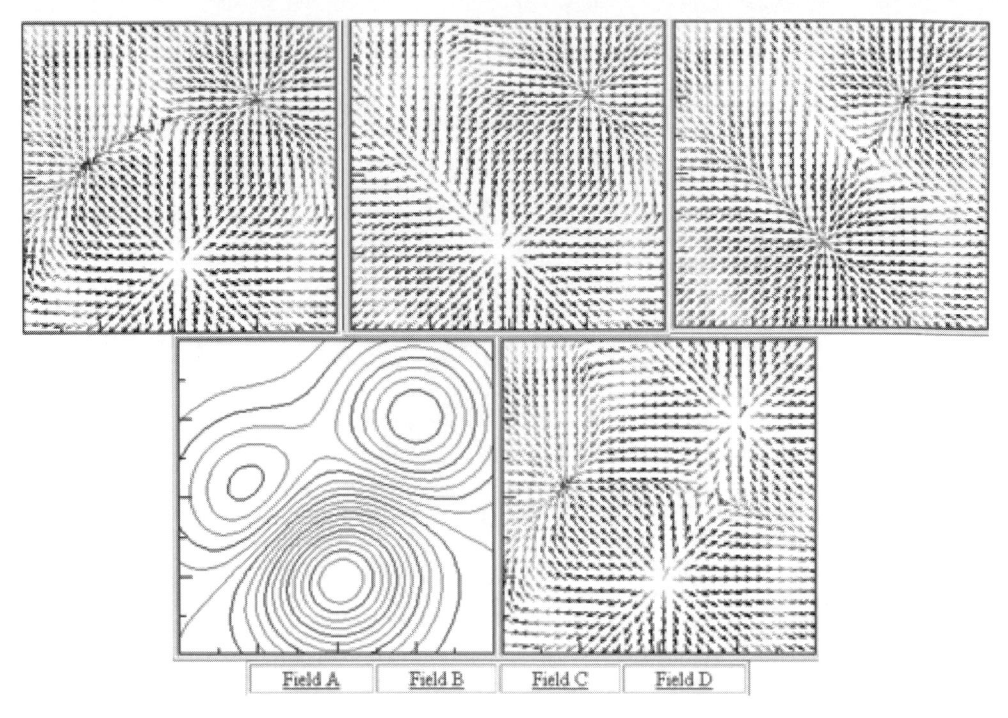

FIGURE 24: *EField* can be used to plot both electric fields and equipotential curves.

charge. This task is more complex than it first seems. Students are comparing each charge with a reference charge of their choice. One strategy is to put one charge in each of the four corners of the screen and then bring one to the middle next to be the test charge. Other strategies can be deployed. This exercise allows you to show some creativity in solving a physics problem [Bonham 1999].

Example 8: *EField* is used in Figure 24 and generates equipotential lines for an electric field. There are also four electric fields described by color-coded unit vectors. Students are expected to determine which field matches that described by the equipotential lines in the illustration, using the mouse to measure potential.

Example 9: *BField* uses vector addition to display the magnetic field from a collection of wires or coils. Figure 25 demonstrates how it can be used to reinforce the importance of symmetry when applying Ampere's law. The animation on the CD is not static. The field plot in each configuration evolves as wires are slowly added. Stopping the calculation after only half a cylinder has been constructed and asking students to predict the outcome is an effective Peer Instruction activity [Mazur 1997].

Example 10: Physlets can be used in many different contexts. *Poisson*, for example, can be used in upper-division E&M courses when discussing the solution of Poisson's second-order partial differential equation for a collection of conductors and dielectrics. Some introductory texts have adopted a molecular approach to electrostatic phenomena, and this Physlet can be used effectively to demonstrate how fields are affected by a material medium. You can even show the appearance of bound charge at the surface of the dielectric if this simulation is run from the CD.

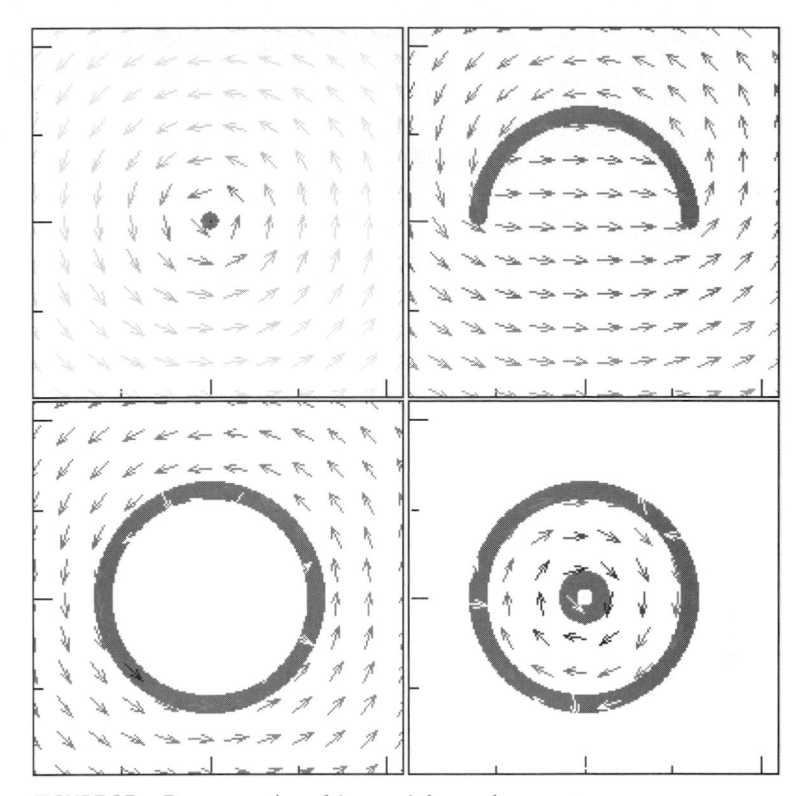

FIGURE 25: Demonstration of Ampere's law and symmetry.

Example 11: The *Optics* Physlet models mirrors, lenses, and changes in index of refraction for various types of light sources. Figure 27 shows how light is trapped inside a fiber optic cable. The light beam can be dragged into and out of the medium and its angle adjusted by click-dragging the white hot spots.

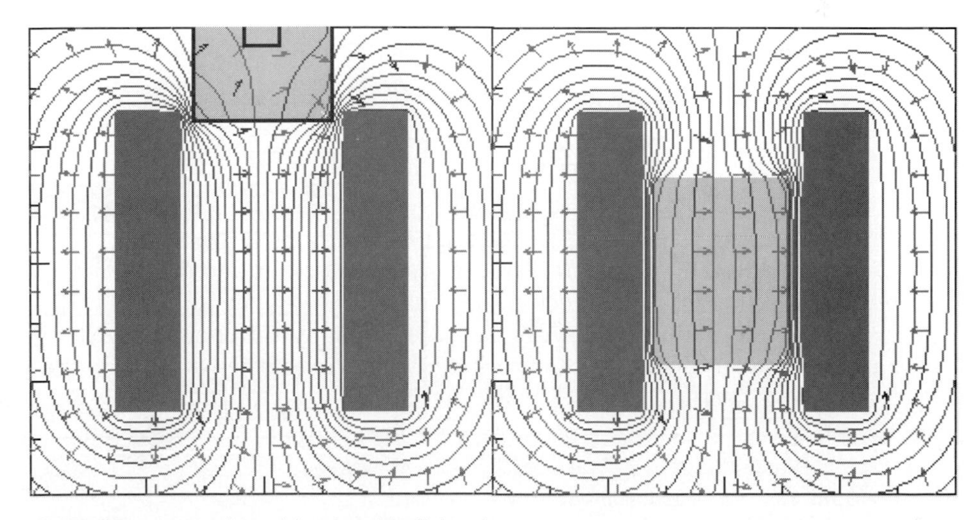

FIGURE 26: A capacitor with a dragable dielectric.

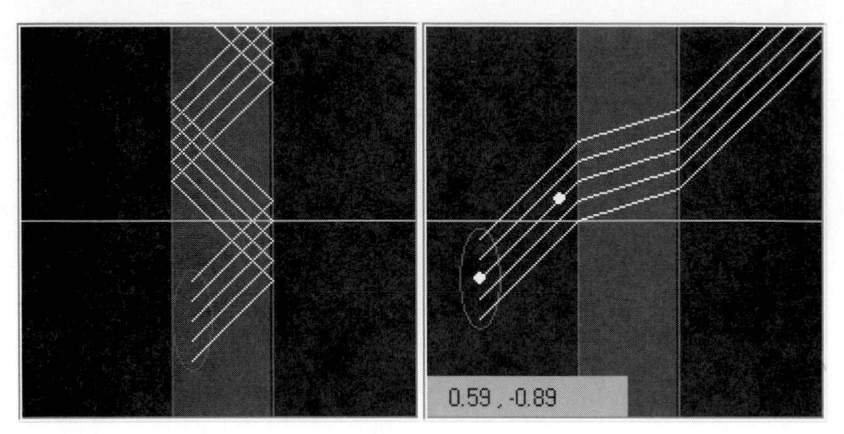

FIGURE 27: Refraction of light.

Example 12: Curriculum authors usually want to process data and to present the data in various formats. Over the past year, the functionality of Physlets has been greatly extended using interapplet communication. This makes it possible to use a modular object-oriented approach for the design of interactive curricular material. Many Physlets, including *Animator*, are now capable of generating data in response to

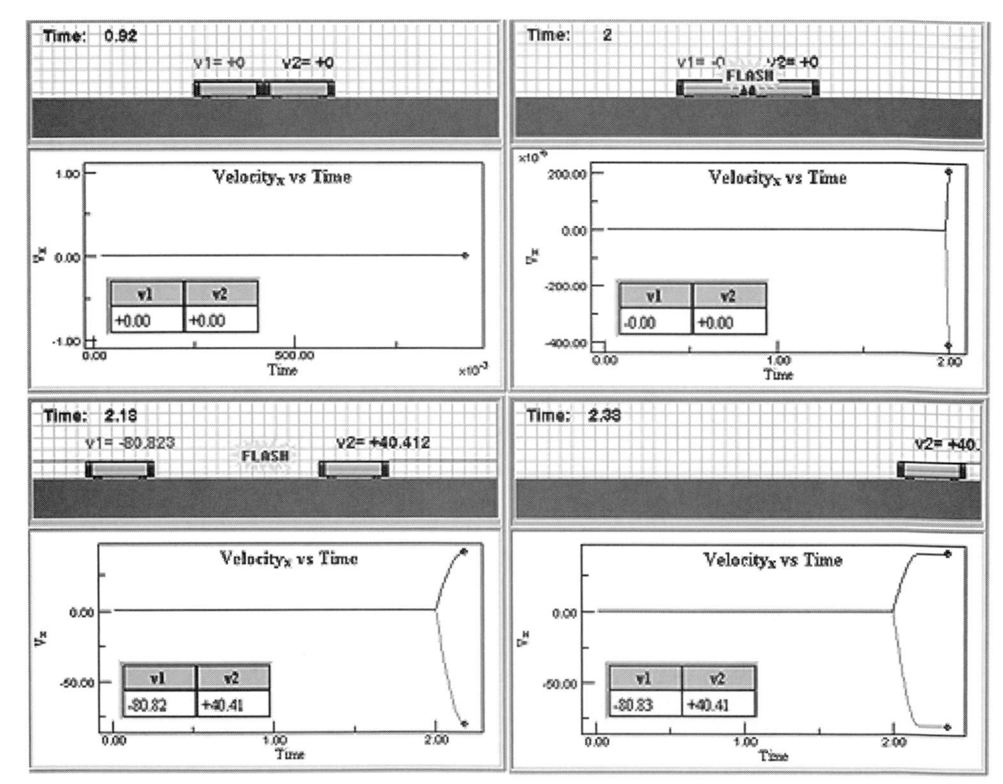

FIGURE 28: An inelastic collision between two carts.

an internal clock or in response to user actions. These data can then be passed to a bar graph, a table of numeric values, or an x-y graph using one line of JavaScript to establish the communication link. This technique is very flexible since the code to process and present the data is written in an interpreted runtime environment. Figure 28 combines *Animator*, *DataGraph*, and *DataTable* to present data in three separate formats. An explosion between two carts sends them moving in opposite directions. The velocity of each cart is detailed in the animation, graph, and table simultaneously to facilitate student understanding.

4.2 INSTALLING PHYSLETS

Although it is possible to access Physlet problems from the CD or from the Davidson College WebPhysics server, teachers will almost certainly want to copy the necessary files to a local hard drive or a local Web server. Not only will this provide faster and more reliable access, but it is essential to authoring your own problems and to modifying existing Physlet-based material.

4.2.1 Addressing

There are two ways to specify the location of a file on a computer disk or on a Web site: absolute and relative addressing. Absolute addressing specifies the location of a file starting from a single directory, known as the root, and progresses down into subdirectories from that location. A file that is specified using this type of addressing starts its name with a slash.[1] For example, we can embed an html anchor—also known as a hyperlink—that loads the tour_1.html file on the CD by specifying its location as

```
<a href = "\physletprob\ch4_tour\tour_1.html">Go to tour 1.</a>
```

You should be able to locate the tour_1.html file on your CD by navigating into the physletprob and then into the ch4_tour subdirectories on the CD file system.

Relative addressing, on the other hand, specifies the location of a file starting from the current location. Current location is a somewhat arbitrary concept since it is application dependent. Browsers treat the directory containing the html file as the current location. Consequently, the tour_1.html page can contain an anchor that loads the tour_2.html page using the following abbreviated syntax:

```
<a href = "tour_2.html">Go to tour 2.</a>.
```

Subdirectories within the current location are accessed by pre-pending the subdirectory name to the file name. If the example.html file were in a subdirectory called examples, we could access it as

```
<a href = "examples/example.html">Go to more examples.</a>
```

[1]The MS Windows operating system uses a backslash character instead of a slash when specifying a file name path. Fortunately, Internet Explorer recognizes the slash when files are referenced within an html page.

Relative addressing can also use *dot* and *dot-dot* to refer to files in the current and parent directories, respectively. For example, the notation codebase = "../examples/" instructs the browser to move up one level from the location of the html file in the directory tree before looking for a directory called examples. If a single *dot* codebase = "./classes/" had been used, the browser would have assumed that the classes subdirectory was located in the same directory as the current html document.

You will notice in the next section that we use relative addressing in order to refer to Java archives on the CD. These archives are located in a subdirectory named classes within the applets directory located at the root of the CD. Although it would have been simpler to have referred to the codebase directory for all Physlet examples as "/applets/classes," a bug in one of the major browsers prevents us from doing so. When Java files are accessed on the CD, we refer to them with relative addressing. There you will see syntax similar to the following as you copy and edit our examples.

```
codebase = "../../../applets/classes/"
```

This notation specifies that java files necessary to run an applet are to be found by moving up three levels from of the current directory and then down two subdirectories.[2]

4.2.2 Code, Codebase, and Archive

Create an empty directory on a local hard drive and, for the sake of consistency, name this subdirectory *applets*. This will be your working directory. Create a subdirectory called *classes* in the working directory. Starting at the root of the CD, navigate into the *applets* subdirectory on the CD. This directory contains the *classes* subdirectory. Copy two files, Animator4_.jar and STools4.jar, from the classes subdirectory on the CD into the classes subdirectory of the working directory as shown in Figure 29. The *Animator* Physlet is now installed on your local hard drive. The next step will be to copy an existing html file to the local drive and modify it so that the browser can locate this applet.

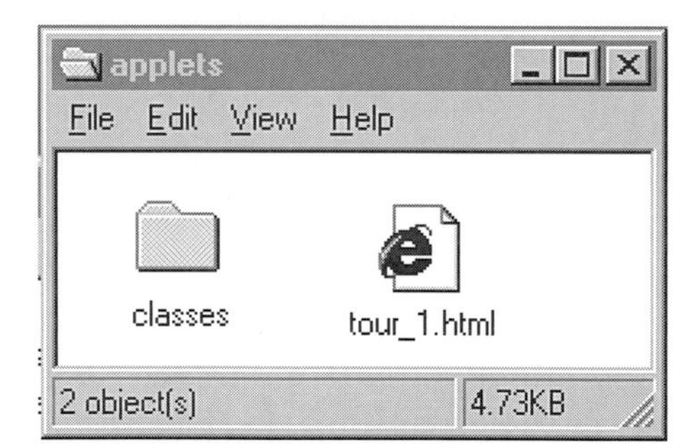

FIGURE 29: A working directory containing an html page and codebase directory.

[2]It is possible to specify files using a complete URL pointing to an entirely different Web server from the one containing the html document. However, due to bandwidth limitations and sluggish Internet response time, it is generally advisable to download and mirror the necessary files on a local hard disk or file server.

Applets are embedded into html pages using the applet tag (See Script 1.). This tag specifies attributes that the browser needs in order to find and load the Java class files. In particular, it has the following format for Ferris Wheel Physlet problems on the tour:

```
<applet
codebase = "../../applets/classes"
archive = "Animator4_.jar,STools4.jar"
code = "animator.Animator4.class"
name = "animator"
width = 300
height = 250
vspace = 0
hspace = 0
>
</applet>
```

Script 1: The applet tag and its attributes.

Physlets are usually packaged and distributed in one or more jar files. These jar files are specified in the archive attribute as shown previously. The version of the *Animator* Physlet that was just copied, for example, requires both the Animator4_.jar and STools4.jar files.

```
archive = "Animator4_.jar,STools4.jar"
```

Notice that when multiple jar files are required, they are separated by commas.

The *codebase* tag specifies where the browser begins to search for the classes and jar files necessary to run the applet.[3] On the CD, the codebase directory is the classes subdirectory located within the applets subdirectory. The codebase attribute is the attribute most likely to require editing when an html file is copied from the CD or from the Internet onto a local hard drive.

Open the CD in a file viewer—not the html browser—and navigate from the CD root into the book subdirectory for chapter 4, called ch4_tour. Copy the file tour_1.html from the tour subdirectory on the CD to the working directory on your drive that you created earlier. Navigate to this working directory and open the tour_1.html file in a text editor. The opened file may look a little strange if you are unfamiliar with html tags. Don't worry about the details just yet. Scan the opened file in order to locate the applet tag and the codebase attribute.

```
codebase = "../../applets/classes".
```

[3]Browsers will begin their search in the directory containing the html file if the codebase attribute is not present.

Replace the reference to the codebase directory with the name of the subdirectory containing the jar files.

```
codebase = "classes"
```

This change tells the browser to move down one level from the working directory, the directory containing the html page, into the classes directory. Since the classes directory is in the working directory, the browser will load the necessary jar files from this subdirectory even if the working directory is renamed or copied to another disk drive or to a Web server. You can now open the edited html file in a Web browser and check to see that the Ferris Wheel example runs.

Finally, there may be other resources, such as images or style sheets, that are referenced within the html page. Either these resources must be copied or their references should be removed. Most documents on the Physlets CD contain a reference to an html style sheet that is located in the root of the CD. Look for the following tag in the head section of the html page:

```
<link rel = "STYLESHEET" type = "text/css"
href = "../../styles_demo.css">.
```

Although you can copy the referenced style sheet to your working directory and adjust the reference path, the simplest solution is to remove the style sheet reference from the document. Although style sheets are a good way to produce a consistent look and feel across hundreds of Web pages, you should be aware that there is a bug in one of the browsers that may keep your html page from loading if a requested style sheet is unavailable.

In summary, three steps are necessary to install Physlet problems onto a local drive. First, the necessary html files and jar files must be copied to a local drive. Second, the codebase attribute for embedded applets must be edited in the html page to properly reference the directory where the jar file is located. Third, check for other resources that are referenced within the html page and either copy these references from the CD or remove them from the page if they are optional.

4.2.3 Other Attributes

Five other attributes are used with the applet tag in Script 1.[4]

- **name** is an optional string attribute. It should be present in order for JavaScript to refer to the applet. Although it can lead to confusion, it is common practice to use the name of the applet as the name attribute. Name attributes must, of course, be different whenever multiple applets are embedded in a page.

---

[4]The name and attributes in a tag are not case sensitive in today's version of html. In other words, WIDTH, width, and Width will work the same as will APPLET and applet. Values for particular attributes may be case sensitive, particularly filenames and paths such as code and codebase values. However, we strongly urge authors to adopt the xhtml standard. This standard requires that all element and attribute names be in lowercase. Attribute values can be mixed case but must be quoted.

- **width** and **height** are required integer attributes. They define the applet's size on the embedding page. These values are sometimes set to zero for applets that perform calculations but do not have an on-screen representation.
- **hspace** and **vspace** are optional integer attributes that specify the horizontal and vertical padding around an applet. These attributes are useful if you embed a small applet directly into the text.

Parameter tags, <param *value* = "*name*">, are often nested within the beginning and ending applet tag. Since parameters are applet specific, they will be discussed later. None of the parameters need to be changed when an applet is copied from the CD or the Internet onto a local disk.

4.2.4 Troubleshooting

If an applet does not load properly, you will most likely encounter a gray box in the html page. The first place to begin debugging is in the applet tag. The codebase attribute is the most common source of error.

- Remember that Java is case sensitive. Class names typically use mixed case, and directories are lowercase. Although many developers use lowercase, we often use mixed case for Physlet jar files. We have noticed that the case of a character can sometimes change when files are copied between computers with different operating systems.
- The code attribute in the applet tag usually consists of the package name followed by the class name, followed by the file extension, class. For example,

  ```
  code = “animator4.Animator.class".
  ```

  tells us that the file Animator.class is located within the animator4 package. You can use a decompression utility, such as WinZip, to open the jar file and observe that the jar file actually contains a file named Animator.class in a subdirectory called animator4.

- The codebase attribute is usually specified relative to the html directory containing the applet. Dot and dot-dot notation are often used to designate the current directory and the parent directory, respectively, relative to the html directory. It is possible, but uncommon, to use a complete URL as the codebase.
- Attributes; are specified as strings. Make sure that the values of the code, codebase, archive, and name tags are enclosed in quotes.[5]
- Finally, the Java Virtual Machine may be disabled—or possibly not even installed—as a matter of school or company policy. It is even possible to enable or disable Java applets that are downloaded from particular servers. Dealing with paranoid administrators and network security is, of course, beyond the scope of this—or any other—book.

[5]This is not strictly true, but consistency is a good idea. If a value is a word or a number that contains only letters or numerals, then it is OK to place it directly after the equals sign. If a value contains several words separated by commas, such as our archive attributes, then it must be enclosed in quotation marks.

4.2.5 Updates

Physlet updates are available on the Davidson College WebPhysics server: http://webphysics.davidson.edu/applets. Look for the download button on the right-hand side. A number of files are usually available for each Physlet.

- *A complete jar file that can be used to embed a Physlet into an html document.* The Physlet should run without further modification, but you must write your own HTML document. Use the following archive attribute if you want to use this file:

```
archive = "Animator4.jar"
```

- *A Physlet-only jar file without the tools package.* This file will not run by itself. You must also download the appropriate tools archive, usually STools4.jar, and include this file in the archive tag.

```
archive = "Animator4_.jar,STools4.jar"
```

- *A template archive.* Templates contain an html page and all necessary jar files in a compressed zip archive. Download the template, unzip the contents into a hard drive, and then open the file in a browser that supports both Java 1.1 and JavaScript to Java communication. The html file should load and run the applet.

Please note that it is possible that the jar files included in a template archive are not the most recent. After you have downloaded the template and determined that it runs in your browser, you should download and replace any jar files contained in the template package. The jar files obtained using the first two options come directly from our production web site and are always the most recent.

CHAPTER 5

TECHNOLOGY OVERVIEW

Java: A simple, object-oriented, distributed, interpreted, robust, secure, architecture neutral, portable, high-performance, multi-threaded, and dynamic language.

Sun Java Language White Paper

Web technology is still evolving, and it is currently not possible to run every Physlet on every computer platform using any browser. The promise of "write once, run everywhere" is unfortunately still a promise. However, there are powerful commercial reasons for adhering to a core set of Internet standards, and it is very likely that these standards will prevail. The Sun-Microsoft court battle over the Java trademark notwithstanding, Microsoft is supporting Java and has adopted html as the new Windows Help file format. Microsoft is producing a scriptable Java product to allow any Java-enabled application to connect to any standard Open Database Connectivity (ODBC) database. (This idea is very similar to Physlets: namely, write a general-purpose Java applet for Internet commerce, and script the applet to provide functionality needed for a particular application.) The European Computer Manufacturers Association (ECMA) has adopted a specification for JavaScript, officially called ECMAScript, which browser vendors will be required to adhere to if they are to sell in Europe. Numerous software packages are available that simplify high-level authoring using JavaScript. These packages also check syntax and browser compatibility. Hewlett Packard, Novell, and IBM are making substantial investments in Java and are putting pressure on Sun to keep the standards process open. Finally, an enterprising group of developers with the Free Software Foundation is reverse engineering Java to provide a freely available public domain version of the product complete with source code. Anyone who has ever written and distributed curricular material using proprietary authoring packages will appreciate the benefits to the education community of supporting and adhering to standards. Java and JavaScript are likely to be an excellent option for producing Web-based curricular material for the foreseeable future.

It is not necessary to become a Web expert in order to use Java applets in html documents. Many public domain Java applets are available on the Web, and the process of downloading and embedding these applets is only slightly more difficult than embedding an image. However, just as there are various image file formats, there are various versions of Java and various ways of packaging the applet. Computers are notoriously unforgiving of almost any syntactic error, so it is important to have a bird's eye view of Java technology in order to understand what can go wrong. You might want to skip ahead to the scripting tutorial, Chapter 6, and refer back to the remainder of this section when you want to learn more about Java.

5.1 A HISTORY OF JAVA

Java has evolved rapidly since its introduction by Sun Microsystems in late 1995, and it has been difficult for software vendors to update their products to keep pace with new language features. Netscape, Microsoft, and other vendors quickly adopted the first version of Java, version 1.0, and applets written for this version will almost certainly work with any application that supports Java. This includes Netscape Navigator, Internet Explorer, and authoring packages such as Microsoft Front Page, Macromedia Dreamweaver, and Adobe Page Mill. Unfortunately, Java 1.0 had serious limitations as a programming language and needed fundamental changes and extensions in order for the language to reach its full potential. Although Java development environments from Borland, Symantec, Sun, and Microsoft quickly adopted the new specification, Internet browsers are only now beginning to support the new version. Running a Java 1.1 applet in a Java 1.0–aware application is likely to produce a gray box and a rather cryptic "Class not found" error message. Furthermore, Sun Microsystems and Microsoft have extended the language in different ways to provide access to native graphical user interface (GUI) components such as dialog boxes, toolbars, and menus. It is unlikely that the promise of platform independence is attainable in the foreseeable future if developers use some of the extensions available from competing vendors. However, the core language, including many version 1.1 features, is very stable and many educational software developers have chosen to stick with the original, albeit more limited, GUI components available since Java 1.0. This seems like a small price to pay, and all new Physlets have been written in Java 1.1 to provide maximum vendor and platform independence.

Although both Java 1.1 and JavaScript are currently implemented on all major platforms, the ability of JavaScript to communicate with a Java applet is still problematic on Apple computers.[1] Physlets will run in interactive mode, that is, without script, using the latest Apple Java Virtual Machine. However, Microsoft Internet Explorer 5.0 for the Mac still does not support JavaScript to Java communication. We believe that this problem will be solved shortly. Both Netscape and iCab, a new browser being developed in Europe, have announced plans to support the necessary features in their next release. In fact, older Physlets written can be scripted on the Mac using Netscape Navigator version 4.5. Unfortunately, this browser has a rather ancient Java 1.0 Virtual Machine (VM) that will not support our recent Physlets.[2]

---

[1]Support on MS Windows machines using older browsers may also be problematic. Netscape Communicator version 4.05 and below, for example, do not support Java 1.1 and will not run most Physlets.

[2]In the interim, it is possible to run Physlets on all operating systems using the Sun Microsystems HotJava browser. This browser is actually a Java application that runs inside the local Java VM. In other words, the Java VM is an operating system within your operating system. In order to use HotJava, you must first install a stand-alone Java VM onto your computer. Then download HotJava and start this application. You will see a browser interface that is very similar to a browser that was written specifically for the operating system that you are using. There is almost no performance penalty using this approach since Physlets are written in Java and the HotJava browser is an optimized Java application.

5.2 JAVA LANGUAGE

It is helpful to understand the process of writing Java code in order to understand the file structure that authors using JavaScript will encounter when using Java applets. Preparing Java code that will eventually run inside a Java aware application is a multistep process. First write the code! Java syntax is similar to C in some respects, but its philosophy is much closer to Object Pascal or Smalltalk. The programmer begins by creating a file that has the applet name as the filename and has the extension "java." This main file often contains the code necessary to generate the user interface. It also contains code for various public methods that can be accessed from outside the applet using a scripting language such as JavaScript. (Methods appear to behave like subroutines or functions in procedure oriented languages such as Basic or FORTRAN, but they are really quite different. See, for example, *Thinking in Java* by Bruce Eckel [Eckel 1998].) Almost any significant software project will require additional source code files. Applets that need additional Java files can access them by specifying their name next to the "import" keyword near the beginning of the file. This process is very similar to accessing a library routine in other languages. A number of files can be grouped together by adding the "package" keyword followed by a name as the first line of a Java program. These files are usually stored in a subdirectory having the same name as the package. Blank spaces and other non-alphanumeric characters in filenames and subdirectory names will almost certainly cause problems. The compiled class files that are necessary for an applet to run will reflect these organizational principles.

5.3 CLASS FILES

After the code is written, it is compiled into an intermediate state called a class file. These class files contain the byte code for the Java virtual machine and have the "class" file extension. A typical programming project can produce dozens of class files. Since most Java programmers organize their projects into packages, there will often be a number of subdirectories each containing one or more class files.

Although applets can be distributed as a collection of individual class files organized into subdirectories, there is a better way. Browsers are now capable of accessing and downloading a single compressed or "zip-ed" file that contains everything necessary to run an applet. Java 1.1 takes this concept one step further and specifies a new file format called a Java Archive (jar) file. Jar files are zip files on steroids. In addition to an applet's class files, a Java archive contains a manifest listing the archive's contents and encrypted security information such as the applet vendor. Figure 30, for cxample, shows a classes directory that has been set up to run *Animator*, *Doppler*, and *EField* Physlets as well as an older applet named balldrop. The balldrop applet's class files are in the balldrop subdirectory. *Doppler*, *Animator* and *EField* class files are packaged in their respective jar files. The STools4.jar file contains a common library of files that can be shared among all Physlets.

When using an applet, you can copy the entire file structure containing an applet's class files and all the associated subdirectories to a new location on a hard

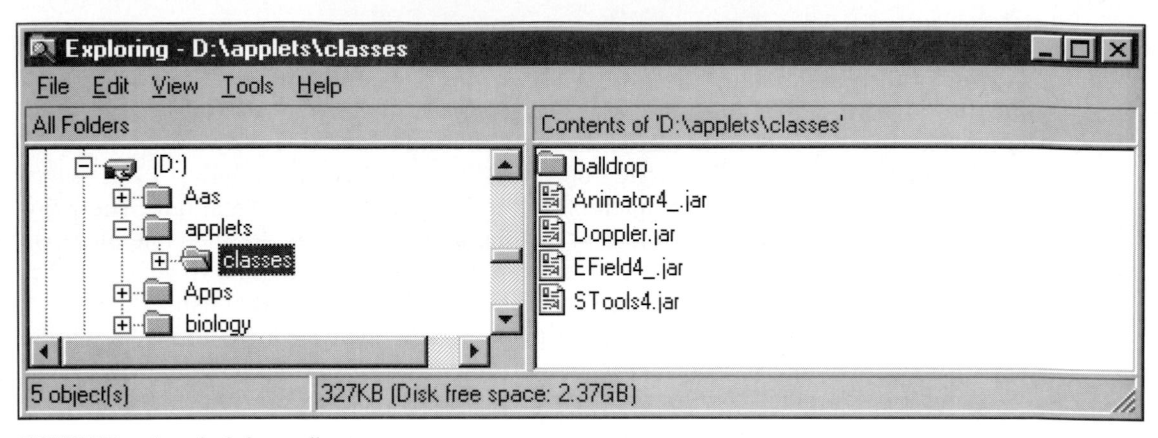

FIGURE 30: A typical classes directory.

disk, but you must not change the names of any files or subdirectories. Since Java has its roots in UNIX, even capitalization is important!

A "Class not found" error message is usually the result of a jar file not being in the correct directory; however, it is also useful to open a jar file with a standard file decompression utility, such as WinZip, to see if the requested file is actually available. Class files are, of course, also distributed as part of a browser package, and it is possible that a "Class not found" error is due to the fact that a browser vendor does not support the necessary functionality.

An applet begins its life cycle inside an html-aware application—typically a Web browser—after the class files have been downloaded. Since a Java machine does not really exist in most desktop computers, Java byte code is interpreted one instruction at a time and converted into one or more native machine language instructions. These instructions are specific to the Pentium, PowerPC, or whatever microprocessor is running the applet. This process is slow; it is the penalty to be paid for processor independence. Early Java Virtual Machines (Java VMs) performed this translation for every line of code as it was being executed. Fortunately, browser vendors have now developed compilers that translate the entire class file into native machine code after downloading. These just-in-time compilers have the potential to make Java almost as fast as C++ code. The shipping versions of Netscape Navigator and Microsoft Internet Explorer include such compilers.

5.4 EMBEDDING

Insertion tags in an html document specify the type and location of multimedia content. For example, the image tag, <img>, can be used to insert a 300 by 250 pixel image of an apparatus into a document using the following syntax:

```
<img src = "apparatus.gif" width = "300" height = "250">.
```

Notice that the image tag starts with a < and ends with a >. It contains attributes that define the source, width, and height of the image. The ending > is important and is easily overlooked, particularly if there are attributes that span multiple lines, as in the case of the applet tag discussed next.

Applets are embedded into an html document using just the applet tag, <applet>. This embedding is similar to adding a graphic to an html page with an image tag. An applet's class files are downloaded into the browser along with other multimedia objects, such as sounds files or GIF images, which are referenced within the containing html document. It is then up to the browser to lay out the page on the screen and to translate the machine-independent class files into native binary code. It is the browser's job to provide the hooks, such as memory management, into the operating system. Many applets can be used effectively by merely embedding them in an html page and asking students to describe the relevant physics. For example, the *Doppler* Physlet shown in Figure 31 can be embedded within a Web page using the following tags:

```
<applet
code = "doppler.Doppler.class"
archive = "Doppler.jar"
codebase = "classes"
width = 320
height = 370 >
</applet>
```

The browser displays the applet, and the user interacts with the applet using the applet's intrinsic controls. Doppler has a slider to set the velocity and an optional radio button to enable relativistic effects. In addition, the user can click-drag the mouse to make position measurements on the wave fronts. The input burden is on the user. The html author assumes the user is knowledgeable in the operation of the applet. This approach is easy and effective for simple applets.

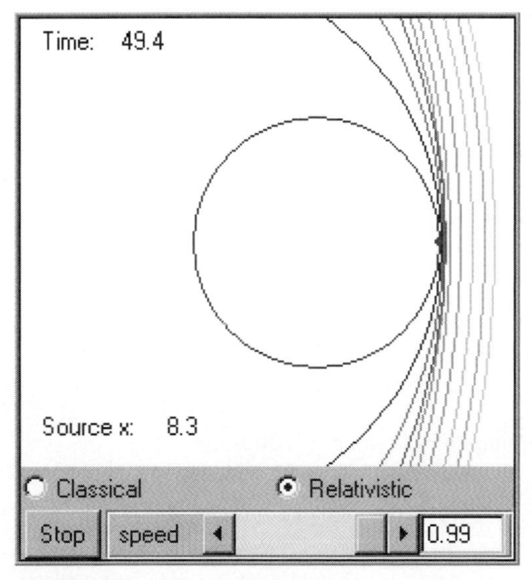

FIGURE 31: Doppler Physlet.

The codebase attribute specifies that the browser should begin its search for the files necessary to run the applet in a subdirectory named "classes" located in the directory containing the html document. Just as with the <img> tag, you can specify a complete URL for the codebase, enabling you to access a Physlet, or any applet for that matter, on an entirely different Web server from the one containing the html document. Due to bandwidth limitations and sluggish Internet response time, it is generally advisable to download and mirror an applet on a local file server, as described in the tutorial, Chapter 6.

Notice the code attribute inside the applet tag. This attribute can be confusing since it consists of two parts, the package name and class filename. The class file is at the end. In this case, it is a file called Doppler.class. This is because the Doppler.class file is part of a collection of java class files known as a java package. The package name in this example is doppler.[3] Appending the package name to the file name gives rise to the fully qualified code attribute, doppler.Doppler.class.

Packages are always located in directories of the same name. Therefore, all files associated with the doppler package will be located in the doppler subdirectory. This subdirectory is hidden inside the Dopper.jar file, as described in Section 5.3. Although it is possible for a package to be distributed as a subdirectory, it is much more likely—and useful—for packages to be distributed using one or more jar files. Most Physlets are distributed this way.

5.5 PARAMETER TAGS AND USER INTERFACES

Even a moderately complex user interface may distract from the physics that a problem author wishes to present. It may invite video gaming or it may result in user error and frustration if too many options are presented on screen. Most authors prefer to add parameter, <param>, tags nested within the beginning and ending tags, <applet> and </applet> respectively, in order to define the applet's behavior. Each applet has a unique set of parameter fields determined by the Java programmer. The programmer should, of course, document them. Applets can now be embedded multiple times with different conditions in order to simulate different physics problems. The Superposition Physlet uses two parameters to set arbitrary functions of position and time, $f(x, t)$ and $g(x, t)$. These functions are each displayed within a panel in the applet and their sum is displayed in a third panel. This Physlet is set up to display a standing wave in an html page with the following tags:

```
<applet code = "superposition.Superposition.class"
codebase = "../classes/"
width = "404"
height = "300">
<param name = "numPoints" value = "100">
<param name = "numGraphs" value = "3">
<param name = "pixPerX" value = "20">
```

---

[3]Using the same name for the applet and the package may seem confusing at first, but it is common practice among Java programmers.

```
<param name = "pixPerY" value = "10">
<param name = "gridX" value = "1">
<param name = "gridY" value = "1">
<param name = "func1" value = "2*sin(pi*x/2-pi*t)">
<param name = "func2" value = "2*sin(pi*x/2+pi*t)">
<param name = "showControls" value = "true">
<param name = "FPS" value = "10">
<param name = "dt" value = "0.1">
</applet>
```

The preceding parameter fields can easily be modified to produce other wave phenomena, such as beats. A good way to learn about the *Superposition* Physlet is to use an html editor to change a parameter and reload the html page to observe the effect. *Superposition* parameter fields are very extensive and allow a high degree of customization. Since applets will provide default values for parameters, it is often not necessary to assign each and every parameter-name pair.

Although parameter fields spare the user from having to worry about many of the more arcane details of an applet, it is often desirable to change an applet's behavior after it has been loaded into a browser. Parameter fields, however, are read only once when the html page is loaded.

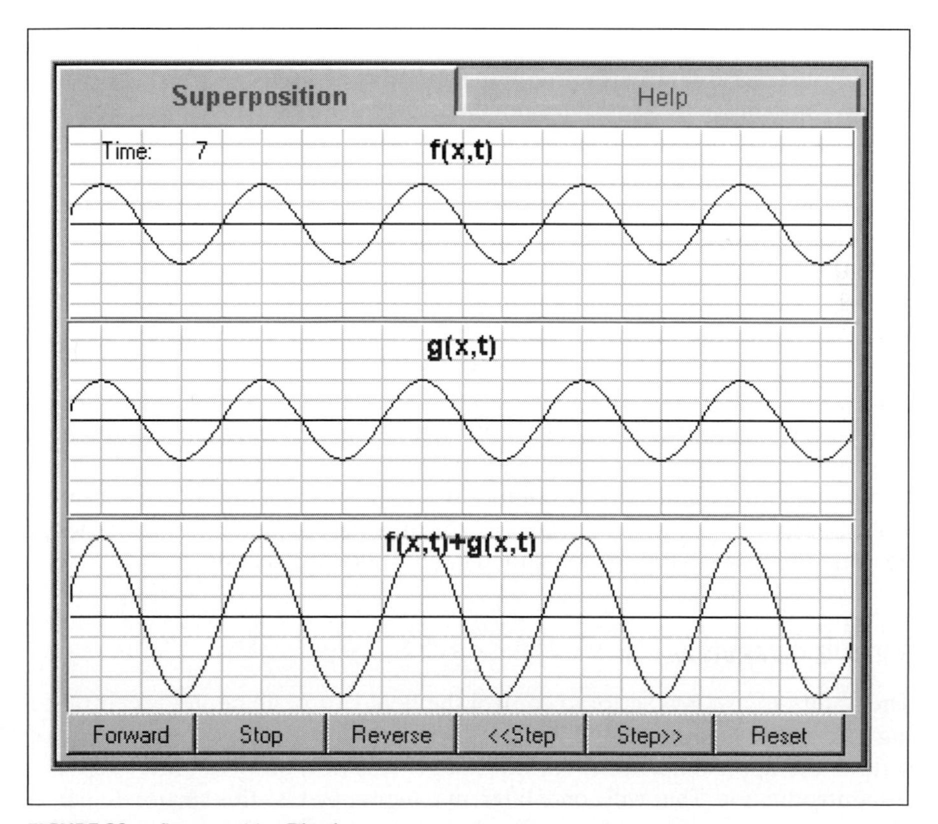

FIGURE 32: *Superposition* Physlet.

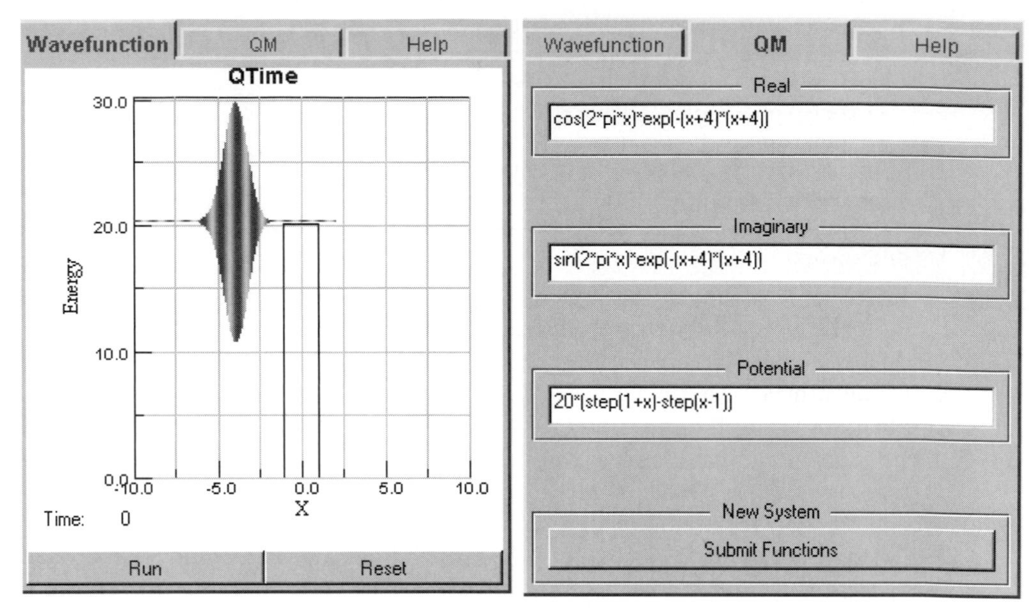

FIGURE 33: *QTime* Physlet.

Therefore, many applet authors provide a graphical user interface (GUI) that enables users to control the applet. Designing a user interface for students often requires a delicate tradeoff between ease of use and control. Even simple operations, such as entering a function string, are far from foolproof and can get in the way of the underlying physics.

Consider the *QTime* Physlet shown in Figure 33 and used to display the time evolution of a quantum wave packet in an arbitrary potential, $U(x)$. This applet has a tabbed panel, named **QM**, that allows a user to access (and change) three function strings. These strings define the real and imaginary parts of the wave function and the potential. Using this interface may be desirable for advanced students, but it is unlikely that a sophomore-level modern physics student needs to be confronted with such detail. Few students would know how to write the real and imaginary parts of a Gaussian wave function in atomic units with the appropriate momentum boost needed to produce a reasonable group velocity. Almost all Physlets have a "showControls" parameter that is designed to simplify the user interface by hiding the interface controls. Changing parameters, as well as starting and stopping the animation, must now be done using JavaScript, as explained in the following section.

5.6 SCRIPTING OVERVIEW

The ability to use JavaScript to control the behavior of an applet is central to developing curricular material with Physlets. This section provides an overview of the way scripts and applets appear in an html page. It is not, however, intended as a step-by-step introduction. That will come later, in Chapter 6.. Use this section to get the general idea of the process of scripting a Physlet before going on to learn what all the methods (commands) mean. Of course, if you understand basic html and Java already, you may wish to go on to the specifics.

The most efficient way to include JavaScript in an html page is to define the script in the <head> section of an html page. A typical blank html page looks like this:

```
<html>
<head>
<title>Describing Motion</title>
</head>
<body>
</body>
</html>
```

A head contains information that is not part of the document text, such as the title between the title tags: <title></title>. It may also contain the JavaScript that controls an applet. Here is an example of a script embedded in the head of an html document that produces the animation shown:

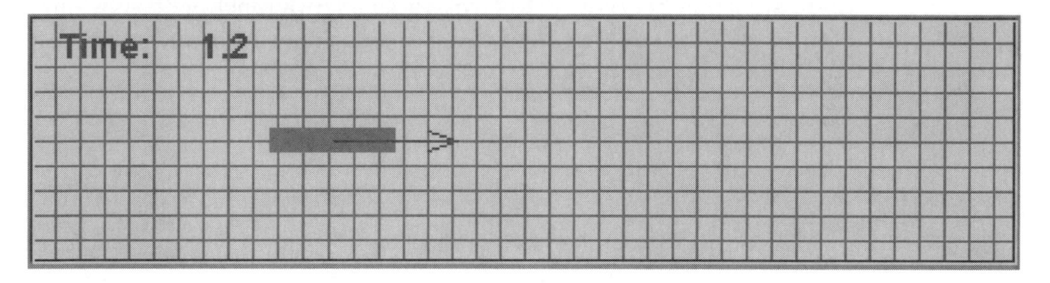


```
<head>
<title>Describing Motion</title>
<script language = "JavaScript">

xDefault = "5*t*t";
yDefault = "0";

function initApplet(xStr,yStr) {
document.animator.setAutoRefresh(false);
document.animator.setDefault();
document.animator.shiftPixOrigin(-150,0);in(-150,0);
id = document.animator.addObject("rectangle", "w = 50,h = 15");
    animator.setTrajectory(id,xStr,yStr);
document.animator.setRGB(id, 255,0,0);
id = document.animator.addObject("arrow","h = 5,v = 0");
document.animator.setTrajectory(id,xStr,yStr);
document.animator.setTimeInterval(0,3);
document.animator.setAutoRefresh(true);
document.animator.forward();
}

</script>
</head>
```

SCRIPT 2: Creating a moving rectangle with a fixed-length arrow.

Script 2 defines two JavaScript variables, xDefault and yDefault, and one JavaScript function, initApplet(xStr,yStr). These JavaScript elements can be referenced later in the document. Notice that string variables are enclosed in single or double quotes.[4] Strings often denote functions that will be evaluated repeatedly while the animation takes place, whereas integer values usually denote fixed values that that do not change while an animation takes place. Evaluating a string function at every time step is computationally expensive, so we use integer values wherever possible, such as in the setRGB method. It is the author's responsibility to ensure that the mathematical functions represented by strings are well formed.

Script 2 is used with Problem 1 in Chapter 2 of the Giancoli Companion WebSite to control the *Animator* Physlet. The Physlet responds to this script and draws a red rectangle and a black arrow. These objects move in the *x*- and *y*-directions according to the analytic functions passed into the initApplet method. A thorough description of the *Animator* Physlet will follow this section, while a complete description of the public methods can be found in Part Three of this book. Documentation for other Physlets can be found on the CD and on the Davidson WebPhysics site.

There are a number of ways to invoke a JavaScript function from within the body of an html page. The easiest method is to use a variant of the anchor tag. The anchor tag, <a>, is most commonly used to link to another html page

```
<a href = "http://webphysics.davidson.edu">
Go to WebPhysics! </a>
```

but it can also be used to send e-mail, connect to an ftp server, or execute JavaScript. In order to execute a JavaScript function, the http link is replaced with the JavaScript as follows:

```
<a href = "JavaScript:initApplet(xDefault,yDefault)"> Problem 1.</a>
```

Clicking on the text bracketed by the <a> and </a> tags will execute the JavaScript function that was previously defined as initApplet (x, y). The script can be called a second time elsewhere on a page using different parameters[5]:

```
<a href = "JavaScript:initApplet('8+4*t','0')">Problem 2.</a>.
```

Notice that the first script uses the *x*- and *y*-variables defined in the head, while the second script makes use of two strings that are defined on the spot. Scripting techniques

---

[4] A string is computer jargon for a sequence of characters. This type of variable is fundamentally different from two other types of variables, integers and floating point numbers. These types of variables are defined without quotation marks—for example, n = 100 or $c = 2.998 \times 10^8$. Numbers are stored internally in a more compact format than strings are, and it is usually more efficient to do arithmetic with numbers. But strings are more flexible and must be used when called for. Notice how the setTrajectory method is used to specify the position of the arrow's origin using two strings representing functions of time.

such as these make it possible for a Physlet to support many different types of problems and to change its behavior while the student is working through an exercise.

In summary, a complete html page containing an interactive Physlet problem will usually contain the following elements: (1) a <head> containing JavaScript functions capable of communicating with the Physlet, (2) a <body> containing text and a Physlet embedded with the <applet> tag, and (3) an anchor tag to invoke the appropriate JavaScript. A complete html page is shown in Script 3 for reference. It is included on the CD in the Chapter 5 subdirectory.

```
<html>
<head>
<title> Describing Motion </title>
<script language = "JavaScript">
x = "5*t*t"; y = "0";
function initApplet(xStr,yStr){
document.animator.setDefault();
document.animator.shiftPixOrigin(-150,0);
id = document.animator.addObject("rectangle","w = 50,h = 10");
document.animator.setRGB(id,255,0,0);
document.animator.setTrajectory(id,xStr,yStr);
id = document.animator.addObject("arrow","h = 5,v = 0");
document.animator.setTrajectory(id,xStr,yStr);
document.animator.setTimeInterval(0,3);
document.animator.forward();
}
</script></head >
<body>
<h3 align = "left"> Describing Motion </h3>
<p align = "center">
<applet code = "animator4.Animator.class"
codebase = "../classes/"
archive = "Animator4_.jar,STools4.jar"
align = "baseline" width = "400"
height = "150"
id = "animator" name = "animator">
<param name = "dt" value = "0.05">
<param name = "FPS" value = "20">
<param name = "gridUnit" value = "1.0">
```

[5]Single quotes must be used when a string is nested within a string. The hypertext reference, href, within the anchor tag is a string. But within this string we define another string that is passed to the initApplet function. Removing the single quotes would cause an error since the browser's JavaScript interpreter would attempt to evaluate 8+4*t, assuming that t was a variable. The second parameter, 0, would also be interpreted as a pure number and not as a character string. The difference between numbers and characters is fundamental and a common cause of script errors. These syntactic pitfalls are exacerbated by the fact that JavaScript does not distinguish between strings and numbers, while Java does. That is, JavaScript will automatically convert a number to a string when necessary while Java will produce an error if a string is passed to a function that is expecting a number.

```
<param name = "pixPerUnit" value = "10">
<param name = "showControls" value = "false">
</applet>
</p>
<p>Physicists use arrows to represent many things in diagrams. What vector quantity is
    being represented by the arrow in this simulation? 
<a href = "JavaScript:initApplet(x,y)">Start simulation</a> </p>
<p><input type = "radio" name = "answer.1" value = "1"> displacement</p>
<p><input type = "radio" name = "answer.1" value = "1"> velocity</p>
<p><input type = "radio" name = "answer.1" value = "1"> acceleration</p>
<p><input type = "radio" name = "answer.1" value = "1"> speed</p>
<p><strong>Instructions</strong></p>
<p>Answer the question after you have run the simulation. </p>
</body>
</html>
```

SCRIPT 3: HTML template with a JavaScript function, Physlet, and text.

CHAPTER 6

SCRIPTING TUTORIAL

6.1 AUTHORING TOOLS

Although the Internet is designed to be platform and operating system independent, content creation takes place in the real world of an author's personal desktop. Prior training, personal preference, school policy, and intended use all play a role in determining the appropriate authoring package. It would be foolish for us to specify a single environment for authoring Physlet problems. Physlet problems can be run on most browsers and can be edited using the simplest ASCII text editor. Nevertheless, we believe that specialized tools are time and cost effective. They simplify routine editing, enforce proper html syntax, and provide Web site management. This section briefly describes the features of the FrontPage® 2000 authoring platform that we have found to be most helpful in writing our own Physlet problems. Other tools have similar features. Whatever tool is chosen, we believe that it is more efficient for authors to master that tool than to dabble with multiple tools.

Good What-You-See-Is-What-You-Get (WYSIWYG) html editors are akin to the best word processors and hide the code from the author. The author simply does routine editing using cut and paste to insert images, or highlighting to apply formatting and font styles. The finished document can then be published on the Web. Unfortunately, high-level integration of advanced interactive Web-based technologies, such as JavaScript and Java, is still sketchy in some authoring packages. A passing knowledge of html and JavaScript syntax is usually required for developing interactive curricular material and for those times when it is absolutely necessary to get under the hood to find out what is going on.

The single most important feature we look for in an authoring package is the ability to switch between html markup and a WYSIWYG editing environment, as shown in Figures 34 and 35, respectively. FrontPage® 2000 performs this switch using tabs at the bottom of the editing window. Integrated packages, as opposed to separate tools, usually maintain the position of the edit/insertion cursor during the switch. For example, editing the heading Projectile Motion in WYSIWYG mode and switching to html mode should display the code

```
<h3 align = "left">Projectile Motion</h3>
```

near the center of the edit window without having to scroll.

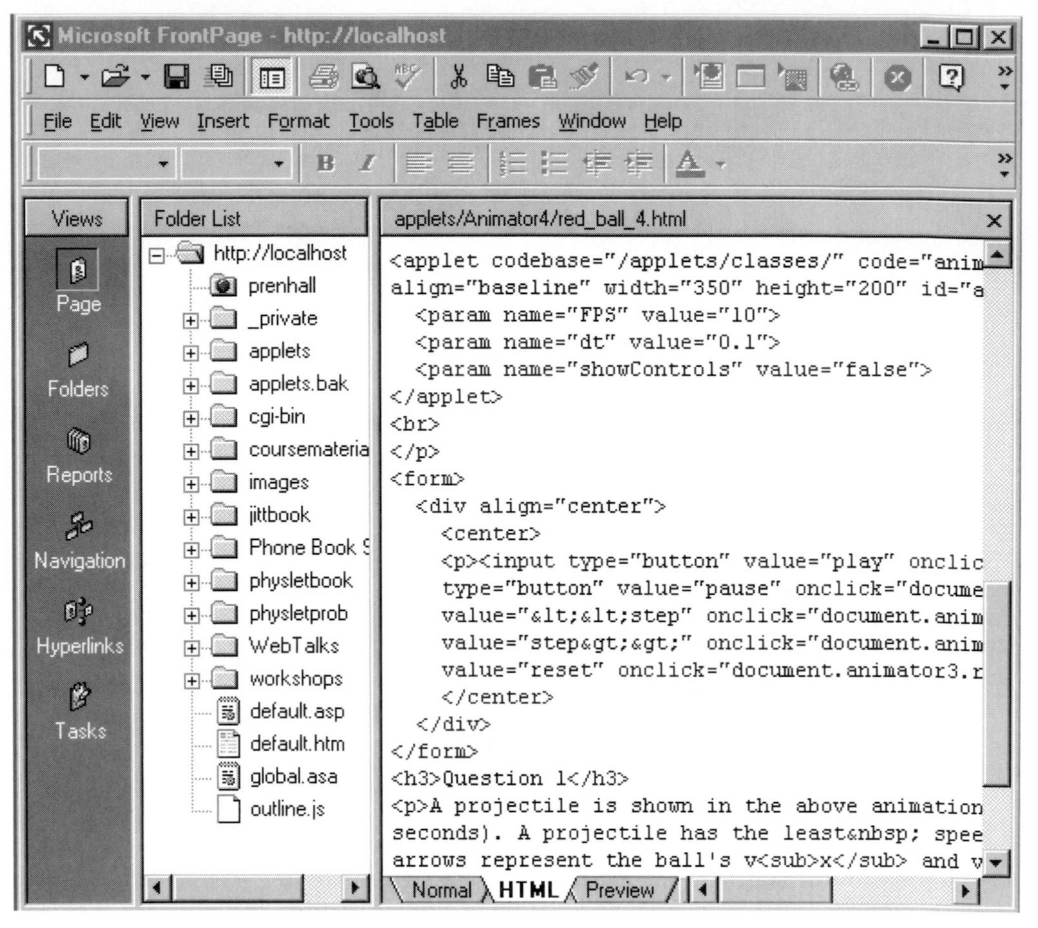

FIGURE 34: FrontPage® authoring tool in html mode.

Another important feature is the ability to drag and drop objects in WYSIWYG mode onto an html page and then access their properties. The html buttons shown in Figure 35 are an excellent example. Few authors would want to type the following code whenever they wished to insert a centered on-screen button to invoke a JavaScript function.[1]

---

[1]Standard html defines a small number of user interface elements: button, text field, check box, and radio button. Proper html syntax requires that these elements be enclosed by <form> and </form> tags. This requirement makes it possible for multiple pieces of information to be sent to another program in a single operation. An online purchase, for example, might contain text fields for name, address, and credit card number; these are processed simultaneously when a button is pressed. Physlet problems will most often process information using embedded JavaScript. This is accomplished by specifying a JavaScript function as the receiver of the element's onclick action.

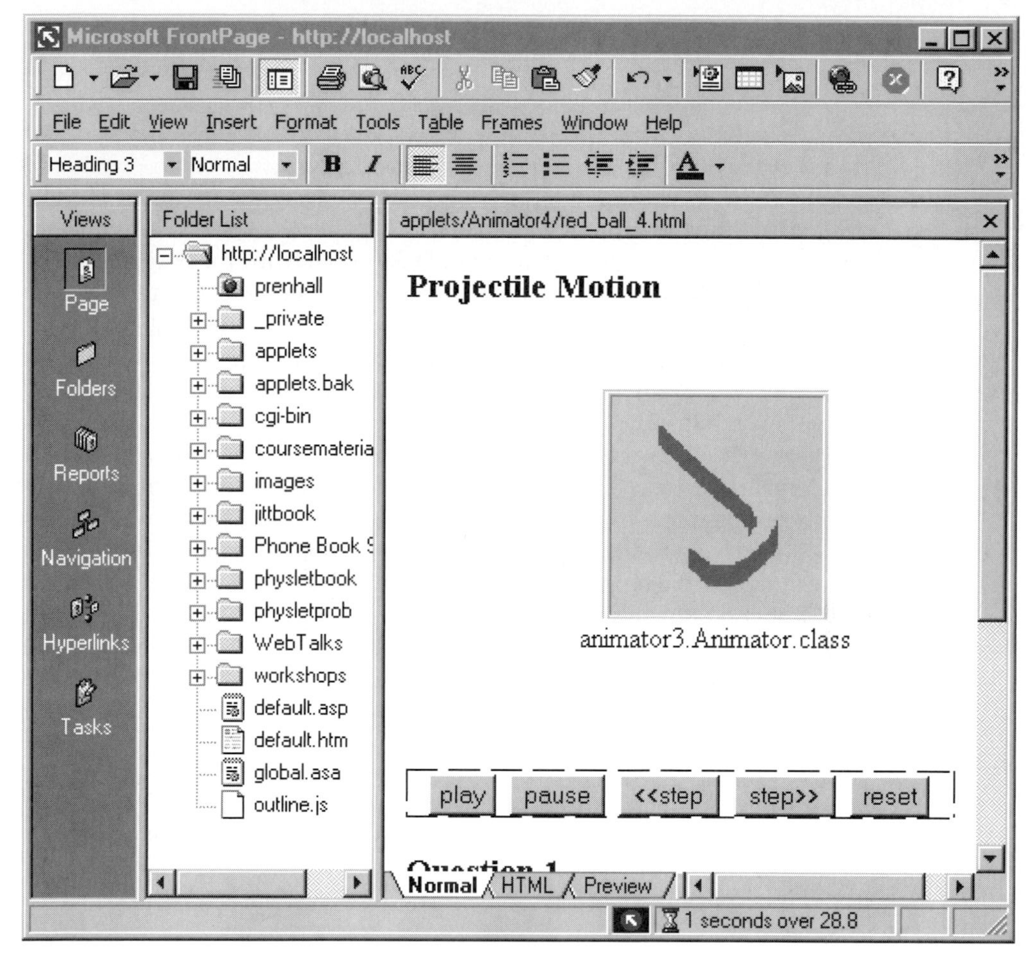

FIGURE 35: FrontPage® authoring environment in normal mode.

```
<form>
<div align = "center">
<p>
<input type = "button" value = "play" onclick = "initApplet()">
</p>
</div>
</form>
```

Drag and drop not only saves typing, but it also reduces syntax errors. Right-clicking on the button pops up a dialog box with text fields for editing the button's label and the action, play and initApplet() in the preceding example. This expert prompting is especially useful for beginning html authors, who may be unfamiliar with all the options and attributes that are available.

A third feature to look for is the ability to check for and correct incorrectly written html. Specifically, the following code is poorly formed since the bold and italic tags are not nested properly:

```
<b>This code is <i>not</b></i>
```

Although most browsers will interpret this code properly, it does not conform to the latest standards.[2] Another common mistake is not to close the paragraph tag, <p>, with the end paragraph tag, </p>. FrontPage® 2000 only does a poor job of identifying improper syntax. For a real eye opener, html authors should check their pages using the World Wide Web Consortium's, W3C, validation service at http://validator.w3.org/.

The fourth feature we look for in an authoring package is document and site management. Site management is a crucial feature for Web masters managing Internet servers, but even beginning authors writing stand-alone html pages will eventually need some structure. Whenever a file is moved from one folder to another folder in the folder list pane, hypertext links that reference the document on other html are updated to reflect the new location.[3] Tools on the left-most pane can be used to check for broken hyperlinks and to automate Web-site navigation. Finally, a graphical display showing all hyperlinks to and from a page is available to help find orphaned and out-of-date files.

In order to show exactly what needs to be entered into an html page, we will show html markup in the following tutorial. It is understood, however, that readers should switch into WYSIWYG mode when convenient.

6.2 *ANIMATOR*

Animator provides an environment for the student to see real-time animations of physical (and nonphysical) situations, an example of which is shown in Figure 36.

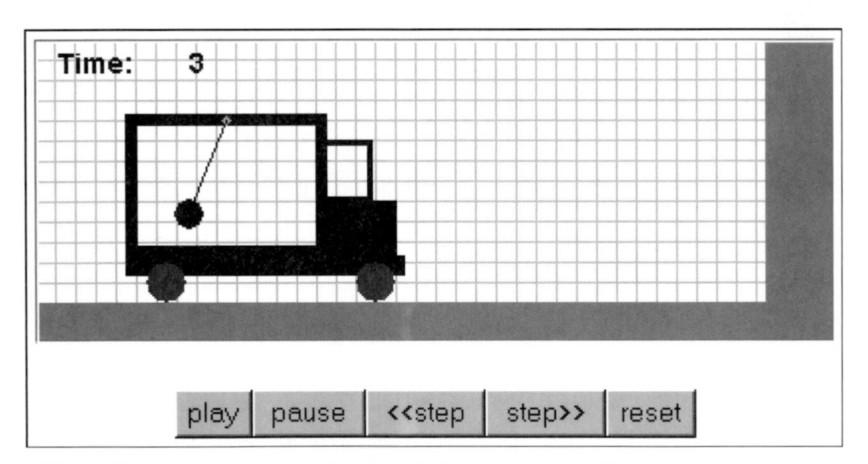

FIGURE 36: Pendulum in a truck scripted with the *Animator* Physlet.

---

[2]See xhtml in the glossary.

[3]Although FrontPage® does a good job renaming hypertext anchors, its Page will incorrectly rename the archive attribute inside an applet tag. FrontPage® prepends a path to the first jar file. This path should be removed after the html file is copied.

Students view these animations and make time, position, velocity, acceleration, and force measurements in order to answer questions about the motion of objects that appear in the Physlet. Because of its versatility, *Animator* is probably the most robust and widely used Physlet currently available. It has been used to create problems ranging in content from kinematics to particle physics to help students to visualize physical situations. (See Part Two for example problems using *Animator*.)

The description *Animator* refers to the current version of the Java code, Animator4, and all of the problems contained in the CD use this version of *Animator*. For early adopters of Physlets who have already written script to control a Physlet, there are some new and interesting features in the latest version, and you should script using this version.[4] There are also a few backward compatibility issues (orphaned methods and new methods) in updating old *Animator* scripts to the new *Animator* that are described in Chapter 13 Section 2. Scripts using version 3 of *Animator* will still run, but they are no longer supported. Scripts using version 1 and 2 may not run if they use orphaned methods. If you have never scripted before, knowledge of the existence of these old scripts is only important if you obtain Physlet problems from others.

The goal of this chapter is to give you an idea of how to script a simulation in *Animator* from the ground up. Many of the methods and techniques that are found in this chapter regarding *Animator* will also appear in other Physlets. This chapter proceeds slowly and methodically to introduce scripting. Subsequent chapters rely heavily on this introduction.

6.2.1 Creating an *Animator* Template

We begin by creating a red ball in the *Animator* Physlet. Once this script is written, getting the ball to move and adding more features is not too difficult. An html page is necessary to deliver the applet to a Web browser, so we begin by creating an empty html page.

Start your favorite text editor and enter the following html code exactly as written:

```
<html>
<head>
<title>Projectile</title>
</head>
<body bgcolor = "#FFFFFF">
<h3 align = "left">Projectile Motion</h3>
</body>
</html>
```

This code creates a Web page entitled **Projectile** that has a single text line and has a white background (*<body bgcolor=*"*#FFFFFF*"*>*). Notice the use of the attribute bgcolor to set the background color. Use the SAVE AS... feature to save the docu-

---

[4]Animator versions 1 and 2 only supported objects following analytic trajectories. These early versions were written for Java 1.0 and were designed to run on computers with 90-MHz microprocessors. *Animator* version 3 was written in Java 1.1 and required a 200-MHz processor in order to solve the differential equations for real-time dynamics. *Animator* version 4 added interapplet communication and introduced a consistent naming convention for adding objects to Physlets, the *addObject* method.

ment as an ASCII text file with the name projectile.html in a directory of your choice. You now have a Web page locally on your computer that can be opened using any standard browser. Make sure you can do this before you proceed.

You will now add script that will create a red ball inside *Animator*. Insert the following few lines of JavaScript into the head of the Web page, that is, between the <head> and </head> tags.[5]

```
<script language = "JavaScript">
function proj1_0(){
document.animator.setAutoRefresh(false);
document.animator.setDefault();
document.animator.setPixPerUnit(5);
document.animator.setGridUnit(2);
id1 = document.animator.addObject("circle", ↵
  "x = 0, y = 0,r = 10");
document.animator.setRGB(id1, 255,0,0);
document.animator.setTimeInterval(0,6);
document.animator.setAutoRefresh(true);
document.animator.forward();
}
</script>
```

Script 4: Adding a circle to *Animator*.

Script 4 defines a function, **proj1_0()**, that will animate a red ball. Note the syntax within the code. Java uses "dot" syntax, which is common in object-oriented programming. Invoking a method of an applet that has been embedded into a document begins by specifying a reference to the embedded object: **document.animator**. Immediately following the object reference is the method to be performed by the *Animator* Physlet. This script uses eight methods. One method is used twice with different arguments:

- setAutoRefresh(false); //turns off the automatic drawing of the applet.
- setDefault(); //clears the Physlet and sets default parameters.
- setPixPerUnit(5); //sets the pixels per unit on the animation.
- setGridUnit(2); //sets the background grid unit for the animation. A value of zero turns off the background grid.
- id1 = addObject("circle","x = 0, y = 0, r = 10"); //adds an object, here a circle with a radius of ten pixels, to the animation at the coordinate x = 0 and y = 0. The addObject method returns a unique integer identifier, id1, which can be used to identify the object.
- setRGB(id1,255,0,0); //sets the color of the object with the given id. Here it makes our red ball red. Specifically, the argument of the command is in the form (id,R,G,B) where R(ed),G(reen), and B(lue) are integers ranging from 0 to 255.[6]

---

[5]Because of width limitation in the text, we will often have to include a line break in the scripts we present where one does not really appear. To minimize confusion, we will use the symbol ↵ to indicate a typographical line break that should *not* appear in the actual script. In addition, we often end lines with periods to be grammatically correct. All lines appearing in the script must end in a semicolon.

[6]Typical RGB colors are red=(255,0,0), green=(0,255,0), blue=(0,0,255), black=(0,0,0), white=(255,255,255), light gray=(155,155,155), and dark gray=(80,80,80). Black is the default color for all objects, and white is colorless on a white background.

- setTimeInterval(0,6); //sets the time interval. Here the time interval of the animation is set to run 0<t<6. Note that the time interval is internal to the applet and may not represent the real time interval that the applet animates. There is also another method that stops an animation. It is **setOneShot(0,6,"End of Animation");** which stops the animation and displays an author defined message, here the words End of Animation, when the applet stops running at t = 6.
- setAutoRefresh(true); //turns on the automatic drawing of the applet.
- forward(); //makes the animation run.

The double slash, //, is a shorthand that is used in many programming languages to insert a short comment. It signifies to the JavaScript compiler that the remainder of the line is text and should not be interpreted as code. The double slash is used here as a comment to separate the methods from their descriptions, just as one would in an actual script.

The Web browser must now be told where to find the *Animator* class, how many pixels the applet should occupy in the html page, and what initial parameters should be used. This is accomplished by adding the following to the body of the html page:

```
<h3 align = "left">Projectile Motion</h3>
<p align = "center">
<applet
codebase = "classes"
code = "animator4.Animator.class"
archive = "Animator4_.jar,STools4.jar"
align = "baseline"
width = "350"
height = "200"
id = "animator"
name = "animator">
<param name = "FPS" value = "10">
<param name = "dt" value = "0.1">
<param name = "showControls" value = "false">
</applet>
</p>
```

This markup does two things: It creates some text and embeds the applet. The text is shown as a heading "**Projectile Motion**." The code in between the <applet> </applet> tag embeds the applet in a portion of the Web page. This applet will be centered and have a size of 350 by 200 pixels. The attributes inside the <applet> tag tell the Web browser where to find the Animator.class file. In this example, it resides in the directory named **classes**, which is in the same directory as the html page. See Section 4.2.1 for more details on relative addressing. In addition, the keywords "id" and "name" give the Java document a name that can be used by scripts in Internet Explorer and Netscape Navigator/Communicator, respectively (both are included for backward compatibility). Note that this is the exact same name that appeared after the word **document** in the preceding script, in this example: **animator**. In order for the applet to run properly, these names **must be exactly the same**. Also noteworthy, the

World Wide Web Consortium, W3C, has recommended that vendors adopt the keyword "object" to replace both the modifiers "id" and "name". Expect to see this standard adopted in the next browser release.

As these applets will vary in complexity and appear on a variety of different computers—all with different processor speeds—the rate at which the screen redraws and the animation time step should be carefully considered. This is done by adding the following **param**eters to the animation right above the end applet </applet> line. The Frames Per Second (**FPS**) parameter controls how often a new image will be calculated and displayed on the screen. Too small a value produces image flicker and too large a value causes jerky motion due to the microprocessor not being able to perform the calculations quickly enough. The parameter **dt** defines the time step for each animation frame. If FPS is set to 10 and dt is set to 0.1, the animation will proceed in real time without noticeable flicker on most microcomputers. Setting the **showControls** parameter to false hides the user interface, thereby allowing the author to add a custom one using html form elements later.

The html page is finished with the addition of a hyperlink to the function defined in the head that will start the animation when the reader of the page clicks on it:

```
<a href = "JavaScript:proj1_0()"> Start</a>
```

Now reload the Web page by clicking the reload button on the browser. The browser is unaware of your changes to the html page unless this update is performed. If all went well the applet shown in Figure 37 should appear on a Web browser.

When the applet is started by clicking the Start hyperlink, the time in the upper-left-hand corner of the applet will cycle from 0-6 seconds, and the ball will remain stationary.

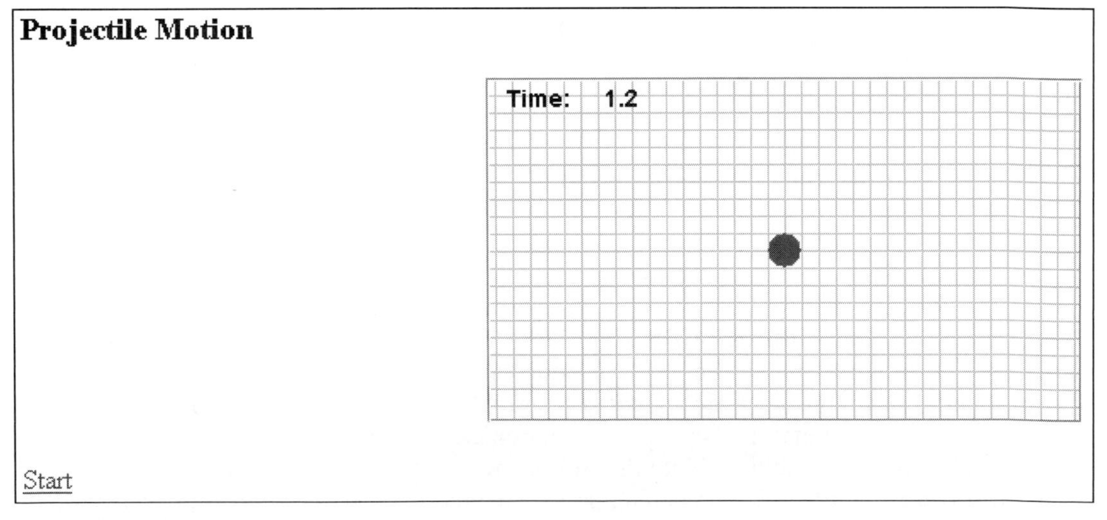

FIGURE 37: Stationary red ball on screen.

6.2.2 Simple Projectile Motion

We are now ready to create the animation. To do so, the position of the red ball must vary with animation time t. If the following line is added after the **addObject** method of the Script 3

```
document.animator.setTrajectory(id1,"(8.0*t)", "(20*t-4.9*t*t)");
```

and the page is reloaded, the ball will move across the screen along the $x(t)$ and $y(t)$ trajectory specified by the second and third arguments, respectively, of the setTrajectory method.[7] The particular values give an initial velocity (in both the x- and y-directions) and the correct acceleration due to gravity (acting solely in the y-direction).

There are now several problems to be addressed. The animation starts at $t = 0$ with the ball appearing on the screen and then moving off screen. For pedagogic reasons the ball should be a little more centered in the applet, appearing in the bottom left corner and disappearing in the bottom right corner some time later. For objects that were in motion at $t = 0$, starting the object off screen allows students to view the motion as intended. Otherwise, the motion appears to begin as soon as the Start button is pressed. This is accomplished by adding the following line to the script directly below the **setDefault** line[8]:

```
document.animator.shiftPixOrigin(-100,-100);
```

The method **shiftPixOrigin** will shift the origin of coordinates down and to the left by 100 pixels in each direction, which is exactly what is needed for the animation of the ball to be centered on the applet grid. There is yet another problem: How is the user supposed to control the applet? Assuming that we will be asking the user to measure something pertaining to the animation, how does he or she do that without starting and stopping the animation? To add this functionality, standard html form buttons are used to create VCR-type controls beneath the animation grid. It is easiest to add these buttons using a WYSIWYG authoring tool, but they can also be created by adding the following lines of html to the page following the end applet tag, </applet>.[9]

---

[7]The argument of setTrajectory can accept any function of t. See Section 6.4.4 for the types of functions available.

[8]The location in a script of scaling methods such as setPixPerUnit, setGridUnit, and shiftPixOrigin is arbitrary. However, we tend to group these methods together near the beginning of the script for consistency.

[9]It is necessary to enter characters that have special meaning, such as $<$ and $>$, using escape sequences. The escape sequences for $<$ and $>$ are < and >, respectively. In this example, escape sequences are used to inform the browser that we intend to render the symbols $<$ and $>$ rather than the beginning or the end of a tag.

```
<form>
<input type = "button" value = "play"
onclick = "document.animator.forward()">
<input type = "button" value = "pause"
onclick = "document.animator.pause()">
<input type = "button" value = "&lt;&lt;step"
onclick = "document.animator.stepBack()">
<input type = "button" value = "step&gt;&gt;"
onclick = "document.animator.stepForward()">
<input type = "button" value = "reset"
onclick = "document.animator.reset()">
</form>
```

Note again that the name appearing in the word document in the preceding html is (and must be) exactly the same as what appeared in the name attribute of the applet tag (id/name) and in the script itself, again for this example: **animator**. You should now see Figure 38 on your browser after the page is reloaded.

The red ball appearing in the lower left of the applet and following a parabolic path before disappearing on the lower right of the applet. The complete html page is as follows:

```
<html>
<head>
<script language = "JavaScript">
function proj1_0(){
    document.animator.setAutoRefresh(false);
```

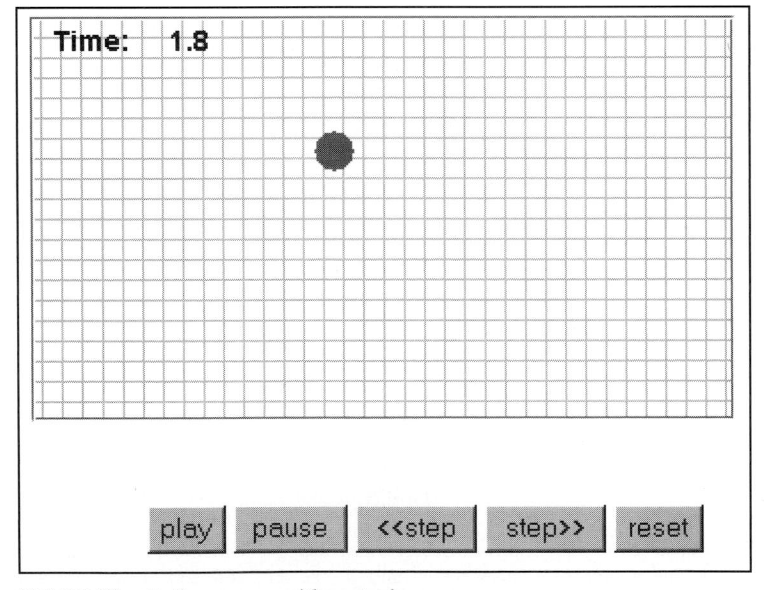

FIGURE 38: Ball on screen with controls.

```
        document.animator.setDefault();
        document.animator.shiftPixOrigin(-100,-100);
        document.animator.setPixPerUnit(5);
        document.animator.setGridUnit(2);
        id1 = document.animator.addObject("circle","r = 10");
        document.animator.setTrajectory(id1, "(8.0*t)","(20*t-4.9*t*t)");
        document.animator.setRGB(id1, 255,0,0);
        document.animator.setTimeInterval(0,6);
        document.animator.setAutoRefresh(true);
        document.animator.forward();
}
</script>
<title>Projectile</title>
</head>

<body bgcolor = "#FFFFFF">
<h3 align = "left">Projectile Motion</h3>
<p align = "center">
<applet codebase = "classes" archive = "Animator4_.jar,STools4.jar"
code = "animator4.Animator.class"
align = "baseline" width = "350" height = "200" id = "animator" name = "animator">
<param name = "FPS" value = "10">
<param name = "dt" value = "0.1">
<param name = "showControls" value = "false">
</applet>
</p>
<form>
<p><input type = "button" value = "play" onclick = "document.animator.forward()">
<input type = "button" value = "pause" onclick = "document.animator.pause()">
<input type = "button" value = "&lt;&lt;step" onclick = "document.animator.stepBack()">
<input type = "button" value = "step&gt;&gt;" onclick = "document.animator.stepForward()">
<input type = "button"value = "reset" onclick = "document.animator.reset()"></p>
</form>
<a href = "JavaScript:proj1_0()">Start</a></p>

</body>
</html>
```

Script 5: Projectile motion scripted using *Animator*.

6.2.3 Projectile Applet Sample Problems

The basic projectile motion applet will be improved with the addition of a few more lines of script. These additions will enable the applet to be more functional for questions the author might want to ask students regarding projectile motion. Text for a standard question in the physics textbook by Giancoli, page 70. [Giancoli 1998] “A projectile has the least speed at what point in its path?” can be easily embedded below the html buttons using standard tags.

```
<h3>Question 1</h3>
<p>A projectile is shown in the above animation (position is given in meters and time is in
    seconds). A projectile has the least speed at what point in its path?</p>
```

In writing the text for these types of problems, the author should be careful to stress units. The applet itself is unitless, since the grid spacing can represent any spatial length. The same is true of the time display. A simple addition to the text of the problem, such as "*position is given in meters and time is in seconds*," is usually sufficient to stress the importance of units.[10]

The current applet correctly depicts projectile motion, but perhaps not in the correct form for most students to visualize the answer to the preceding question (and subsequent questions). Two successive additions to the script will help students answer this question correctly (and in the process better understand projectile motion). First, add the line

```
document.animator.setTrail(id1,150);
```

below the **setRGB** line. (What is crucial in the line placement is that the id must have already been instantiated. Hence, any line after the **addObject** line will suffice.) This addition turns on the trail of the object in question, namely id1, the red ball. This will leave a parabolic curve on the screen where the ball has traveled during the last 150 time steps, an effective visual aid to answer the question. Note that the trail will inherit the color of the original object; in this example the trail is red.

Depending on the questions asked of the student, footprints marking where the ball has traveled may make a better visual aid than the solid trajectory line. Since these footprints will be evenly spaced in time, this technique gives students a visual aid when looking at constant velocity or constant acceleration animations. To add these footprints, add the following two lines beneath the **setTrail** line:

```
document.animator.setFootPrints(id1,5);
document.animator.setGhost(id1,true);
```

The argument of the **setFootPrints** method places a marker where the object with identifier id1 appears at an interval of five time steps (the **dt** parameter that was set in the <applet> tag). If instead we had **setFootPrints(id1,1),** the footprints would be left after every time step. There are other ways to mark the path of an object. The Ghost feature in the second line of the preceding script paints a pale ghost image of the

---

[10]Units are important to physicists. However, computer simulations store numbers and these numbers do not have units. Calculations are performed just as they are on a pocket calculator. This can cause confusion since the time and distance units shown on the computer display do not have an a priori relationship to the real world. In other words, we can assign the relationship to be anything we want it to be. For example, an animation modeling the motion of an electron about the nucleus of an atom will define the distance unit to be 10^{-9} meter (i.e., a nm) and the time unit to be 10^{-6} second (i.e., a μs). Another simulation might define distance and time units to be kilometers and hours, respectively.

object on screen with the same frequency as the FootPrints. This is exactly the effect depicted in Figure 2 of Chapter 1. As with the Trail and FootPrints, the Ghosts inherit the color of the object in question.

This applet's flexibility can be exploited by asking questions regarding the motion of the red ball, such as

- What is the red ball's horizontal speed?
- What is the red ball's velocity at the top of its motion?
- What is the red ball's vertical acceleration?
- What is the speed of the projectile at $t = 3.7$ s? What is the velocity of the particle?
- What is the minimum speed of the projectile?

Calculations can be made by mousing-down in the animation. When this procedure is performed, a yellow text box appears in the lower left-hand corner of the applet that gives a continuous reading of the x- and y-coordinates of the cursor as it is moved around the applet.

Students often see projectile motion and automatically respond that all accelerations are 9.8 m/s^2. To counter this tendency, state in the problem that the motion is, say, on the moon, and then change the acceleration in the animation to match this assertion. You may do so by changing the line

```
document.animator.setTrajectory(id1,"(8.0*t)", "(20*t-4.9*t*t)");
```

to

```
document.animator.setTrajectory(id1,"(8.0*t)", "(20*t-0.816*t*t)");
```

since the acceleration due to gravity on the moon is about one-sixth of the acceleration due to gravity on the earth. This points out one key feature of these animations: The author manipulates the JavaScript to make the situation depicted in the animation physical. Hence, the author may also make the animation nonphysical to test student understanding of the material.

Again, since the applet itself is unitless, the author should be careful to stress units since the grid spacing can represent any spatial length.

6.2.4 Adding Arrows

With a few more modifications to the script, the applet can include velocity vectors. First, comment off the **setFootPrints** and **setGhost** lines using the double slash at the beginning of the lines:

```
//document.animator.setFootPrints(id1,1);
//document.animator.setGhost(id1,true);
```

since we now want just the solid red trail to follow the red ball. Should we want these two lines back in the script, simply remove the double slashes. Consider the following additions to the applet below the **document.animator.setTrail(id1,150)** line:

```
id2 = document.animator.addObject("arrow", "h = 0,v = 20-9.8*t");
document.animator.setAnimationSlave(id1,id2);
id3 = document.animator.addObject("arrow", "h = 8.0,v = 0");
document.animator.setAnimationSlave (id1,id3);
id4 = document.animator.addObject("arrow", "h = 8.0,v = 20-9.8*t ");
document.animator.setAnimationSlave (id1,id4);
document.animator.setRGB(id4, 14,49,188);
```

There are now three new objects that follow the red ball around the screen. This addition to the script places black arrows corresponding to the ball's velocity in the *x*- and *y*-directions and a pale blue arrow representing its total velocity, all placed with their tails at the center of the ball. This is accomplished by a new method, **setAnimationSlave,** which enables the one object to follow another. In this example, the arrows follow the red ball since the red ball is its master. The signature of this method is

setAnimationSlave(*int master_id, int slave_id*).

Upon reloading the page, your browser should look like Figure 39.

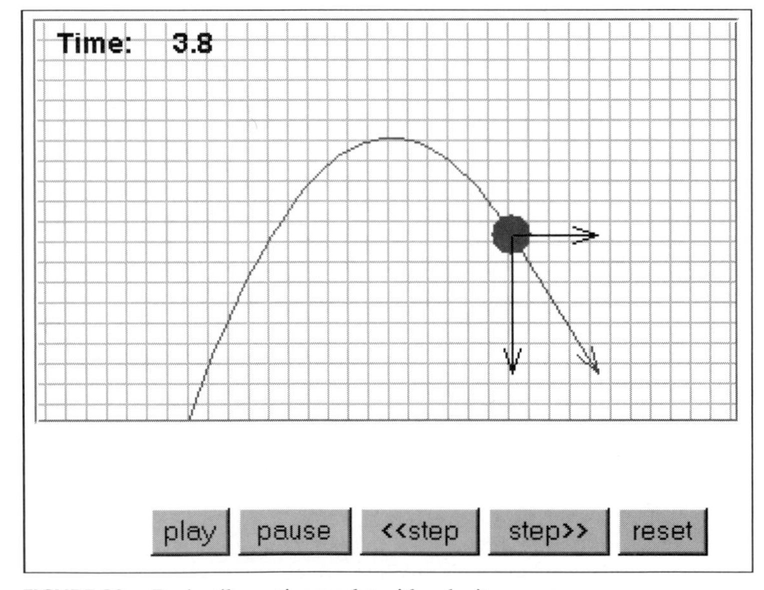

FIGURE 39: Projectile motion applet with velocity arrows.

To avoid student confusion regarding the arrows and what they mean, in the text of the problem, state that the black arrows represent the velocity components and the blue arrow represents the total velocity vector.

6.2.5 Adding Text

Two more optional features will finish off this example. Sometimes it is helpful to give students a continuous display of coordinates instead of having students mouse-down to read coordinates. A caption or text also can also be added in cases where information for students is best given in the applet itself as opposed to in the text of the problem. In this case add the following lines of code to the existing script beneath the **setTrail** line:

```
document.animator.setShowCoordinates(id1,true);
document.animator.setCoordinateOffset(id1,-35,0);
document.animator.setFont(id1,"Helvetica",1,12);
capid = document.animator.addObject("caption", "text = My First Physlet");
document.animator.setFont(capid,"Helvetica",1,16);
```

These additions draw the x-y coordinates of the object with identifier id1—the red ball—offset to the left so as not to be right on top of the ball. The coordinates are drawn with the font Helvetica bold in 12pt. The second argument of setFont, 1, bolds the font in question (0 draws the default font, 2 draws the italic font, and 3 draws the bold, italic font). Also added is a caption, "My First Physlet", in Helvetica bold 16pt that is slightly offset from its default position. Boldfacing and changing font size makes it easier to see text on individual computer screens and in the classroom. The final version of this *Animator* applet is shown in Figure 40, which is exactly Physlet problem 8.1.4 (see Chapter 8) with a caption.

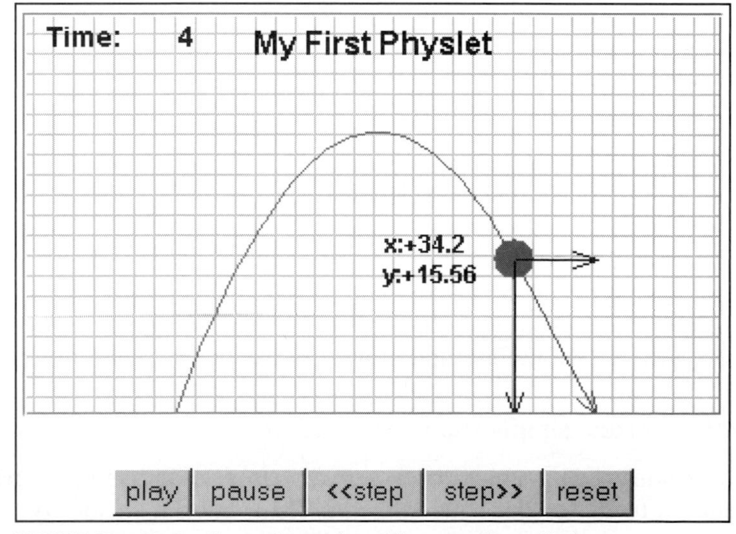

FIGURE 40: Projectile motion with position coordinate display.

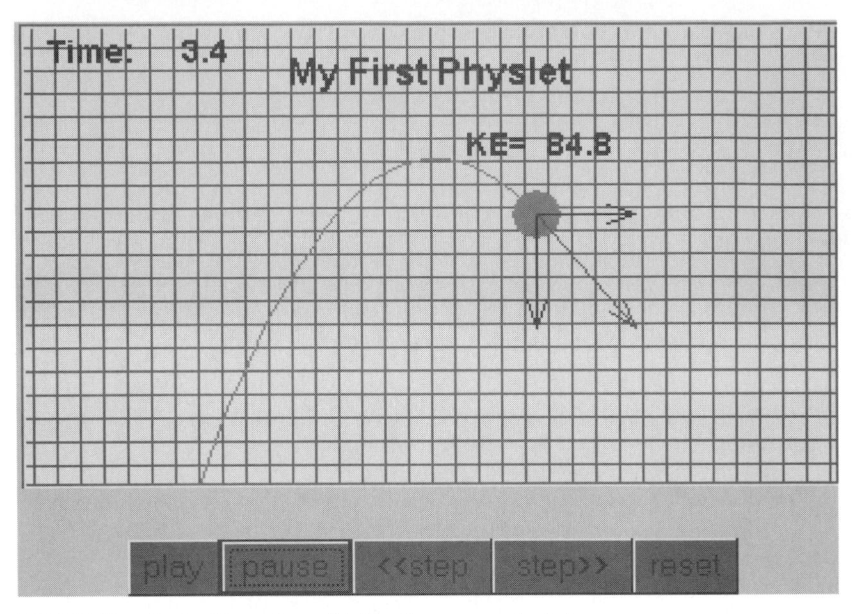

FIGURE 41: Projectile motion with kinetic energy calculation.

However, there is one other way to add the coordinates of the object. This time we will add text that is slaved to the red ball. A text object can contain both a text string and a calculation. In this example the kinetic energy is calculated. We do so by first commenting off the setShowCoordinates and setCoordinateOffset lines we just added. Now add the following lines:

```
tid1 = document.animator.addObject("text","text = KE = ,calc = (vx*vx+vy*vy)*m/2");
document.animator.setAnimationSlave(id1,tid1);
document.animator.setDisplayOffset(tid1,-70,4);
document.animator.setFormat(tid1,"%-+5.2f");
document.animator.setFont(tid1,"Helvetica",1,12);
```

and reload the Web page. You should now see Figure 41 in your browser. What we have done is to add a text object labeled tid1. The text object will follow the red ball around (setAnimationSlave), will have an offset for readability (setDisplayOffset), and will be displayed in the Helvetica, bold, 12 point font (setFont). The calculation could have been any analytic function of the state variables: $m, t, x, y, vx, vy, ax, ay, fx$, and fy. Since the calculation will show three digits to the right of the decimal place, we format the text with the setFormat method to show two digits to the right of the decimal place.

6.2.6 Scripting Forces: Simple Harmonic Motion

The previous section animated a red ball based on knowing the ball's trajectory. This amounted to the author solving the problem in order to specify the ball's position at any given time. In this section a red ball will again move, but now subject to a force scripted by the author. Objects that move under the action of forces is said to be dynamic. In order to make an object dynamic, replace the setTrajectory statement in the projectile script with the following statement:

```
document.animator.setForce(id1,"0","-9.8", 0,0,8.0,20);
```

The argument of **setForce** is has the following signature:

setForce(*int id, String Fx, String Fy, double x0, double y0, double vx0, double vy0*);.

Here the object upon which the force acts is again the red ball, which is identified by id1. **Fx** and **Fy** are strings representing the components of the force in the x- and y-directions, respectively. These forces must appear in the setForce method with quotes around the function, as they are strings. The force can depend on x, y, vx, vy, and t, but in all cases must be a string. The following four arguments set the initial position of the object in x- and y-coordinates followed by the initial velocities in the x- and y-directions, respectively. These values are not strings but can be any number. In this example, the red ball will experience a linear restoring force, -9.8, solely in the y-direction, starting out at the origin and having initial velocity components of 8.0 and 20.0. Therefore, despite this change in the script, the red ball should have the exact same trajectory as in the previous example.

It is easy to change the type of problem being displayed by changing the force. In order to study simple harmonic motion, start with the html page used in the previous example and change the script as follows:

```
<html>
<head>
<script language = "JavaScript">
function shm1_0(){
document.animator.setAutoRefresh(false);
document.animator.setDefault();
document.animator.setPixPerUnit(5);
document.animator.setGridUnit(2);
id1 = document.animator.addObject("circle","r = 10,m = 1")
document.animator.setForce(id1,"-2*x","0",20,0,0,0);
document.animator.setRGB(id1, 255,0,0);
document.animator.setOneShot(0,2.3, "End of Animation");
document.animator.setTrail(id1,150);
document.animator.setFootPrints(id1,2);
document.animator.setGhost(id1,true);
document.animator.setAutoRefresh(true);
document.animator.forward();
}
</script>

<title>Projectile</title>
</head>
<body bgcolor = "#FFFFFF">
<h3 align = "left">Simple Harmonic Motion</h3>
<p align = "center">
```

```
<applet codebase = "classes" code = "animator4.Animator.class" archive =
"Animator4_.jar,STools4.jar"
align = "baseline" width = "400" height = "100"
id = "animator" name = "animator">
<param name = "FPS" value = "10">
<param name = "dt" value = "0.1">
<param name = "showControls" value = "false">
</applet>
</p>
<p align = "left"><a href = "JavaScript: shm1_0()">Start</a> </p>
.
.
</body>
</html>
```

Script 6: Simple harmonic motion.

You should also add the standard VCR buttons as in the previous examples. The ball will move in one dimension, so in the applet tag the **width** and **height** of the applet window have been changed and the **shiftPixOrigin** line has been removed to better show the simple harmonic motion soon to be scripted. To distinguish this script from the projectile script, the name of the function has been changed to **shm1_0()**. Finally, the animation is set to run for a half cycle of the motion (2.3 seconds) and will stop and display "End of Animation" with the **setOneShot** method.

The line

```
document.animator.setForce(id1,"-2*x","0",20,0,0,0);
```

sets the force that the ball will experience to be a linear restoring force, $F = -2x$. For some visual feedback, add ghosts:

```
document.animator.setTrail(id1,150);
document.animator.setFootPrints(id1,2);
document.animator.setGhost(id1,true);
```

below the **setRGB** line. Figure 42 should now appear on your Web browser after the page is reloaded.

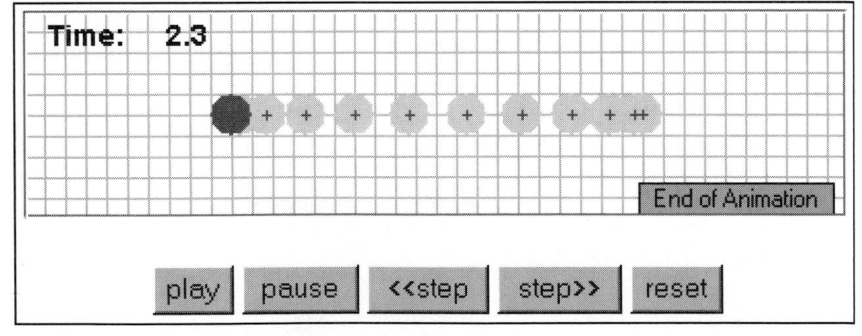

FIGURE 42: Simple harmonic motion applet with ghosts.

Thus far, the animation shows simple harmonic motion (the ghosts show that the velocity of the ball is greatest near $x = 0$ and the least at maximum compression/elongation of the spring), but there is no spring. To add a spring in this context, three additional lines must be added to the script. First, comment off or remove the last three lines added since the trail and the ghosts will get in the way of the spring. Then add

```
id2 = document.animator.addObject("rectangle", "w = 10,h = 40,x = -30");
document.animator.setRGB(id2,0,0,0);
document.animator.addConnectorSpring(id1,id2);
```

to the script. The first addition adds a rectangular anchor for the spring. The second line makes the anchor black, and the final line connects a spring from the ball to the anchor. Finally, change the animation time in the **setOneShot** command to 13.4 so the animation will run through three cycles and look like Figure 43.

Note that since only the red ball knows about the force, the anchor rectangle is indeed an anchor and does not move in response to the **setForce** method.

6.2.7 Harmonic Motion Sample Problems

The basic simple harmonic motion applet is now complete. New scripts can be written by varying the spring constant and the mass of the red ball. One can ask questions involving mass, velocity, acceleration, work, kinetic and potential energy, spring constant, and period.

By changing the arguments of the **setForce** and **setOneShot** commands to

```
document.animator.setForce(id1,"-2*x-0.2*vx", "0",20,0,0,0);
document.animator.setOneShot(0,40, "End of Animation");
```

damped harmonic motion arises from the original simple harmonic motion applet. Questions regarding kinetic energy, potential energy, and energy loss can now be asked. For example,

- If the equilibrium position is at $x = 0$ and the spring constant is $k = 5$ N/m, what is the force required to hold the mass in place at $x = 8$.

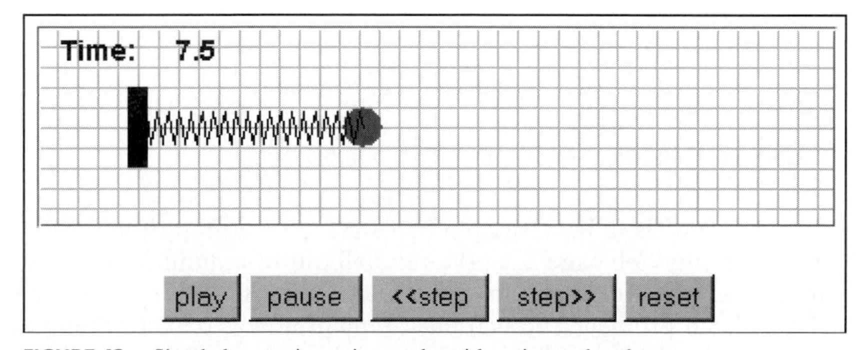

FIGURE 43: Simple harmonic motion applet with spring and anchor.

- If the equilibrium position is at $x = 0$ and the spring constant is $k = 5$ N/m, what is the initial energy of the system?
- What is the period for this motion? Does it depend on the amplitude?
- When and where are the kinetic energy and potential energy a maximum?
- Determine the percent of the initial energy lost during the time intervals: (0–4.5 s) and (4.5–13.5 s).

6.2.8 Mass

The mass of a particle is the ratio of the force to the acceleration in whatever units the script author has chosen. It can be set in one of two ways. The easiest way is to specify the mass as a parameter when the object is created using the addObject method.

```
id1 = document.animator.addObject("circle","r = 10, m = 0.5");
```

The default value for mass is one for geometric objects such as circles, boxes, and rectangles. It is zero for lines, text, and connectors.

The mass can also be set later in a script using the setMass method.

```
document.animator.setMass(id1, 2.5);
```

6.2.9 Dragging, Constraints, and Compound Objects

Another set of additions can make the reader of the page interact even more closely with the applet. Consider the following additions to the script below the **document.animator.setRGB(id1, 255,0,0);** line:

```
document.animator.setDragable(id1,true);
document.animator.setConstrainY(id1,0,-18,18);
document.animator.setShowConstraintPath(id1,false);
tid = document.animator.addObject("text", "text = drag me");
document.animator.setAnimationSlave(id1,tid);
document.animator.setDisplayOffset(tid,-20,15);
document.animator.setFont(tid,"Helvetica",1,12);
```

and the following change to the **setForce** command:

```
document.animator.setForce(id1,"-2*x","0",0,0,0,0);
```

The idea is to have the red ball start at its equilibrium position and have the reader of the Web page drag the red ball out of equilibrium, let go, and watch the animation. The first line allows the reader to drag the red ball. Since this is a one-dimensional problem, the ball must remain at $y = 0$ even when the student drags the ball. The next command **setConstrainY** does just that by limiting the red ball to

$-18 < x < 18$ so as to keep the animation physical (no over stretching or bunching up of the spring). Since the constraint is a necessary feature, but will get in the way of the animation, use **setShowConstraintPath** to hide the path. The next four lines place the text "drag me" in the animation. The **setAnimationSlave** is used to make the text to follow the red ball. In fact, since the slave will be placed at the center of the master object, make the text, "drag me," more readable by using **setDisplayOffset** and **setFont** to offset and format the text. The ball should start off in equilibrium, so the **setForce** command is used to place the red ball at the origin with no initial velocity.

This applet is now an excellent "laboratory" to test, for example, how the period of simple harmonic motion depends on the amplitude. Instead of hearing the independence of the period, students can test this idea in the virtual spring lab by setting initial amplitudes and then measure the period of motion. Sample questions could include the following:

- Drag the ball from its equilibrium position. Determine the period.
- Does the period depend on the amplitude?
- Determine the spring constant of the spring.
- Where is the potential energy the greatest? The least?
- Where is the kinetic energy the greatest? The least?
- When you drag the ball from its equilibrium position, where does the energy come from?
- How much work must you do on the system to move the ball from its equilibrium position to $x = 22$ cm? What is the potential energy at that point?

The two examples discussed in this section, projectile motion and simple harmonic motion, have stressed the primary methods and syntax involved in writing a script to control *Animator*. Other Physlets, such as *EField*, *DataGraph*, and *Poisson*, are scripted similarly. Of course, not all of the methods in *Animator* were discussed. (See the *Animator* documentation in Part Three and on the CD.) In fact, one of the most interesting and useful features of *Animator*, the fact that data (position, velocity, acceleration, and combinations of these) from an object can be sent to a graph, was not

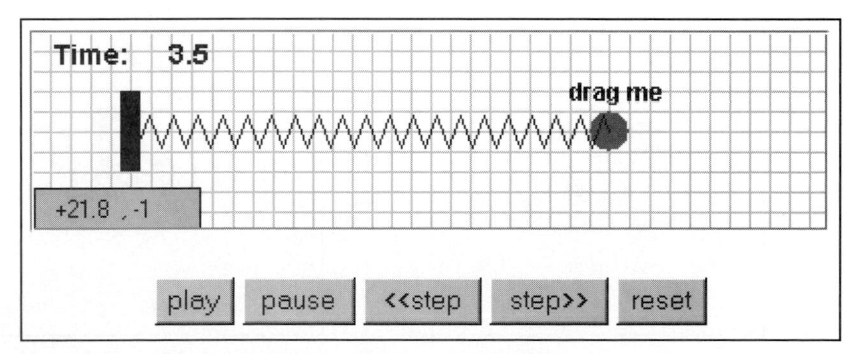

FIGURE 44: Final dragable simple harmonic motion applet.

discussed at all. For a discussion of these details, see Section 6.5, "Data Connections." For a full description of all the methods and their descriptions, see the *Animator* reference section in Part Three. For more examples, see Part Two and the problems on the CD.

6.3 *EFIELD*

6.3.1 Introduction

In *EField*, point charges interact with each other by a Coulombic force, with an external electrostatic potential, and with a constant magnetic field. Charges come in two varieties: (1) fixed charges, which feel and exert forces on all other particles but cannot move in response to forces; and (2) test charges, which move under the influence of the potential and the fixed charges but do not exert forces or influence the fields. Both types of charges can be moved if their dragable property is set to true, both can show their force vectors, and both can be labeled with colors or text. Just as in the *Animator* Physlet, geometric shapes such as circles, boxes, and arrows can be added to the Physlet and scripted to follow analytic trajectories. Objects may also be interconnected to follow each other across the screen using the setAnimationSlave method.

The external field may be any two-dimensional electrostatic potential accompanied by a constant magnetic field normal to the computer screen. (See Section 18.2, "*BField*" for a more complete simulation of magnetic fields.) Fields in *EField* can be represented with field vectors, field lines, and equipotential lines, or they may not be evidenced at all. Click dragging with the mouse can display values for position, field, force, and/or potential.

6.3.2 Forces on charges

A typical *EField* animation usually creates one or more charges using the addObject method. Our first script will create three dragable charges and their associated force vectors as shown in Figure 45. But before we begin scripting we must embed *EField* into an html page using the appropriate parameter tags.

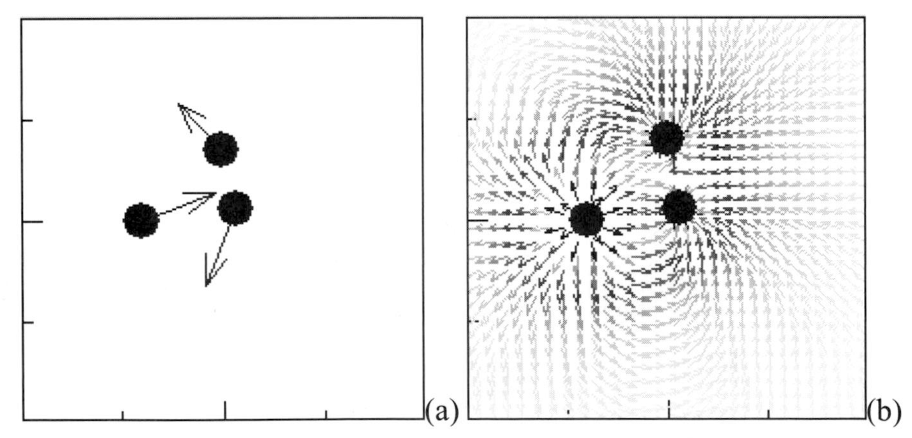

FIGURE 45: Three charged objects displaying (a) their net force and (b) their electric field.

```
<applet codebase = "../classes" archive = "EField4_.jar,STools4.jar" code =
    "eField4.EField.class" name = "efield" id = "efield" width = "350" height = "300">
<param name = "FPS" value = "10">
<param name = "dt" value = "0.02">
<param name = "ShowControls" value = "false">
<param name = "ShowContours" value = "false">
<param name = "ShowFieldLines" value = "false">
<param name = "ShowFieldVectors" value = "false">
<param name = "ShowLabels" value = "false">
<param name = "ShowCharge" value = "true">
<param name = "Potential" value = "0">
<param name = "Range" value = "-5,5,-5,5">
<param name = "Gridsize" value = "64">
</applet>
```

Enter the preceding lines into the body of an html page and check to make sure that the *EField* Physlet loads properly. The codebase attribute should, of course, be adjusted to refer to the correct classes directory on your hard disk or Web server. You may also want to set the ShowControls parameter to true and explore the applets capabilities via the user interface. You should also set the potential parameter to a nontrivial value and the ShowFieldVectors parameter to true and observe the effect.

Now add the following script to the head of the html document.

```
<script language = "JavaScript">
function initApplet(){
document.efield.setAutoRefresh(false);
document.efield.setDefault();
document.efield.setShowTime(false);
document.efield.setShowFieldVectors(false);
id = document.efield.addObject("charge", "x = -2,y = 0,q = 3");
document.efield.setDragable(id,true);
document.efield.setShowFVector(id,true);
document.efield.setRGB(id,0,0,0);
id = document.efield.addObject("charge", "x = 2,y = 0,q = -2");
document.efield.setDragable(id,true);
document.efield.setShowFVector(id,true);
document.efield.setRGB(id,0,0,0);
id = document.efield.addObject("charge", "x = 0,y = 2,q = -2");
document.efield.setDragable(id,true);
document.efield.setShowFVector(id,true);
document.efield.setRGB(id,0,0,0);
document.efield.setAutoRefresh(true);
}
</script>
```

Script 7 Create three dragable charges using the *EField* Physlet.

EField scripts follow the same pattern as *Animator* scripts. First, drawing is disabled using the setAutoRefresh method. Next, objects are created and their properties set to the appropriate values. In this example each charge is created, made dragable, asked to display its net force, and has its color set. Finally, drawing is enabled to update the screen with the newly created objects. Execute this script by adding the following anchor tag to the body of the document.

```
<a href = "JavaScript:initApplet()">
Initalize the applet.</a>
```

You should obtain a close approximation to Figure 45(a) when Script 7 is executed. Drag the charges and observe the change in the force vectors. Next, change the showFieldVectors boolean to true and the three showFVector booleans to false and reload the html page. The electric field produced by the charge distribution is now displayed as shown in Figure 45(b) using color-coded unit vectors.

The complete html page used to generate Figure 45 is shown in Script 8 for reference. It will be used as a template for other *EField* examples. Notice the use of the *onload* event handler in the <body>. This is yet another way to execute a script. The onload event is generated after the html page and its content, such as Java applets, have been loaded and initialized.[11] You may want to use an alternate method, such as an anchor tag or html button, to run this script if more than one script is being used.

```
<html>
<head>
<scriptlanguage = "JavaScript">
function initApplet (){
document.efield.setAutoRefresh(false);
document.efield.setDefault();
document.efield.setShowTime(false);
document.efield.setShowFieldVectors(true);
id = document.efield.addObject("charge", "x = -2,y = 0,q = 3");
document.efield.setDragable(id,true);
document.efield.setShowFVector(id,showForce);
document.efield.setRGB(id,0,0,0);
iid = document.efield.addObject("charge", "x = 0.28,y = 0.28,q = -2");
document.efield.setDragable(id,true);
document.efield.setShowFVector(id,showForce);
document.efield.setRGB(id,0,0,0);
id = document.efield.addObject("charge", "x = 0,y = 2,q = -2");
document.efield.setDragable(id,true);
document.efield.setShowFVector(id,showForce);
document.efield.setRGB(id,0,0,0);
document.efield.setAutoRefresh(true);
}
</script>
```

[11]There was a bug in early versions of Netscape Communicator that caused the onload event to be generated before JavaApplets were fully initialized, thereby freezing the browser. Browsers bugs are becoming less common but they are, unfortunately, a possibility.

```
<title>EField Example</title>
</head>
<body bgcolor = "#FFEFCE" onload = "initApplet()">
<h3 align = "left">EField4 Example</h3>
<p align = "center">
<applet codebase = "../classes" archive = "EField4_.jar,STools4.jar" code =
    "eField4.EField.class"
name = "efield " id = "efield " width = "200" height = "200">
<param name = "FPS" value = "10">
<param name = "dt" value = "0.02">
<param name = "ShowControls" value = "false">
<param name = "ShowContours" value = "false">
<param name = "ShowFieldLines" value = "false">
<param name = "ShowFieldVectors" value = "false">
<param name = "ShowLabels" value = "false">
<param name = "ShowCharge" value = "true">
<param name = "Potential" value = "0">
<param name = "Range" value = "-5,5,-5,5">
<param name = "GridSize" value = "64">
</applet>
</p>
</body>
</html>
```

Script 8: *EField* html page used to generate the three charges in Figure 45.

6.3.3 Potential Energy

Unlike *Animator*, which can be used to model any force, *EField* was designed to model only conservative forces. This restriction enables *EField* to calculate and display potential energy. Remove the line containing the setShowfieldVectors method in Script 8 and replace it with

```
document.efield.setShowContours(true);
```

Reload the page and observe the appearance of contour lines as shown in Figure 46(a). Drag a charge and notice that the contours are recalculated only after the mouse button is released. Unfortunately, it is still not possible to calculate contours in real time as the dragging is being performed since this calculation is very processor intensive.

We will now modify Script 8 to display a charged object's potential energy by slaving a text object to each charge. Set the show contours variable to false and add a text object to the Physlet immediately after each charge has been created. Make this text object an animation slave of the charge. Each charge requires the addition of three lines of script, as follows:

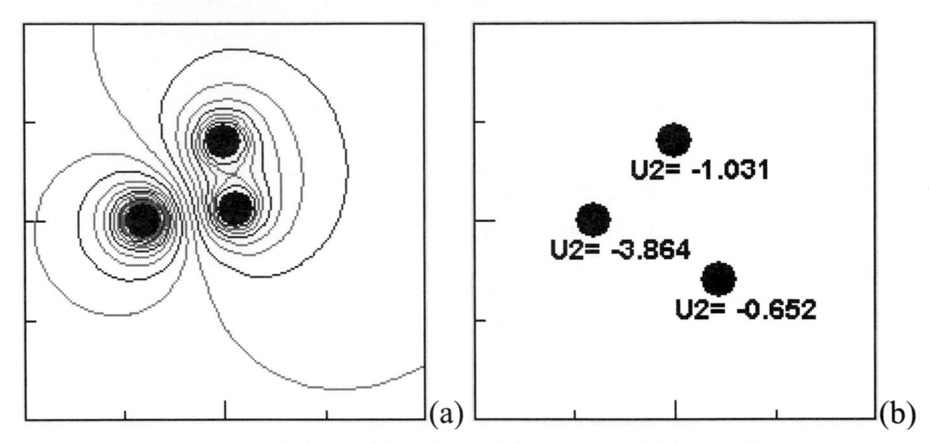

FIGURE 46: Three charged objects: (a) equipotential contours and (b) potential.

```
id = document.efield.addObject("charge","x = -2,y = 0,q = 3");
document.efield.setRGB(id,0,0,0);
tid = document.efield.addObject("text", "text = U1 = ,calc = p");
document.efield.setDisplayOffset(tid,-22,-20);
document.efield.setAnimationSlave(id,tid);
```

Drag any of the charges and observe the change in the display.

Since animation slaves take on the state of the master object, we can specify the potential, p, of the charge as the function to be shown in the text object's calculation field. Text objects that are animation slaves in *EField* can use any of the following state variables when defining the calculation: t, x, y, vx, vy, ax, ay, m, p, and f. The two variables p and f are only defined for certain objects and return zero otherwise. The variable p is the potential at the geometric center of point charges and test charges. The variable f is the electric flux per unit length passing through a box or shell object. An example of its use will be presented later in this tutorial. All other dynamic variables have their usual meaning.

The ideas in the forces and potential scripts that we have just presented can be used for a number of exercises. Three examples can be found on the accompanying CD:

- Create a number of charges on the screen with distinctive labels and ask the students to rank them from most negative to most positive.
- Ask students to determine which charges are of like sign.
- Create charges and a dragable test charge and find the equilibrium points.

6.3.4 Particle and Field Lines

The previous *EField* examples used static charges that do not move in repose to their net force. This example creates a dipole and places a dynamic test charge within it, as

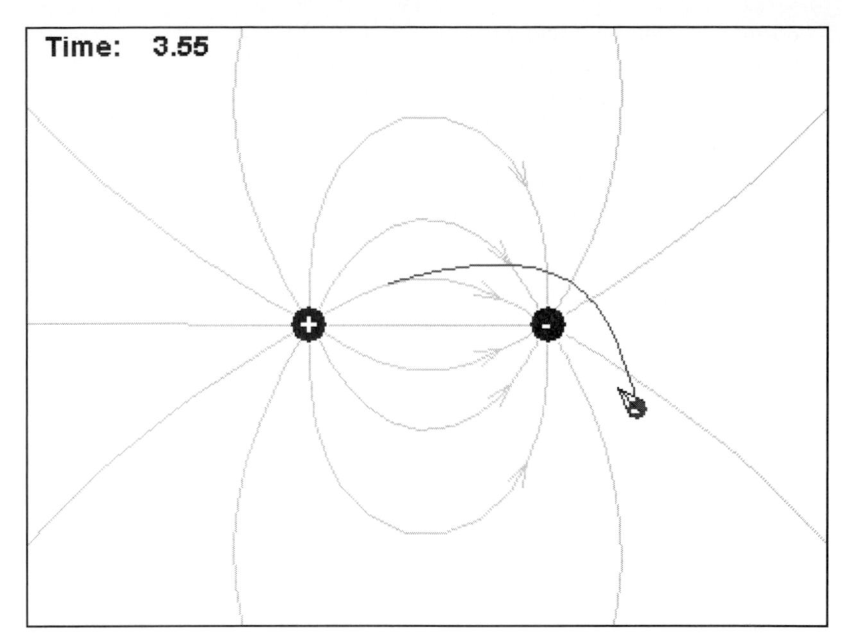

FIGURE 47: *EField* dipole field.

shown in Figure 47. The two regular charges that form the dipole are black and are labeled '+' and '−'. The user cannot drag these charges. We use another method of *EField* in order to draw electric field lines.[12] The small test charge can be moved if the user click-drags using the computer's mouse. After the mouse is released, the charge accelerates and leaves behind a permanent trail.

```
<script language = "JavaScript">
function initApplet(){
document.efield.setDefault();
document.efield.setAutoRefresh(false);
document.efield.setShowFieldLines(true);
id = document.efield.addObject("charge", "x = -1.5,y = 0,q = 1");
document.efield.setRGB(id,0,0,0);
document.efield.setDragable(id,false);
document.efield.setLabel(id,"+");
id = document.efield.addObject("charge", "x = 1.5,y = 0,q = -1");
```

[12]Script authors should use the electric field lines option cautiously. A canonical algorithm to generate electric field lines from a distribution of charges does not exist. In fact, most textbook figures finesse the problem of how to represent three-dimensional field lines on a two-dimensional page. Does the straight line connecting the two charges in Figure 47 project into or out of the page? Does this line hide other lines that have the same projection? You may want to script a half-dozen dragable charges of varying magnitudes and observe the field lines. Decide for yourself how helpful the resulting field line diagrams are likely to be for beginning students. We believe that field vectors are a more robust and instructive visualization of electric fields.

```
document.efield.setRGB(id,0,0,0);
document.efield.setDragable(id,false);
document.efield.setLabel(id,"-");
id = document.efield.addObject("testcharge", "x = -0.5,y = 0.5,r = 5");
document.efield.setDragable(id,true);
document.efield.setShowFVector(id,true);
document.efield.setRGB(id,255,0,0);
document.efield.setTrail(id,255);
document.efield.setAutoRefresh(true);
document.efield.forward();
}
</script>
```

Script 9: A dipole field with a dynamic test charge.

Since the name of the embedded applet has not changed, we can enter Script 9 into the *EField* template that was used previously. You should also adjust the width and height of the embedded applet to 400 and 300 pixels, respectively. Finish the html page by adding the usual play, pause, step, and reset buttons. A quick way to do this is to copy the html button code fragment from the template developed in the *Animator* tutorial. Be sure and replace all occurrences of "animator" with "efield" after you have copied the code. The method names, such as forward() and pause(), that control the animation are, after all, the same.

The trajectory of the test charge clearly cuts across the field lines as the charge moves. This example can be used to test the common misconception that charges follow field lines or that field lines somehow represent trajectories. Have students start the test charge at various locations and ask them to explain the relationship between the trajectory and the field line.

6.3.5 Modeling a Velocity Selector with Magnetic Fields

The following example creates a test charge moving in crossed electric and magnetic fields. The magnetic field points out of the screen, and the electric field points in the vertical direction. The electric field is set using the setPotential function in the script. The test charge displays a force vector. The script and the embedding commands are shown.

```
<script language = "JavaScript">
function initApplet(ey,bz){
potStr = "y*"+ey;
document.efield.setAutoRefresh(false);
document.efield.setDefault();
document.efield.setPotential(potStr,-30,30,-20,20);
document.efield.setBz(eval(bz));
id = document.efield.addobject("testcharge", "x = -20.0,vx = 5.0");
document.efield.setShowFVector(id,true);
document.efield.setShowVVector(id,false);
```

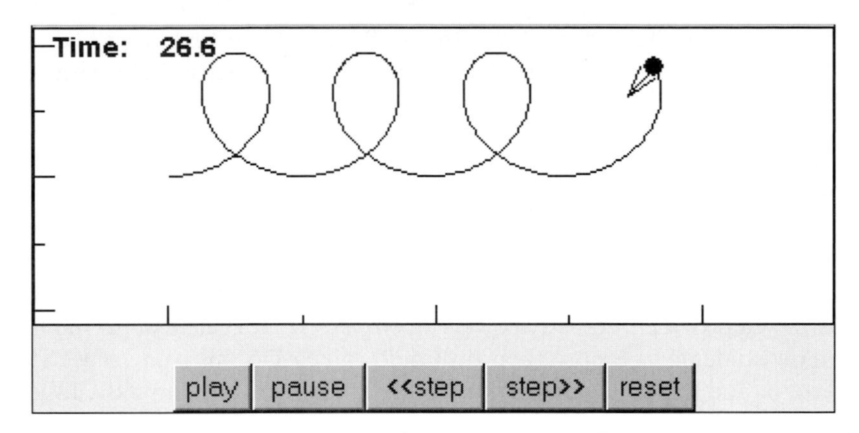

FIGURE 48: *EField* Physlet—crossed electric and magnetic fields.

```
document.efield.setDragable(id,false);
document.efield.setTrail(id,800);
document.efield.setMaxTime(80,"Animation Stopped");
document.efield.setAutoRefresh(true);
document.efield.forward();
}
</script>
.
.

<applet codebase = "../classes/" archive = "EField4_.jar, STools4.jar" code =
    "eField4.EField.class" name = "efield" id = "efield" width = "400" height = "200">
<param name = "FPS" VALUE = "10">
<param name = "dt" VALUE = "0.05">
<param name = "ShowControls" value = "false">
<param name = "ShowContours" value = "false">
<param name = "ShowFieldLines" value = "false">
<param name = "ShowFieldVectors" value = "true">
<param name = "ShowLabels" value = "false">
<param name = "ShowCharge" value = "true">
<param name = "Potential" value = "-20*y">
<param name = "Range" value = "-40,40,-20,20">
<param name = "GridSize" value = "64">
</applet>
```

Script 10: Charged particle in crossed electric and magnetic fields.

Pay close attention to the use of two parameters, ey and bz, that are used to set the electric and magnetic field, respectively. Since JavaScript does not enforce data types, these parameters could be passed to the script either as numbers or as strings. Defensive programming is necessary to ensure that the correct data type is passed to the Physlet's methods.[13] Since the setPotential method requires that the potential energy function be a string, we concatenate the electric field with the string "y*" to form the electric potential function. JavaScript will convert operands of a plus operator into strings before concatenation if either argument is a string. The setBz method, however, requires a number, and we must ensure that the argument, bz in this case, is a number. The JavaScript function eval() performs this conversion.

Perpendicular E and B fields are commonly used as velocity selectors in ion beams. One possible use of Script 10 is in a pre-laboratory exercise for an introductory modern physics laboratory, where students measure the *e*/*m* ratio of the electron. In order to provide interactivity, we can embed two text fields into the html document to allow students to choose their own field values. Do this using a WYSIWYG authoring tool or add the following markup to the html page containing the following:

```
<form name = "dataForm">
<p>
Ey = <input type = "text" name = "E" size = "10" value = "0"
Bz = <input type = "text" name = "B" size = "10" value = "0">
<input type = "button" value = "Set" onclick =
"initApplet(document.dataForm.E.value,
document.dataForm.B.value)">
</p>
</form>
```

The preceeding markup should produce two text fields and a button as shown:

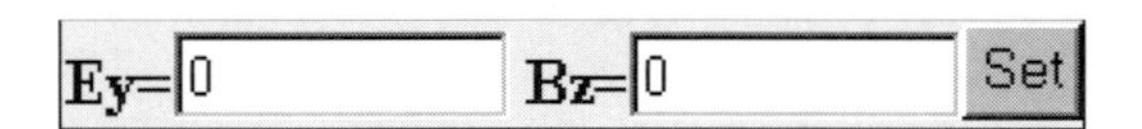

The dot notation for accessing html text fields inside a named form field is consistent with the notation for scripting an applet. The name of the form containing the text fields is dataForm, and the first text field is named E. Therefore, we access the value in this text field as document.dataForm.E.value. The second text field is accessed in a similar manner. These values are passed to the initApplet function in order to initialize the appropriate conditions whenever the button is clicked.[14]

6.3.6 Geometry and Gauss's Law

Whenever three-dimensional space is rendered in two dimensions, a decision is made as to the geometry of on-screen objects. Does a circle represent the cross section of a

---

[13]Browsers have the nasty habit of converting numbers to strings whenever they see fit. Calling the initApplet function from within an html page with numeric arguments, initApplet(1,0.5), does not guarantee that these parameters will be passed at numbers.

[14]A more robust script would check to see if the parameters that were entered into the text field are numeric and within a predetermined range. JavaScript was, in fact, designed to perform this type of data processing before the information was passed on to common gateway interface (cgi) programs on the Web server. Consult any good book on JavaScript programming to learn more about JavaScript's capabilities.

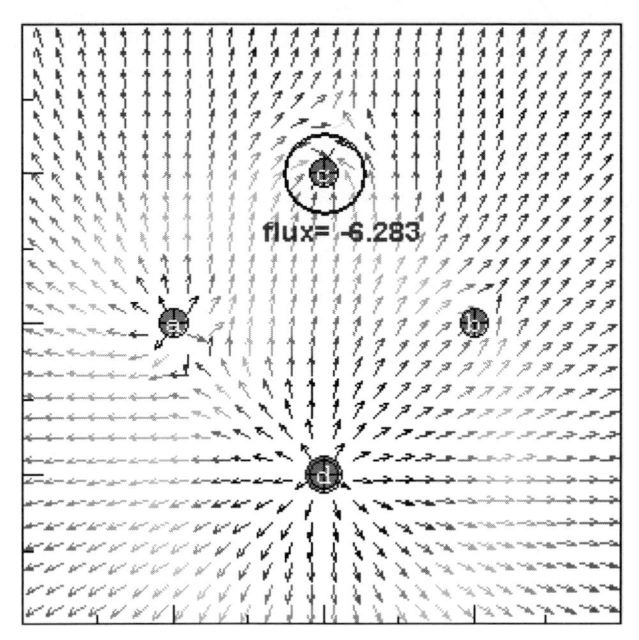

FIGURE 49: Flux measurement using *EField*.

sphere or a cylinder? The script author can, in fact, choose either interpretation by setting a parameter tag when *EField* is embedded.

```
<param name = "PointChargeMode" value = "false">
```

The choice of mode is fundamental since true or false values determine if two charges interact via a $1/r^2$ or a $1/r$ force law, respectively. Add this embedding parameter tag to the *EField* template that was developed at the beginning of the *EField* tutorial and observe the effect of changing the PointChargeMode parameter. The visual effect is subtle and will go unnoticed by users unless it is brought to their attention in the text. Although charges are drawn with a slightly different fill pattern depending on the mode, the definitive test of the geometry is to measure the radial dependence of the potential or force between two interacting charges. Whenever PointChargeMode is set to false, the charge parameter, q, in the addObject method represents the charge per unit length of a line charge. The simulation will do the right thing and show vectors representing force per unit length. Since the external potential function, $U(x, y)$, does not depend on z, it is unaffected by the choice of geometry. The potential energy is always calculated as $q*U$ and therefore represents either the potential energy of a point charge or the potential energy per unit length.

In order for *EField* to model Gauss's law, geometric objects, such as box and circle, are able to calculate the flux passing through their Gaussian surface. This calculation is easy to perform for static charges. The simulation assumes that a shell is a hollow sphere in point charge mode and a unit cylinder in line charge mode. A script to create four charges and a dragable shell is easy to write using the *EField* template.

```
<script language = "JavaScript">
function initEField(){
document.efield.setAutoRefresh(false);
document.efield.setDefault();
doucment.efield.setShowFieldVectors(true);
document.efield.setShowTime(false);
id = document.efield.addObject("charge", "x = -2,y = 0,q = 0.75,r = 7");
document.efield.setRGB(id,0,0,0);
document.efield.setLabel(id,"a");
id = document.efield.addObject("charge", "x = 2,y = 0,q = -0.25,r = 7");
document.efield.setRGB(id,0,0,0);
document.efield.setLabel(id,"b");
id = document.efield.addObject("charge", "x = 0,y = 2,q = -0.5,r = 7");
document.efield.setRGB(id,0,0,0);
document.efield.setLabel(id,"c");
id = document.efield.addObject("charge", "x = 0,y = -2,q = 0.625,r = 7");
document.efield.setRGB(id,0,0,0);
document.efield.setLabel(id,"d");
bid = document.efield.addObject("shell", "w = 30,h = 30");
document.efield.setDragable(bid,true);
tid = document.efield.addObject("text", "text = flux = ,calc = f");
document.efield.setRGB(tid,255,0,0);
document.efield.setDisplayOffset(tid,-30,-35);
document.efield.setAnimationSlave(bid,tid);
document.efield.setAutoRefresh(true);
}
</script>
```

Script 11: Gauss's law simulation with point charges.

A more advanced variation of this script uses a potential function with a nonzero divergence in order to simulate a continuous charge distribution. Remove the addObject methods that created the charges in Script 11 and add a setPotential statement as shown.

```
<script language = "JavaScript">
function initEField(){
document.efield.setAutoRefresh(false);
document.efield.setDefault();
document.efield.setShowFieldVectors(true);
document.efield.setShowTime(false);
document.efield.setPotential("x*y*x/10", -4,4,-4,4);
bid = document.efield.addObject("shell", "w = 30,h = 30");
document.efield.setDragable(bid,true);
tid = document.efield.addObject("text", "text = flux = ,calc = f");
document.efield.setRGB(tid,255,0,0);
```

```
document.efield.setDisplayOffset(tid,-30,-35);
document.efield.field.setAnimationSlave(bid, tid);
document.efield.setAutoRefresh(true);
}
</script>
```

Script 12: Gauss's law simulation with a continuous charge distribution.

This simulation is only accurate if PointChargeMode is false since cylindrical geometry is assumed. The potential function cannot have a z-dependence and hence the charge distribution will be independent of z.

6.4 *DATAGRAPH*

6.4.1 Introduction

DataGraph is a general-purpose graphing tool designed to collect and display data that have been generated within other Physlets. These data can then be relayed to *DataGraph* using the interapplet communication mechanism described in the next section. This section introduces *DataGraph* as a stand-alone applet and presents a brief tutorial describing its most common methods.

DataGraph organizes data points into one or more data sets called a series. Each series has a drawing style that specifies color and marker shape. Multiple color-coded data sets can, therefore, be displayed on a single graph. In addition, *DataGraph* can display one or more analytic functions. Just as in *Animator* and *EField*, display properties

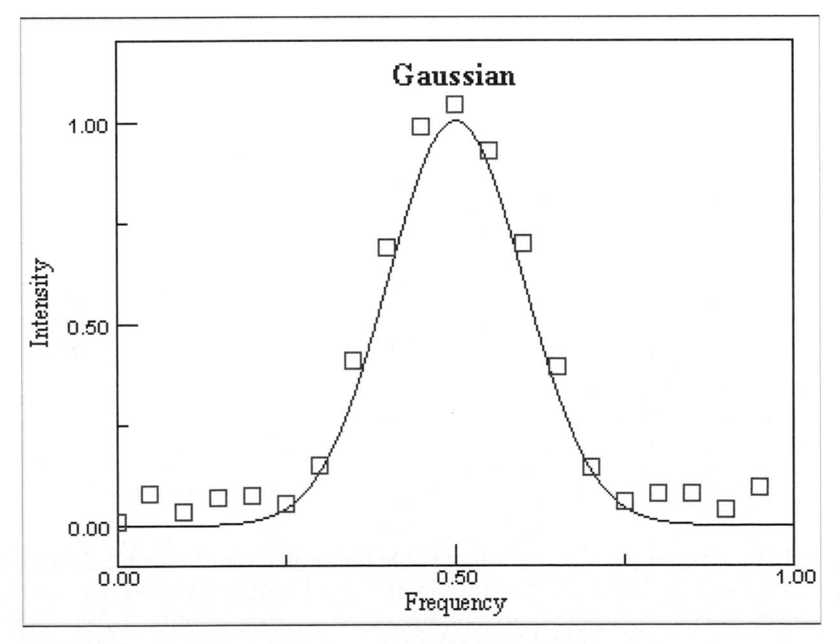

FIGURE 50: *DataGraph* displaying a data series and a function.

of functions can be accessed using the object identifier that is returned when the function is added to the graph.

6.4.2 Parameter Tags

DataGraph is embedded into an html page using the following parameter tags:

```
<applet codebase = "../classes"
archive = "DataGraph4_.jar,STools4.jar"
code = "dataGraph.DataGraph.class" name = "dataGraph" id = "dataGraph"
width = "300" height = "300">
<param name = "Function" value = "sin(1/x)">
<param name="XMin" value=-5">
<param name = "XMax" value = "5">
<param name = "YMin" value = "-1.1">
<param name = "YMax" value = "1.1">
<param name = "AutoScaleX" value = "false">
<param name = "AutoScaleY" value = "false">
<param name = "ShowControls" value = "true">
<\applet>
```

These parameter tags are, as always, optional. The range parameters are self-explanatory. They are ignored if the autoscale property of an axis is set to true. The Function parameter defines an analytic function of x that will be drawn inside the graph. Additional analytic functions can be added with JavaScript using the addFunction method.

6.4.3 Series Styles

Although JavaScript may not be suitable for large programming projects, it is ideal for quick and simple tasks, such as generating artificial data. Consider the following script fragment:

```
x = 0;
dx = 1.0/20.0;
for(i = 0; i<20;i++){
     y = 2*x-4+Math.random()*0.2;
     document.dataGraph.addDatum(1,x,y);
     x = x+dx;
}
```

This fragment starts by defining three variables, x and dx, and assigning values to these variables. The *for* loop, starting in the third line, is used to add 20 data points to the Physlet. Loops in JavaScript follow a syntax made popular by the C programming language. The first argument defines a counter, i, and sets its value to zero. The second argument specifies a condition that must be met in order for the loop to continue running. In this example, the counter i must be less than 20 or the loop will exit after the closing curly bracket. The final argument specifies an action to be performed at the end of the loop. This action usually increments the counter. Our loop contains three statements. The first calculates the y-value. The second adds the data point to the graph. The third increases the value of x in preparation for the next data point. The addDatum method specifies not only the (x, y) coordinate pair, but also the series number.

DataGraph uses series numbers in order to organize data sets. These numbers serve the same purpose as object identifiers, except that they are usually sequential starting at 1. They are assigned by the script author and not the Java VM (Virtual Machine). Almost all series display properties, such as style and color, are set using series numbers. Series style is set using the setStyle method with the following signature:

```
setSeriesStyle(int id, boolean connected,int marker);.
```

The first integer identifies the series number. The second determines whether data points are to be connected, and the third determines the type of marker to be placed at the data point. Marker styles are as follows:

0—one-pixel dot
1—circle
2—square
3—cross

Negative marker styles are reserved for special display modes. Styles of -1 and -2 are used to display areas under curves. Style -1 fills the area between the graph and the x-axis with a solid fill color. Style -2 is similar except that the colors above and below the axis can be different. See problem 8.4.3 for an example.

Marker style -3 is useful for drawing histograms or bar charts. Each coordinate pair, (x, y), is interpreted as a bar drawn from the x-axis to a height y. A good example of its use is the display of the ensemble velocity distribution in molecular dynamics simulations (see Section 18.7).

Setting the connected boolean to true and the marker style to 0 will connect the data points with a solid line without markers

```
document.dataGraph.setSeriesStyle(1,true,0);
```

while setting connected to false and marker style to zero will produce a one pixel dot at each data point.

The setSeriesRGB method is often used in conjunction with setSeriesStyle to set display color. The first integer again identifies the series. The following three integers are restricted to the range [0,255] and specify the amount of red, green, and blue color. For example, the following statement sets series 1 in the applet named dataGraph to red.

```
document.dataGraph.setSeriesRGB(1,255,0,0);
```

Using series numbers to access a data set is a convenience. Data series are, in fact, stored as Java Objects and can be accessed using the object identifier. This identifier can be obtained from the series number using the getSeriesID method. Therefore, it is also possible to set the color using the setRGB method introduced in the *EField* and *Animator* Physlets.

```
id = document.dataGraph.getSeriesID(1);
document.dataGraph.setRGB(id,255,0,0);
```

The following listing shows how a typical *DataGraph* script that generates and displays a data set representing a linear function with slope of 2 and y-intercept of -4, along with some data points having simulated random scatter.

```
<script language = "JavaScript">
function plotData(){
document.dataGraph.setAutoRefresh(false);
document.dataGraph.setDefault();
document.dataGraph.setTitle("Simulated Data");
document.dataGraph.setLabelY("Y");
document.dataGraph.setLabelX("X");
document.dataGraph.clearSeries(1);
document.dataGraph.setSeriesStyle(1,false,2);
document.dataGraph.setSeriesRGB(1,255,0,0);
x = 0;
dx = 1.0/20.0;
for(i = 0; i<20;i++){
  y = 2*x-4+Math.random()*0.2 ;↵
  document.dataGraph.addDatum(1,x,y);
  x = x+dx;
  }
document.dataGraph.setAutoRefresh(true);
}
</script>
```

Script 13: Script used to generate a data set with simulated experimental data.

The setDefault method in the preceding script is important. It removes all series, functions, and other on-screen objects. Without this method, data points would accumulate in the graph every time the script runs.[15] Other methods, such as clearSeries or setDefault, could have been used to clear out old data. These methods are useful when a single data set must be removed while leaving other data unchanged.

---

[15]In order to keep users from shooting themselves in the foot, we have set the maximum allowable number of data points to 10,000.

6.4.4 Math Parser

DataGraph is an excellent tool for evaluating parsed functions for use in other scripts. The parser understands the following single parameter functions $f(a)$:

abs(a)	acos(a)	acosh(a)	asin(a)	asinh(a)	atan(a)	atanh(a)	ceil(a)
cos(a)	cosh(a)	exp(a)	frac(a)	floor(a)	int(a)	log(a)	random(a)
round(a)	sign(a)	sin(a)	sinh(a)	sqr(a)	sqr(a)	sqrt(a)	step(a)
tan(a)	tanh(a)						

as well as the following two-parameter functions $f(a, b)$:

atan2(a,b) max(a,b) min(a,b) mod(a,b)

A good way to test the behavior of an analytic function is to embed *DataGraph* with the showControls parameter set to true. Enter the function into the $F(x)$ text field and see if it evaluates correctly. The y-axis will autoscale.

The four two-parameter functions are particularly useful for generating periodic behavior. Try the following functions and observe the resulting graphs.

```
int(x)
floor(x)
sign(sin(x))
mod(x,1)
frac(2*sin(x))
max(0,sin(x))
```

The atan2(a,b) function is far more useful than the atan(a) function since it properly resolves angles in the second and third Cartesian coordinate quadrants. Note, however, that the y-coordinate should be the first argument when this function is called, atan(y,x).

6.4.5 Clocking Data

DataGraph has an animation clock even though this applet was not designed to perform animation. Before we begin using the clock, we should explain where the animation clock comes from. This clock was not added to the *DataGraph* program; it came for free when we decided to build this applet using the same codebase as *Animator* and *EField*. But the clock does come in handy when we wish to create data. Rather than writing a JavaScript loop to generate a data set as was done in Script 13, we can use the clock. This technique has the advantage that the graph does not appear all at once but is drawn at a predetermined rate.

One of the hallmarks of object-oriented programming is the ability of objects to inherit functionality from other objects. This inheritance is, however, strictly controlled through a parent → child relationship. In Java, a child is derived from a single parent.[16] Since a child knows its parent, a child can perform any of its parent's methods, but a parent cannot perform its child's methods. This inheritance has important consequences. Since many recently written Physlets derive from the parent class SApplet, script authors have access to SApplet's methods.[17] These methods include the following:

```
forward();
pause)();
getClockID();
```

Since scripts can access any public method in the Physlet's superclass, it is important that script authors look at methods documented in SApplet in addition to methods that are documented as part of the Physlet itself. Besides the clock methods just listed, there is a superclass method called makeDataConnection. A data connection takes data from a source and passes it to a listener. We will use it extensively in the next section for interapplet communication. The following script uses it to make a connection between two objects inside DataGraph, the clock and a data set series.

The following script should be copied into the *DataGraph* template and run.

```
<script language = "JavaScript">
function initGraph(){
document.dataGraph.setAutoRefresh(false);
document.dataGraph.setDefault()
document.dataGraph.addFunction("t","exp(-t/5)");
document.dataGraph.addFunction("t","-exp(-t/5)");
document.dataGraph.setSeriesStyle(1,true,0);
document.dataGraph.setLastPointMarker(1,true);
document.dataGraph.setLabelX("Time");
document.dataGraph.setLabelY("Position");
document.dataGraph.setTitle("DataGraph Clock Example");
document.dataGraph.setClockOneShot(0,15);
cid = document.dataGraph.getClockID();
gid = document.dataGraph.getGraphID();
document.dataGraph.deleteDataConnections();
document.dataGraph.makeDataConnection(cid,gid,1,"t","exp(-t/5)*cos(2*t)");
document.dataGraph.setAutoRefresh(true);
document.dataGraph.startClock();
}
</script>
```

Script 14: Use an animation clock to generate data at a predetermined rate.

---

[16]The process of creating a child class from a parent class is also known as subclassing. A child is said to be a subclass of its parent and the parent of a child is called its superclass. This scheme was, of course, first developed for the taxonomy of plants and animals. A dog, for example, is a mammal and has all the properties of mammals. Dogs are therefore a subclass of mammals. You can subclass dogs into collies, terriers, and other breeds to produce a treelike data structure. Physlets are usually subclassed from SApplet, which is itself subclassed from the Applet class provided by Sun Microsystems.

[17]See Chapter 12 for a discussion of SApplet's inherited methods.

The makeDataConnection method takes five variables. The first argument is an integer that identifies the data source. In this example, the data source is a clock, and its id is returned by the getClockID method. The second argument is an identifier for the data listener. The data listener is the graph. The third argument is the series number that will contain the data. The last two arguments are functions for processing the data before they are passed to the source. These functions can use variables that are predefined in the data source. The animation clock only defines a single variable, *t*. The two functions are evaluated at every time step before the data are passed on to the data listener. Script 14 specifies these functions to be time for the horizontal axis and a decaying oscillation for the vertical axis. The following section will deal extensively with data connections and how they can be used for interapplet communication.

6.5 DATA CONNECTIONS

6.5.1 Introduction

A persistent problem facing educational software authors is that programming has no top. In other words, the moment a program is finished a potential user will ask for an enhancement or modification to use the program in a new context. Physlets provide a partial solution since the user can specify the applet's behavior if the applet author has anticipated the need. Some behaviors are sufficiently general that they occur in many different physics problems. Data recording and visualization are examples of such behaviors. Curriculum authors would certainly expect to graph relevant physical quantities as an object moves, but it would be difficult to anticipate the various combinations called for. Will an author want to plot position, energy, or force? What if there are a dozen objects? Physlets implement a programming interface that allows data to pass between applets. Many Physlets, including *Animator*, *EField*, *BField*, and *Faraday*, are capable of generating data in response to an internal clock or in response to user actions. Interapplet communication makes it possible to display these data as a bar graph, a table of numeric values, or an *x*-*y* graph.

In order to implement this interapplet communication, an object in the sending applet must implement the SDataSource interface, and the receiving applet must implement the SDataListener interface.[18] JavaScript can then be used to set up a data connection between the source and the listener using a one-line JavaScript statement with the following signature:

makeDataConnection(*int sid, int lid, int series, String xfuncion, String yfunction*);.

The first two parameters, sid and lid, are integer identifiers for the data source and data listener objects, respectively. The third parameter is a series identifier that the data listener can use to keep track of multiple data sets. The last two parameters are strings representing mathematical functions of the data source variables. These functions are evaluated by the data connection to produce either a data point or an entire data set that is then passed on to the data listener. This scheme is very flexible since the only

[18]Interface is a Java technical term. An interface is a specification for one or more methods that a class must implement. For example, any class that implements the SDataListener interface must have a method called registerDatum.

change that needs to be made to a script in order to change the data being passed is to edit the last two parameters. For example, to change the connection from position, $x(t)$, to kinetic energy, ke(t), requires that the data connection be changed from

```
makeDataConnection(sid, lid, 1, "t", "x");
```

to

```
makeDataConnection(sid, lid, 1, "t", "vx*vx*m/2");.
```

6.5.2 Object Identifiers

As was discussed earlier, an object identifier, id, is an integer that uniquely identifies a *Java Object*. Interapplet communication among Physlets is based on these identifiers. Integers are used since JavaScript does not support the passing of entire Objects as parameters in a browser-independent manner. The documentation for each Physlet lists the available data sources and data listeners and the most efficient way of obtaining their ids. Identifiers are often returned when an on-screen object is created using an *add* method. Animator, for example, returns an id whenever an on-screen circle or rectangle is created using

```
id = document.animator.addObject("circle", "r = 6,x = 1, y = 2");.
```

Likewise, *EField* returns an identifier when a rectangle is created.

```
id = document.efield.addObject("rectangle", "w = 20, h = 10, x = 1, y = 2");.
```

Some objects are not created with JavaScript since they are already present when a Physlet is embedded into a page. Since these objects already exist, their id is obtained using the appropriate *get* method. The *Animator* Physlet contains an ensemble of shapes that move on the screen and an internal clock that drives the animation. These ids are obtained through the methods

```
id = document.animator.getEnsembleID();
```

and

```
id = document.animator.getClockID();,
```

respectively. Another example of a preexisting object is the graph object inside the *DataGraph* Physlet. The id of this object is obtained using the method

```
id = document.dataGraph.getGraphID();.
```

6.5.3 Source Variables

Data sources define variables that are designed to communicate the physics of the object to other Physlets. Rectangles and circles in the *Animator* Physlet, for example, keep track of eight variables: t, x, y, vx, vy, ax, ay, and m. Test charges in *EField* keep track of additional variables such as the charge, q, and the potential energy of that charge, p. Objects in *EField* that have a perimeter, that is, rectangles and circles, track the electric flux, *f*, that is passing through the object's surface. The clock object that is available in every SApplet, on the other hand, only keeps track of a single time variable, *t*.

Data source variables are dynamic. That is, they change whenever the configuration of the system changes in response to actions such as click-dragging on-screen or running a simulation clock. Each of these actions will cause the applet containing the data source to notify its data connections that new data are available. Data connections will also be notified if JavaScript is used to change an object's state. It is the applet's responsibility to calculate new state variables and to pass these values to the data connection. It is the data connection's responsibility to process this information using the functions that were defined in the makeDataConnection method and pass on these processed values to a data listener.

6.5.4 Listener Behaviors

A common data listener is the *DataGraph* Physlet. It was designed to enable multiple data sources to connect to different data sets in a single graph. The following code fragment shows how to establish a connection between two *Animator* objects and a graph.

```
cid1 = document.animator.addObject("circle",↵
"r = 9,x = 0,y = 0");
cid2 = document.animator.addObject("circle",↵
"r = 9,x = 0,y = 0");
// physics goes here
gid = document.dataGraph.getGraphID();
document.dataGraph.setSeriesRGB(1,255,0,0);
document.dataGraph.setSeriesRGB(2,0,0,255);
// set style parameters for dataGraph
document.animator.deleteDataConnections();
document.animator.makeDataConnection(cid1,gid,1,"t","x");
document.animator.makeDataConnection(cid2,gid,2,"t","x");
```

Script 15: Code fragment establishing two data connections.

Connections are always made from the source applet to the listener applet. Since the clock in the *Animator* Physlet drives the simulation, *Animator* makes the data connection. Each datum is processed by the connection and passed on to the graph, where it is appended to the specificd scrics.

Some data sources send entire data sets rather than data points. The *Pipes* Physlet, a simulation of sound waves in organ pipes, sends an array containing the pressure readings along the pipe after every time step. The data connection does the right thing when presented with an array of data: It processes the entire array using the connection functions and then passes an array of points to the listener. Data listeners should react differently if they receive more than one datum in a single event. DataGraph, for example, is programmed to clear the existing series and replace it with the new data whenever it receives an array, but it appends a datum to an existing series when it receives a single data point. Consult the documentation for each data listener to determine its behavior.

6.5.5 Trust Between Applets

It is important to realize that a browser only has one Java VM and that all applets share the resources of this virtual machine. (A Java VM behaves very much like a real microprocessor with resources, I/O, and memory.) When an object is created in the address space of this VM, memory is allocated for the variables for that object. *EField*, for instance, sets aside memory for the magnitude, position, and display color of each charge in addition to references to its methods (i.e., functions) that enable that charge to move under the action of the local fields at its current position. It is also possible for the Java programmer to set aside memory for variables that will be shared by all objects created by an applet, and it is even possible to set aside memory for variables that are shared by all Physlets. Such variables are known as *static variables*. The Java security model does, however, place strict restrictions on access to these static variables. These restrictions are necessary to prevent an applet embedded in an html page written by ShadyCompany.com from accessing an applet in a Web page from MyBank.com. Static variables can be shared between applets only if the applets were downloaded from the same Web site and share code contained in common archive (i.e., jar) files. In addition, both applets must derive (i.e., inherit) from a common class. As stated previously, all Physlets that implement interapplet communication must be derived from the common SApplet superclass.[19] SApplet itself derives from the Applet superclass and therefore inherits all the methods necessary for embedding into an html browser. Consequently, Physlets can communicate with one another but not with other applets that you may have downloaded from the net.

The use of static variables is key to our implementation of interapplet communication. The Physlet superclass, SApplet, has static data structures that keep track of data sources and data listeners that have been created by any Physlet. These data structures do not belong to any one Physlet but are available to all Physlets running on a given VM. A data source in Physlet A can access this static data structure and retrieve a reference to a data listener in Physlet B. Data will then be sent directly from the data source to the data listener without an intermediary. This data passing mechanism is very fast. All that is required is that the script author establishes a connection between the two Physlets, as described in the previous sections.

The key to establishing the required trust relationship between applets is to have *exactly* the same codebase and archive files when each applet is embedded. If the three Physlets (*Animator*, *DataGraph*, and *DataTable*, for example) are to communicate, they should be embedded as follows:

```
<applet
codebase = "../classes/"
code = "dataGraph.DataGraph.class"
archive = "Animator4_.jar,DataGraph4_.jar,DataTable4_.jar, STools4.jar">
```

[19]Older Physlets that do not yet derive from this class are being rewritten.

and

```
<applet
codebase = "../classes/"
code = "animator4.Animator.class"
archive = "Animator4_.jar,DataGraph4_.jar,DataTable4_.jar, STools4.jar">
```

and

```
<applet
codebase = "../classes/"
code = "dataTable.DataTable.class"
archive = "Animator4_.jar,DataGraph4_.jar,DataTable4_.jar, STools4.jar">
```

Note that the *code* (i.e., the applet that will appear on the page) is the only difference in the embedding tags. **Read the last sentence again!** In order for a data connection to be established, codebase and archive tags must be identical for all applets on a page. (Height and width attributes have been omitted for clarity. These attributes can, of course, be different for each applet.) Every detail is important. We have spent hours debugging a script only to discover that one applet specified "../Classes" and another applet specified "../classes" in the codebase attribute. Both applets loaded properly from a Windows NT server since directory names are not case sensitive. However, interapplet communication did not function. **Even the order of the jar files must match!** (Netscape and Sun are more forgiving on the order of the jar files than Microsoft but less forgiving on capitalization.)

Finally, we should explain our notation. We usually break up our jar files to remove common class libraries. Almost all Physlets require a numerical methods and graphics toolkit that has been packaged in a jar file called STools4.jar. Rather than including this library with every Physlet, we prefer to provide Physlet jar files without this common library. These abbreviated jar files have an underscore-appended archive name.

```
Animator4.jar —> Animator4_.jar, STools4.jar
DataGraph4.jar —> DataGraph4_.jar, STools4.jar
DataTable4.jar —> DataTable4_.jar, STools4.jar
```

Embedding a Physlet by specifying multiple jar files does not increase download time, since jar files are only downloaded once no matter how many times they appear on a page.

6.5.6 Ball with a Data Connection

The following kinematics example shows how three data connections can be used to populate a table with position and velocity data as a ball moves across the screen in *Animator*. It will serve as a guide for other data connection examples.

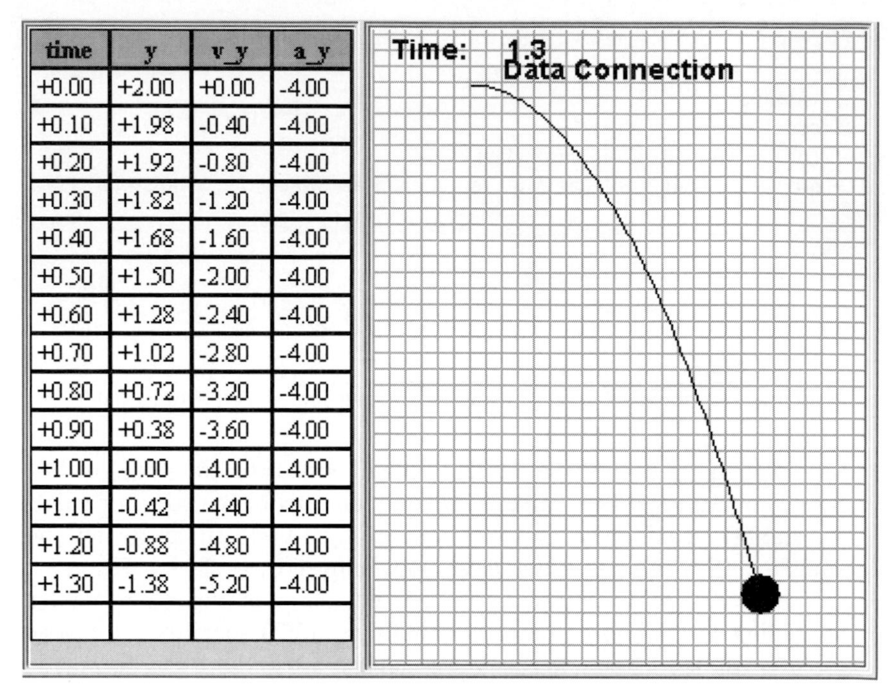

time	y	v_y	a_y
+0.00	+2.00	+0.00	-4.00
+0.10	+1.98	-0.40	-4.00
+0.20	+1.92	-0.80	-4.00
+0.30	+1.82	-1.20	-4.00
+0.40	+1.68	-1.60	-4.00
+0.50	+1.50	-2.00	-4.00
+0.60	+1.28	-2.40	-4.00
+0.70	+1.02	-2.80	-4.00
+0.80	+0.72	-3.20	-4.00
+0.90	+0.38	-3.60	-4.00
+1.00	-0.00	-4.00	-4.00
+1.10	-0.42	-4.40	-4.00
+1.20	-0.88	-4.80	-4.00
+1.30	-1.38	-5.20	-4.00

FIGURE 51: Data connection between *Animator* and *DataTable*.

```
<script language = "JavaScript">
function initApplet(){
document.animator.setAutoRefresh(false);
document.animator.setDefault();
document.animator.setTimeInterval(0,1.4);
document.animator.setPixPerUnit(75);
document.animator.setGridUnit(0.1);
document.animator.shiftPixOrigin(-75,-20);
id=document.animator.addObject("circle","r=10, x=0, y=0");
document.animator.setTrajectory(id, "1.5*t", "2-2*t*t");
document.animator.setTrail(id,300);
document.animator.addObject("caption","text = Data Connection");
document.animator.setAutoRefresh(true);

// initialize the data table
document.dataTable.setAutoRefresh(false);
document.dataTable.setDefault();
document.dataTable.setSeriesLabel(1,"time");
document.dataTable.setSeriesLabel(2,"y");
document.dataTable.setSeriesLabel(3,"v_y");
document.dataTable.setSeriesLabel(4,"a_y");
document.dataTable.setAutoRefresh(true);
tid = document.dataTable.getTableID();l(4,"a_y");
document.dataTable.setAutoRefresh(true);
tid = document.dataTable.getTableID();
```

```
//establish the data connections
document.animator.deleteDataConnections();
cid = document.animator.makeDataConnection(id,tid,1, "t","0");
document.animator.setConnectionStride(cid,10);
cid = document.animator.makeDataConnection(id,tid,2, "y","0");
document.animator.setConnectionStride(cid,10);
cid = document.animator.makeDataConnection(id,tid,3, "vy","0");
document.animator.setConnectionStride(cid,10);
cid = document.animator.makeDataConnection(id,tid,4, "ay","0");
document.animator.setConnectionStride(cid,10);
document.animator.updateDataConnections();
document.animator.forward();
}
</script>
```

Script 16: Data connection between *Animator* and *DataTable* Physlets.

The first ten lines of script create a projectile as described in the *Animator* tutorial. The next nine lines initialize the four columns of a *DataTable*. Each column will be accessed as a *DataTable* series using a data connection. The remaining lines establish the data connection. Notice that the makeDataconnection method returns an object identifier and that this identifier is used to set the data stride property. Setting the data stride to 10 tells the connection to send every tenth datum to the table. Finally, the updateDataConnections() statement sends the initial values to the table before the *Animator*'s clock is started using the forward() command.

The complete html page contains the embedding markup for both applets together with the usual tags needed to produce a well-formed document. It is shown in its entirety and will be our data connection template.

```
<html>
<head>
<script language = "JavaScript">
function initApplet(){
document.animator.setAutoRefresh(false);
document.animator.setDefault();
document.animator.setTimeInterval(0,1.4);
document.animator.setPixPerUnit(75);
document.animator.setGridUnit(0.1);
document.animator.shiftPixOrigin(-75,-20);
id = document.animator.addObject("circle", "r = 10, x = 0, y = 0");
document.animator.setTrajectory(id,"1.5*t", "2-2*t*t");
document.animator.setTrail(id,300);
document.animator.addObject("caption","text = Data Connection");
document.animator.setAutoRefresh(true);

// initialize the data table
document.dataTable.setAutoRefresh(false);
document.dataTable.setDefault();
```

```
document.dataTable.setSeriesLabel(1,"time");
document.dataTable.setSeriesLabel(2,"y");
document.dataTable.setSeriesLabel(3,"v_y");
document.dataTable.setSeriesLabel(4,"a_y");
document.dataTable.setAutoRefresh(true);
tid = document.dataTable.getTableID();

//establish the data connections
document.animator.deleteDataConnections();
cid = document.animator.makeDataConnection(id,tid,1, "t","0");
document.animator.setConnectionStride(cid,10);
cid = document.animator.makeDataConnection(id,tid,2, "y","0");
document.animator.setConnectionStride(cid,10);
cid = document.animator.makeDataConnection(id,tid,3, "vy","0");
document.animator.setConnectionStride(cid,10);
cid = document.animator.makeDataConnection(id,tid,4, "ay","0");
document.animator.setConnectionStride(cid,10);
document.animator.updateDataConnections();
document.animator.forward();
}
</script>
<title>Example 3: Animator-DataTable Template</title>
</head>

<body onload = "initApplet()">
<h2 align = "left">animator-DataTable Template</h2>

<applet codebase = "../classes/" archive = "DataTable4_.jar,Animator4_.jar,STools4.jar"
code = "dataTable.DataTable.class" name = "dataTable" id = "dataTable" width = "164" height
    = "320">
<param name = "ShowScrollBars" value = "false">
<param name = "LastOnTop" value = "false">
<param name = "CellWidth" value = "40">
<param name = "NumRows" value = "15">
<param name = "NumCols" value = "4">
<param name = "ShowControls" value = "false">
<param name = "ShowRowHeader" value = "false">
<param name = "ShowColHeader" value = "true">
<param name = "SizeToFit" value = "true">
</applet>
<br>
<applet code = "Animator4.Animator.class" codebase = "../classes" id = "animator"
archive = "DataTable4_.jar,Animator4_.jar,STools4.jar" name = "Animator" width = "250"
    height = "320">
<param name = "FPS" value = "10">
<param name = "ShowControls" value = "false">
<param name = "dt" value = "0.01">
<param name = "PixPerUnit" value = "10">
<param name = "GridUnit" value = "1.0">
</applet>
```

```
<form>
<p align = "center"> <input type = "BUTTON" value = "play" onclick =
    "document.Animator.forward()"><input
  type = "BUTTON" value = "pause" onclick = "document.Animator.pause()"><input type =
    "BUTTON"
  value = "step&gt;&gt;" onclick = "document.Animator.stepForward()"><input type =
    "BUTTON" value = "reset"
  onclick = "document.Animator.reset()"></p>
</form>
</body>
</html>
```

Script 17: *Animator-DataTable* data connection template.

6.5.7 Pendulum

Pendulum motion is an important problem in physics and lends itself to many different types of data analysis. We will use *Animator* to model this system and a data connection to process the data prior to graphical display.

The first part of the pendulum script, Script 18, creates a mass that is constrained to follow a circle of radius 10. This mass is acted on by an external force in the $-y$-direction, -9.8*m. The second part of the script initializes the *DataGraph*, and the third portion of the script creates the data connection.

```
<script LANGUAGE = "JavaScript">
function initApplet(){
 // part 1
document.animator.setAutoRefresh(false);
document.dataGraph.setAutoRefresh(false);
document.animator.setDefault();
sid = document.animator.addObject("circle","x = 10, r = 10, m = 2");
document.animator.setForce(sid,"0","-9.8*m",10,0,0,0);
document.animator.setShowVVector(sid,true);
document.animator.setConstrainR(sid,10.0,0,0);
document.animator.setShowConstraintPath(sid,false);
document.animator.setDragable(sid,true);
id2 = document.animator.addObject("circle","x = 0, y = 0, r = 2");
document.animator.addConnectorLine(sid,id2);
    document.animator.addObject("caption","text = Pendulum");
document.animator.setAutoRefresh(true);
 // part 2
document.dataGraph.setAutoscaleX(true);
document.dataGraph.setAutoscaleY(true);
document.dataGraph.clearSeries(1);
document.dataGraph.setSeriesStyle(1,false,0);
document.dataGraph.setLastPointMarker(1,true);
document.dataGraph.setLabelY("Omega");
document.dataGraph.setLabelX("Theta");
document.dataGraph.setTitle("Omega vs Theta");
lid = document.dataGraph.getGraphID();
```

```
document.dataGraph.setAutoRefresh(true);
 //part 3
document.animator.deleteDataConnections();
omega = "(x*vy-y*vx)/100";
theta = "atan2(x,-y)";
document.animator.makeDataConnection(sid,lid,1,theta, omega);
document.animator.updateDataConnections();
document.animator.forward();
}
</script>
```

Script 18: Create a pendulum with a data connection to a *DataGraph*.

Constraints were introduced in version 3 of *Animator* in order to restrict the motion of an object along a path. We have previously introduced the setConstrainY and setConstrainX methods to restrict a problem to one dimension. In this example, the setConstrainR method is introduced to restrict the motion to a circle of radius 10 centered on the origin. Since we want the pendulum to appear to be connected to a pivot, we hide the constraint path. The script then creates a circle object to represent the pivot and adds a connector line between the pendulum bob and the pivot. The *DataGraph* initialization, part 2, is similar to previous *DataGraph* examples except that the setSeriesStyle method has been set so as not to connect the data points. This is necessary because of the discontinuity in the tangent function at $+/-\pi/2$.

Part 3 of the script begins by deleting all previous data connections. This is a precaution in case the script has been run multiple times. The makeDataConnections method connects the pendulum bob to series 1 of the *DataGraph*. Since the animation was halted when we invoked the setDefault method, data have yet to be generated or sent to the graph. Calling updateDataConnections forces the first data point, that is, the initial conditions, to be sent through connection to the graph. Additional data will be sent at the end of every clock tick or mouse drag.

Since the data source does not define the angular variables, their definitions must be passed to the data connection.

$$\Theta = \tan^{-1}(-y,x) = \text{atan2}(-y,x)$$
$$\Omega = |\mathbf{r} \times \mathbf{v}| / r^2 = (x \cdot v_y - y \cdot v_x) / r^2 = (x*vy - y*vx)/100$$

In this example, we have chosen to plot omega as a function theta. But we could just as easily have plotted kinetic energy or potential energy by defining different variables.

$$ke = m \cdot v^2/2 = m*(vx*vx + vy*vy)/2$$
$$pe = m \cdot g \cdot h = m*9.8*y;$$

6.5.8 Controlling the Connection

Data connections are Java objects and therefore have properties that can be set. Connections are created and an object identifier, id, is returned when the makeDataConnection method is invoked. The object identifier can be used to set the following properties:

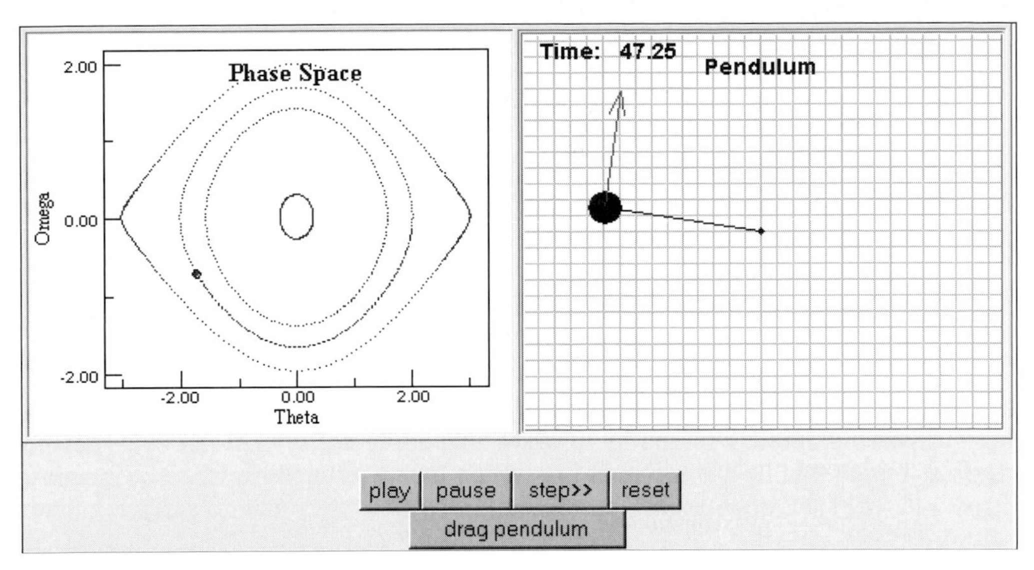

FIGURE 52: Pendulum phase space plot.

```
setConnectionBlock(int id, boolean should block)
setConnectionStride(int id, int num)
setConnectionSmoothing(int id ,int num)
```

The first argument in each method is the object identifier.

The setConnectionBlock method allows a script to temporally disable the data flow. This method was used to generate Figure 52 when the bob was dragged to keep from registering spurious data on the phase space plot.

The setConnectionStride method is often used when generating table data. A time step that produces smooth animation invariably produces data at too fine an interval for a table. The stride of a connection, the second parameter, determines the number of time steps between data points. If the stride is set to ten, every tenth data point will be passed through the connection.

The second parameter in setConnectionSmoothing specifies the number of points that should be included in a running average. If the data source generates an array rather than a single data point, the smoothing is performed on a moving window of *y*-values.

6.6 SCRIPTING TIPS

6.6.1 Data Types

Scripting languages (such as JavaScript, Basic, and Perl) make it easy to program by not requiring that the programmer specify a data type before a variable is used. The script interpreter will try to determine the variable's data type from its context. JavaScript, for example, will execute either of the following sequences of statements:

```
X = 5;
Y = 10;
Z = x+y;
```

or

```
X = 5;
Y = "10";
Z = x+y;.
```

In the first sequence, *Z* is assigned the number 15. It could be passed to any Java method requiring a number. In the second sequence, *Z* is assigned the value of the concatenated strings, 510. This is probably not what the author intended, but it could be passed to any Java function requiring a string. Passing an incorrect data type to a Java function will cause unpredictable results. Some browsers will hang, some will give an error message, and others will ignore the error.

Browsers sometimes change the data type as a variable is being passed from one function to another. Consider, for example, an anchor tag that is being used to call the initApplet function with a particle mass, *m*, as parameter.

```
<a href = "JavaScript:initApplet(2)"> Start</a>.
```

Although the argument of the initApplet function is the number 2, there is no guarantee that this variable will be interpreted as a number inside the function body. Different browsers will handle the following code fragment differently.

```
<script LANGUAGE = "JavaScript">
function initApplet(m){
.
.
document.applet.setMass(id,m)
.
```

Some browsers interpret the variable *m* as a number, while others interpret it as a string.

The only safe approach for script authors is to practice defensive programming. When passing variables into JavaScript functions, always assume that the variable is a string. First check to see if the string is a valid number, then convert the string to a number, and finally check the numerical range. This sequence can be scripted as follows:

```
function initApplet(mStr){
  if(!isFinite(mStr) ){
     alert("Mass is not a number.");
     return;
  }
  m = eval(mStr);
  if(m<0){
     alert("Mass cannot be negative.");
     return;
  }
.
.
```

Script 19: Variable checking and validation with JavaScript.

The *isFinite* function in Script 19 returns true if the argument, mStr in this case, represents a valid number. If the string is valid, we use the *eval* function to convert it to a pure number, *m*. Finally, we test to see if the mass is positive. If either test fails, the script pops an html *alert* message onto the screen and exits the function.

6.6.2 Script Debugging

Even good programmers goof. When a script doesn't run or gives incorrect results, authors can always resort to the time-honored method of printing messages at key points in the script. The JavaScript *alert* function that was introduced in Script 19 stops the execution of a script and displays a message in a dialog box. This message can, of course, be used to examine the value of a variable.

```
alert("The x value is = "+x);
```

The data type of the variable, *x*, is irrelevant since the plus operator concatenates whenever either operand is a string.

A more elegant approach, but with a slightly higher learning curve, is to install a script debugger. The Microsoft Script debugger is a free Internet Explorer add-on that allows an author to stop the execution of a script using a breakpoint. As with most visual debuggers, the author can then single step the script while examining variables.

Some errors manifest themselves within the Physlet itself. Physlets are often able to trap these errors and to write an appropriate message to the Java Console. The Java Console is a small window that can be accessed from the browser's pull-down menu. It is available on most browsers, although it is not enabled using Microsoft's default configuration.

The error shown in Figure 53 was produced by the following code fragment.

```
id = document.animator.addObject("circle","x = 0,y = 0");
document.animator.setTrajectory(id,"1.5t","2-2*t*t");
```

Notice the missing multiplication symbol in the analytic function for the *x*-coordinate.

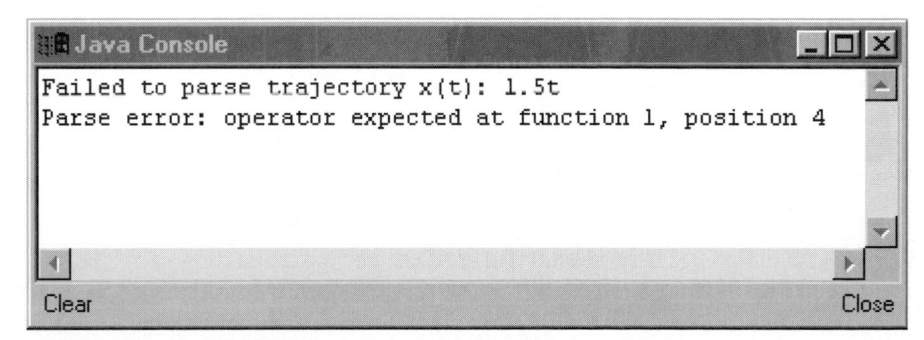

FIGURE 53: Java Console in Internet Explorer showing a script error.

Discovering the source of a Java error can be more challenging than finding script errors. For example, the console message

```
DataConnection not made.
Listener:null Source:animator4.Circle@4ab
```

was generated while trying to establish a data connection between a circle object in *Animator* and a *DataGraph*. The data listener could not be found because the jar files were not in the same order in the applet's respective archive attributes. Since Physlets usually do not halt program execution for syntactic mistakes, it is a good idea to check the Java Console for messages from time to time while authoring

Finally, we urge curriculum authors to test their scripts on all major browsers. JavaScript is a World Wide Web Consortium (W3C) standard, and both browsers do a reasonable job of adhering to the basic standard. However, some browsers are more forgiving of syntactic errors. A script may run on one browser and fail on another not because the second browser's JavaScript implementation is flawed, but because the second browser is a stricter grammarian. You are urged to test the following script fragment on a number of browsers to determine which browsers catch the missing semicolon and which browsers execute the script without the semicolon.

```
id = document.animator.addObject("circle","x = 0,y = 0");
document.animator.setTrajectory(id,"1.5*t","2-2*t*t")
document.animator.setTrail(id,300);
```

Then decide if a forgiving or a strict browser is preferred.

PART TWO
CURRICULAR MATERIAL

CHAPTER 7
IN-CLASS ACTIVITIES

These examples of curricular material are from introductory physics (calculus and algebra based) and selected examples of advanced material. These Physlet exercises can be used at the beginning of class to introduce topics, during the middle of a topic to test whether students are ready to cover additional material, or at the end of a topic to test students' knowledge of the material just covered. Instructors may choose from several different presentation methods depending on the type of animation shown. One method is the Peer Instruction technique developed by Eric Mazur [Mazur 1997]: show the animation(s), ask students to think about their answers and discuss their answers with other students, and then ask for a show of hands. Once this polling is completed, ask students to justify their answers. Despite the relatively few examples in this chapter, any Physlet problem from Chapters 8–10 can be used in this fashion with only slight modifications.

7.1 MECHANICS

Exercise 7.1.1: An animation of a red car moving to the right, hitting a wall, and then moving backward to the left is shown. Also shown is one of four possible graphs depicting the car's position as a function of time. The middle panel shows a cursor on a position versus time graph. This graph is initially blank until the user moves the cursor.

Question 7.1.1: Draw a graph of position versus time for the car.

Answer: Animation 4. The figure also shows the crudely sketched answer drawn by dragging the cursor in the middle panel. Since this is a problem designed for inside the classroom instruction, the instructor can use the middle panel in many different ways. Given Animation 1, the instructor can either show the remaining graphs or solicit a volunteer from the class to use the cursor in the middle panel to "draw" the correct position versus time graph. Once this is completed, the instructor can interact with the class to

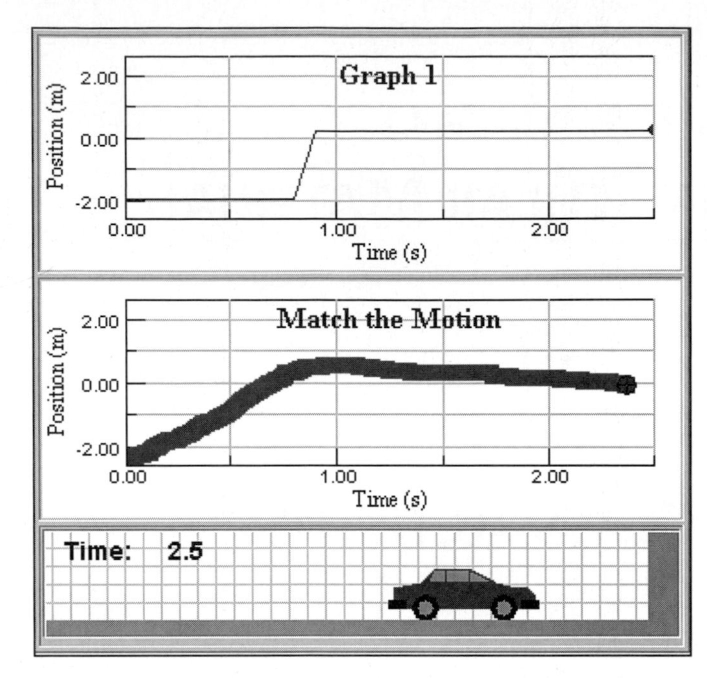

determine whether the volunteer has drawn the correct curve. Another possible use for the middle panel is to have students draw the velocity versus time graph when the correct position versus time graph is shown. This allows students to see an animation and to see the correct graphical representation of that motion.

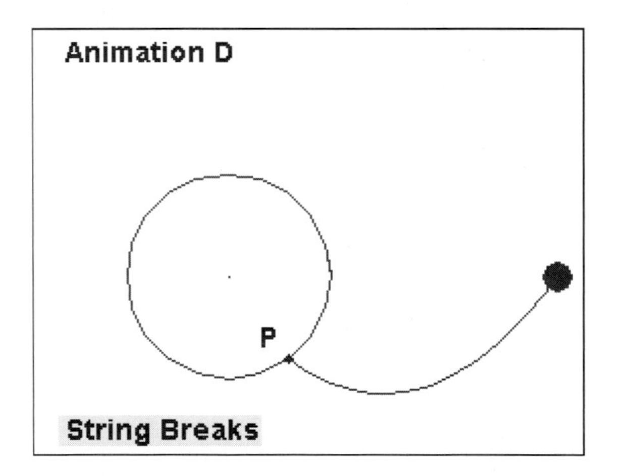

Exercise 7.1.2: A heavy ball is attached to a string and swung in a circular path in a horizontal plane as shown in the animation. At the point indicated the string suddenly breaks at the ball. Four animations represent possible results when the string breaks at point *P*.

Question 7.1.2: Which of the four animations properly represents the path of the ball if the string breaks at point *P*?

Answer: Animation B; the ball moves tangent to the circle at point P *(not shown). This problem is an excellent warm-up for the discussion of uniform circular motion. Students often have a hard time with the cause of uniform circular motion. Here, the string is the cause of the ball's acceleration toward the center of the circle. In fact, without it, there is no longer an acceleration and the ball continues at a constant velocity that is tangent to the circle at point P. Students will often laugh at animations such as Animation D, even though they will answer this question incorrectly (choosing this answer) on the FCI.*

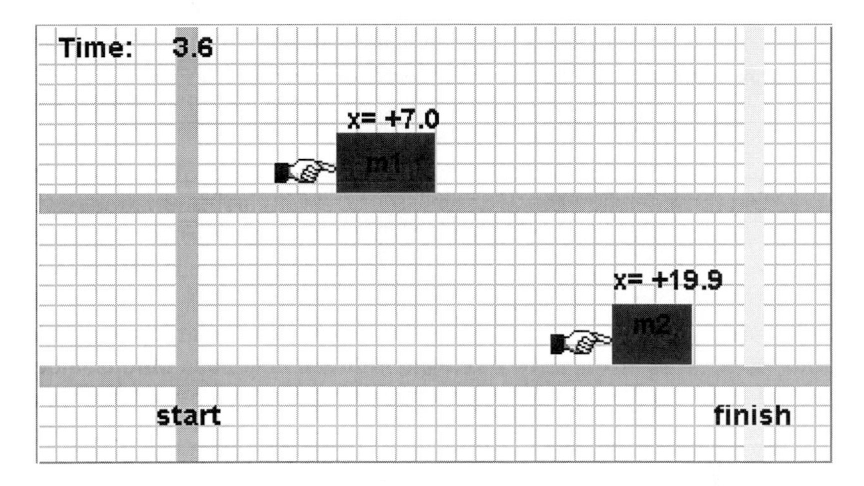

Exercise 7.1.3: Two blocks are pushed by identical forces, each starting at rest at the first vertical rectangle (start). The mass $m_1 = 2m_2$.

Question 7.1.3: Which object has the greater kinetic energy when it reaches the second vertical rectangle (finish)?

Answer: The kinetic energies are identical. This is a good problem to ask in the middle of discussing the work-energy theorem. Typical student reaction is that the second mass has a greater velocity and therefore has the greater kinetic energy. While the second mass does have the greater velocity at the finish, it is not double the velocity of m_1. *Students can see this from Newton's second law and kinematics, although it is a little tedious. The correct rationale is to recognize that the work done by the forces on each mass is identical as the displacements are identical.*

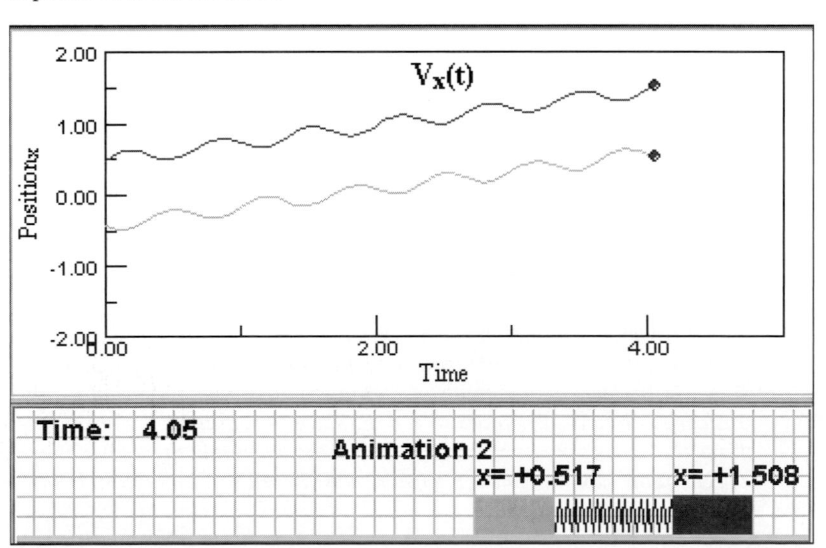

Exercise 7.1.4: A spring is attached to the end of two carts and is compressed. The carts are placed on a low-friction track. The spring is released such that the two carts are "pushed" apart as shown in the animations (position is in meters and time is in seconds). In Animation 1, the system remains at the center of the animation; in Animation 2 (depicted), the system moves to the right; and in Animation 3, the system moves to the left. The graphs depict the motion in the x-direction of the individual masses as a function of time.

Question 7.1.4: Describe the motion of the center of mass in each animation.

Answer: *Animation 1: The velocity of the center of mass is zero. Animation 2: The velocity of the center of mass is constant and positive (to the right). Animation 3: The velocity of the center of mass is constant and negative (to the left). Once these questions are answered, several other questions can be asked. For example, is energy conserved in any of the animations? Energy is of course conserved, but the actual calculation may be difficult depending on how students decide to undertake the calculation. It is apparent in the first animation that* $V_{cm} = 0$ *and that the total energy is* KE_{cm} *plus the energy stored in the spring at maximum compression/extension.*

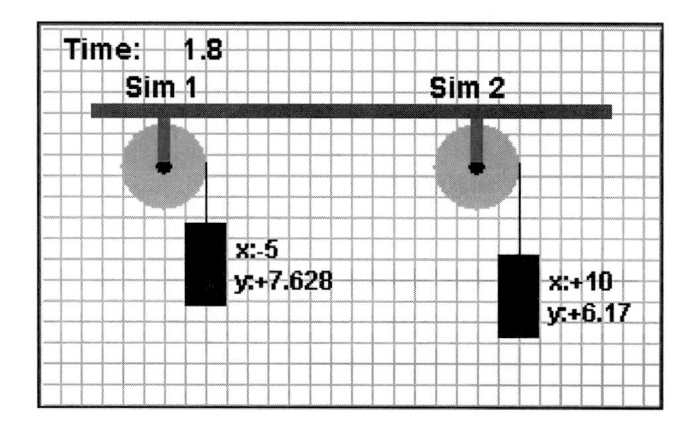

Exercise 7.1.5: Two identical black masses, m, are hung via massless strings over two pulleys of identical mass M and radius R, but different mass distributions. The bearings in the pulleys are frictionless and the strings do not slip as they unwind from their pulleys. The masses fall with different accelerations as shown in the animation (position is in meters and time is in seconds). The mass in the second simulation hits the floor first.

Question 7.1.5a: Which pulley has the greater moment of inertia?

Answer: *Simulation 1. Students will get this one correct. Since torque is* $I\alpha$, *and the acceleration of the black masses is related to the angular acceleration of the pulley through the radius, the object with the least acceleration has the pulley with the greatest moment of inertia.*

Question 7.1.5b: In which simulation is the torque the greatest?

Answer: *Simulation 1. This one is not so obvious. Since we have just asked a question using the definition of torque being* $I\alpha$, *most students will again relate the torque to* $I\alpha$, *however because the moments of inertia are unknown, this relationship is not very useful. Instead, torque is also equal to* $\mathbf{r} \times \mathbf{F}$. *Since the radii are identical, we must consider the force that causes the acceleration, namely the tension in the string. The mass with the larger acceleration has the smaller tension and therefore Simulation 1 has the greater torque applied to the pulley.*

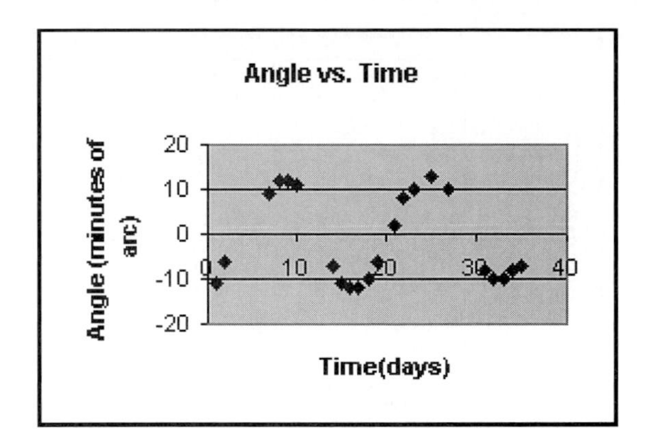

Exercise 7.1.6: Shown is an adaptation of Galileo's data of the angle between Jupiter and its moon Callisto.

Question 7.1.6a: Given what you know about the orbit of planets and satellites, why does Galileo's data look like it does?

Answer: The data suggest a negative cosine curve, yet planets and satellites move in mostly circular orbits. Therefore, Galileo was seeing circular motion edge on.

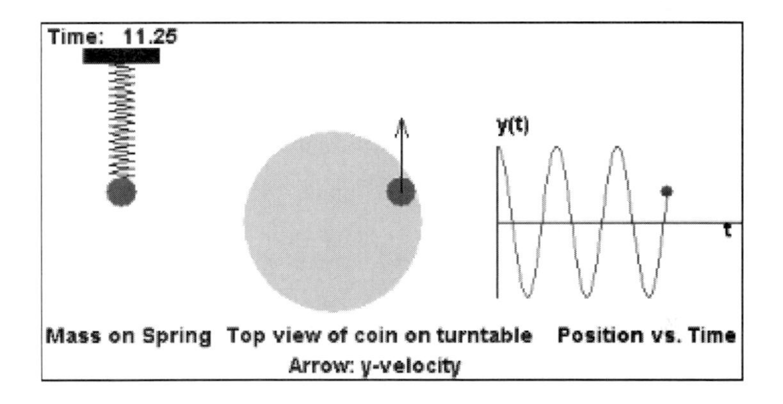

Question 7.1.6b: Given your finding in Question 7.1.6a, explain the possible relationships depicted in the animation shown here.

Answer: As suggested in Galileo's data, circular motion viewed edge on looks like a sine or a cosine curve. Here there is a direct relationship between the position, velocity, and acceleration in the y-direction for the coin on the turntable and the position, velocity, and acceleration in the y-direction for the mass on the spring. Since the magnitude of the velocity for the coin on the turntable is constant, we can determine that when $y = +/- y_{max}$, $v_y = 0$ *and when* $y = 0$, $v_y = +/- v_{max}$. *These relationships are also true for the mass on the spring.*

Exercise 7.1.7: A ball on an air track is attached to a compressed spring as shown in the animations (position is in meters and time is in seconds). Each of the five graphs *correctly* shows a different property of the motion of the ball. Shown is the acceleration versus position graph.

Question 7.1.7: Determine whether the green ball undergoes simple harmonic motion and state which graph(s) tell you this.

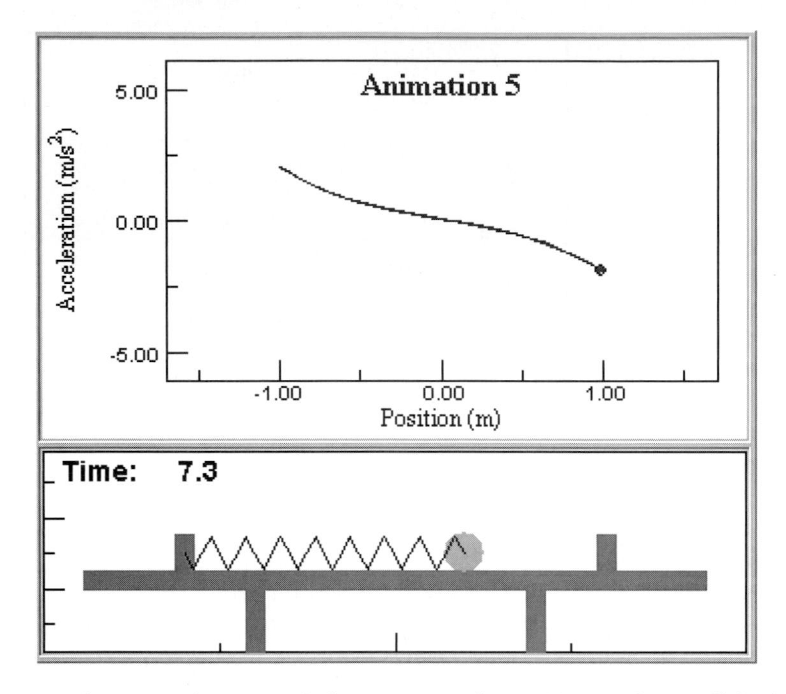

Answer: *Animation 5, the animation shown. Most students will think that this motion is simple harmonic motion when they see the sinusoidal position versus time graph (students often perceive that a sinusoidal* x *versus* t *graph proves simple harmonic motion. Instead a linear restoring force has a sinusoidal* x *versus* t *graph). Students are still relatively convinced of simple harmonic motion when they see the* v *versus* t *graph, which still looks sinusoidal. However, the force versus* x *graph (here the acceleration versus position graph) is what tells us whether the force is a linear restoring force and whether we have simple harmonic motion. Here we do not have a linear restoring force with the amplitude of motion shown. However, with a smaller initial amplitude, this motion is simple harmonic.*

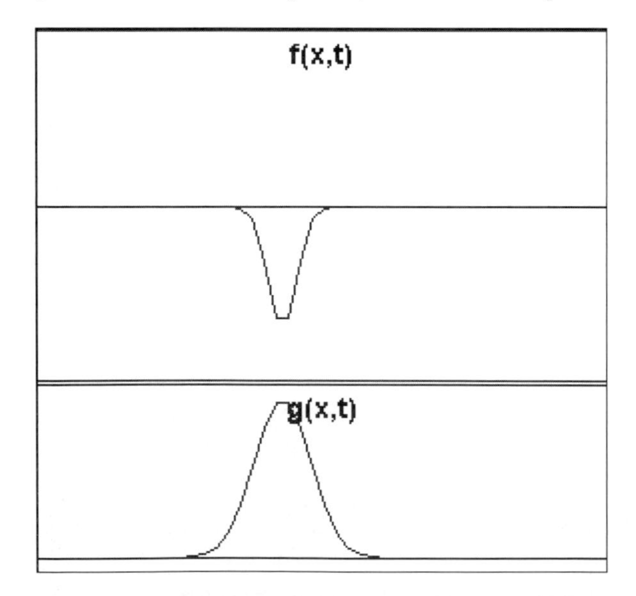

Exercise 7.1.8: The simulations provided here show disturbances on two identical strings (position is in centimeters and time is in seconds). Simulation 1 shows two Gaussian-shaped waves of identical amplitude and width; Simulation 2 shows two Gaussian-shaped waves of equal and opposite amplitude and equal width; and the third simulation is shown.

Question 7.1.8: At $t = 2.0$ seconds, describe the superposition of each of the three pairs of waves.

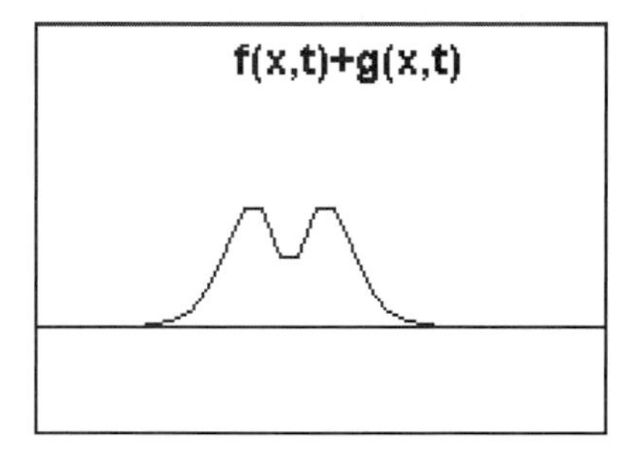

Answer: *Simulation 1: The waves add with double the amplitude. Simulation 2: The waves destructively interfere and at that instant, the string has zero amplitude. Simulation 3: Their sum has a large peak, a depression, and then another large peak, as shown in the preceding figure.*

7.2 ELECTROMAGNETISM

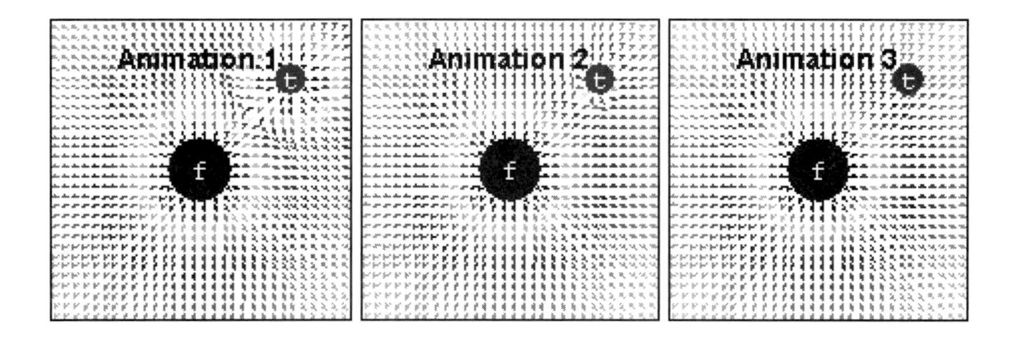

Exercise 7.2.1: One fixed charge and one "test" charge are shown in each animation. The fixed charge is labeled by an *f* and the test charge is labeled by a *t*. You can click-drag on the test charge to change its position. There are three animations depicted.

Question 7.2.1: In which animation is the "test" charge a true "test" charge?

Answer: *Animation 3. As the test charge is moved in each animation, the field changes or remains the same. The animation where the test charge leaves the original field unaffected is the correct animation. This is, of course, the definition of the test charge to begin with. It is important for students to grasp the fact that in the course of measuring the field of the fixed charge, one may disturb the field if one uses a real charge. Note that this is also Problem 9.1.2, used here without major modification as an in-class exercise.*

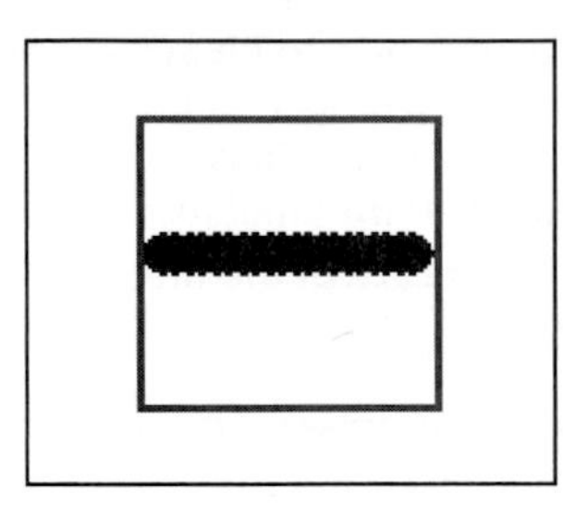

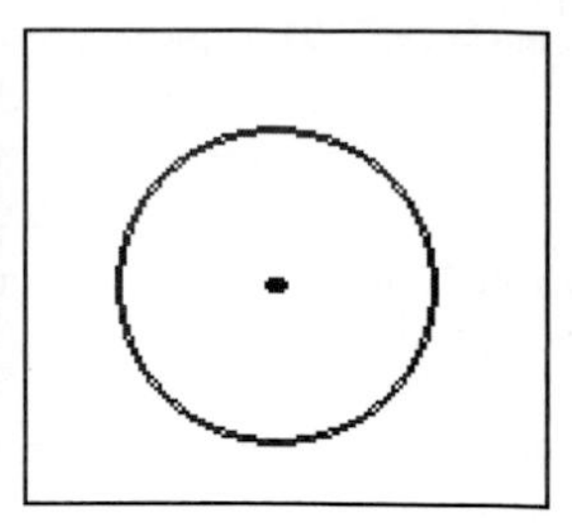

Exercise 7.2.2: You are asked to calculate the electric field due to a charge filament as shown in the animation. You are given three different detectors and three different viewpoints: intermediate distance, very close, and very far from the filament. The bar graph (not shown) displays the electric flux passing through the cubical and spherical Gaussian surfaces. The option of showing the electric field vectors is also given.

Question 7.2.2: From what viewpoint(s) can you safely calculate the electric field using the flux detector's reading and Gauss's law? You may drag around each detector.

Answer: The near and far views have a symmetry that can be exploited. In the classroom, there are several ways to have students come to this conclusion. Primarily, is there symmetry? If so, what is it? In the near view, the charge filament looks long and approximates an infinite line charge. For the far view, the charge looks like a point charge. In both these cases, there is an appropriate symmetry and Gauss's law can be used to determine the electric field on the surface of the two detectors. In the intermediate situation, Gauss's law cannot be used to determine the electric field, as the symmetry is not exact. These conclusions can be further emphasized by turning on the option of showing the electric field vectors.

In addition to discussing electric field calculations, this problem can be used to discuss the meaning of Gauss's law and electric flux. For example, in the far view, as long as the charge filament is enclosed, the electric flux does not change as the detector is moved around. Gauss's law states that this should be the case, and it always is. Now, if the spherical Gaussian surface is not centered on the charge filament, can Gauss's law be used to calculate easily the electric field? No. The symmetry is gone even though the flux calculation remains the same. Hence only when the Gaussian surface is centered on the filament can E *be safely calculated from the flux,* $EA = \Phi$.

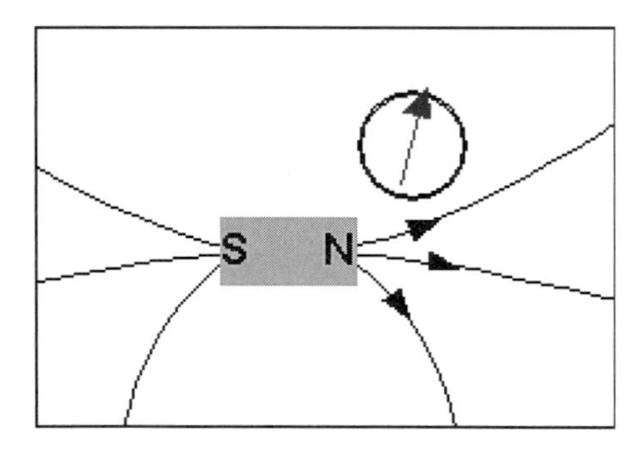

Exercise 7.2.3: The animation shown here represents typical bar magnet with a North and South pole. You may double click anywhere inside the animation to add a magnetic field line. You may also add a compass by clicking the "Add Compass" link.

Question 7.2.3: Which animation correctly depicts a properly labeled magnet?

Answer: *Animation 1, the animation shown. In this animation, there are two ways to determine whether the magnet is properly labeled. One involves double-clicking in the applet to add a magnetic field line. Magnetic field lines leave North poles and enter South poles, and therefore this animation is correct. The second method involves a depiction of an actual in-class demonstration or lab experiment. By adding a compass, the animation depicts moving a compass around a permanent magnet. Since multiple compasses can be added, ask students what they think the relationship between the compass needles and the magnetic field lines will be. The compass needles should point tangent to the magnetic field lines. This can also be shown with one compass by dragging it to follow a field line.*

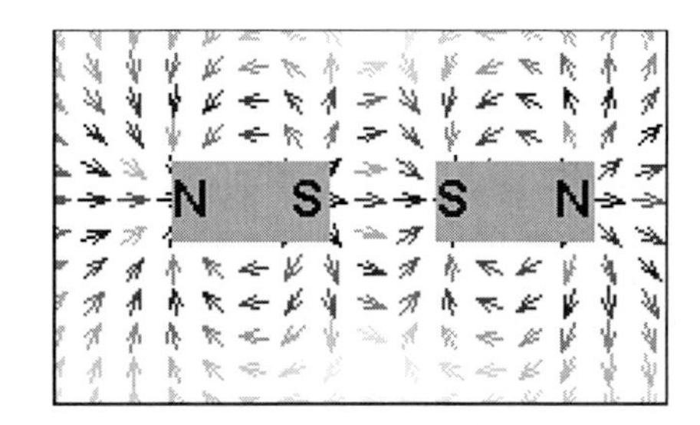

Exercise 7.2.4: The animation shown here represents two typical bar magnets each with a North and South pole. The arrows represent the direction of the magnetic field. The color of the arrows represents the magnitude of the field, with magnitude increasing as the color changes from blue to green to red to black. You may drag either magnet, double-click anywhere inside the animation to add a magnetic field line, and mouse-down to read the magnitude of the magnetic field at that point.

Question 7.2.4: Which animation correctly depicts properly labeled magnets?

Answer: *Animation 2, the animation not depicted. This problem is a slightly more difficult extension of Exercise 7.2.3.*

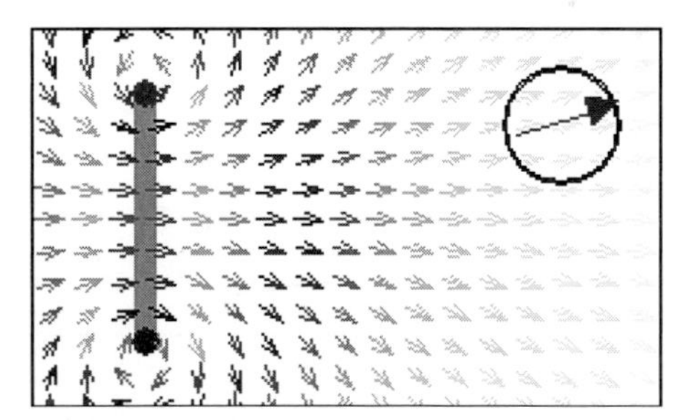

Exercise 7.2.5: A cross section of a circular wire loop carrying an unknown current is shown. The arrows represent the direction of the magnetic field. The color of the arrows represents the magnitude of the field with magnitude increasing as the color changes from blue to green to red to black. You can double-click in the animation to add magnetic field lines, click-drag the center of the loop to reposition it, and click-drag the top or bottom of the loop to change its size. You may also add a compass by clicking the "Add Compass" link.

Question 7.2.5: Does current flow out of the red (top) end or the blue (bottom) end?

Answer: *The current comes out of the red (the top) end and into the blue (the bottom) end. Use the right-hand rule on either cross section and compare the curl of your fingers with the magnetic field vectors. Again, the use of the compass can be used in conjunction with the magnetic field vectors. Another interesting thing to do here is to drag the end of the wire to shrink the loop down to a zero radius. What should the compass needle do? Try it.*

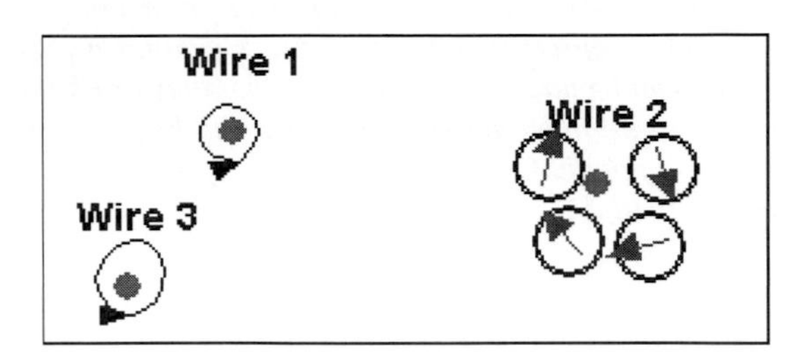

Exercise 7.2.6: A cross section of three wires carrying unknown currents is shown. You can double-click anywhere inside the animation to draw a magnetic field line. You can also click-drag the wires, but this will erase any field line that you have drawn. You may also add a compass by clicking the "Add Compass" link.

Question 7.2.6: Which wires are carrying current out of the plane of the simulation, that is, out of the computer monitor?

Answer: *Wire 1 and Wire 3. Again, students must use the right-hand rule to compare the magnetic field lines on either cross section of wire to the curl of their fingers and the direction that their thumb points. Here is another interesting example of what can be done depending on how students attack the problem. To determine the direction of the magnetic field, double-clicking gives immediate feedback, as shown for the first and third wires. However, the exact same result can be determined by the careful placing of one or several compasses around a wire, as shown around the second wire. With only one compass, getting the same result requires dragging the compass around the wire.*

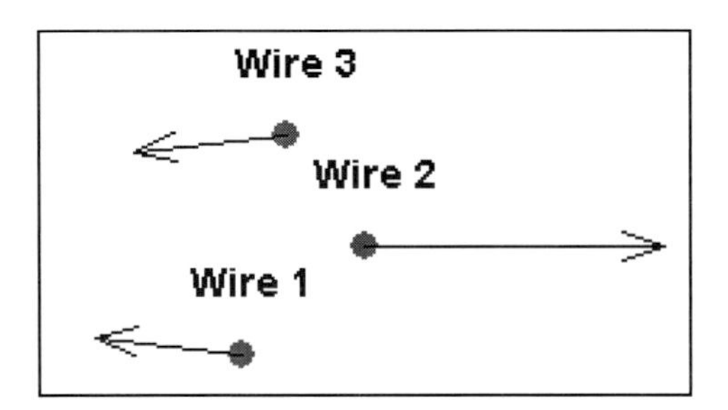

Exercise 7.2.7: A cross section of three wires carrying unknown currents is shown. You can click-drag the wires and the black arrow represents the force experienced by each wire.

Question 7.2.7: How many wires have a current that points out of the page?

Answer: *Cannot determine, not enough information. There are three wires, so how to begin? Students first need to realize that a systematic approach will serve them well in this exercise. Consider two wires at a time with the third as far from the two as possible. Given that like currents (currents pointing in the same direction) attract and opposite currents repel, students can figure out that two of the wires have currents pointing in the same direction. However, this is not enough to determine whether these currents are into or out of the page.*

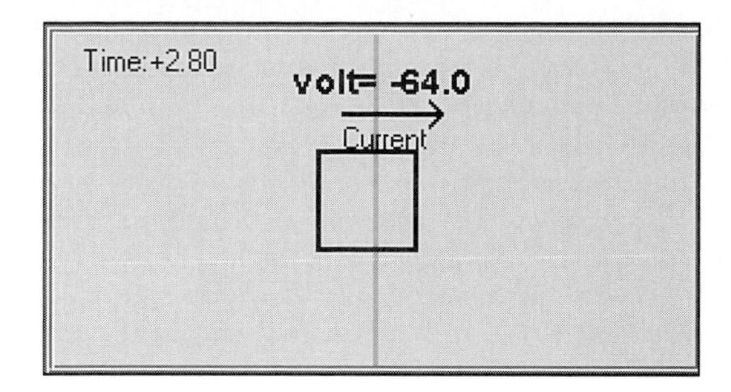

Exercise 7.2.8: A loop of wire travels from the left to the right through an inhomogeneous magnetic field, as shown in the animation. The green vertical line at $x = 0$ is for your reference. The induced *emf* in the loop is shown in Volts, as is the direction of the induced current.

Question 7.2.8: Describe the magnetic field perpendicular to the computer screen.

Answer: $B_z(x < 0)$ = *constant,* $B_z(x > 0)$ = *constant, and* $B_z(x < 0) < B_z(x > 0)$*. From Lenz's law, we know that the magnetic field must be increasing out of the page across the green line. Note that some students will assume that the magnetic field on the left of the green line is zero (because the voltage shown is zero) and the magnetic field on the right is positive (out of the page). However, this is not necessarily the case. The only thing we do know is that the magnetic field to the right of the line must be constant and greater than the constant magnetic field to the left of the line.*

7.3 ADVANCED

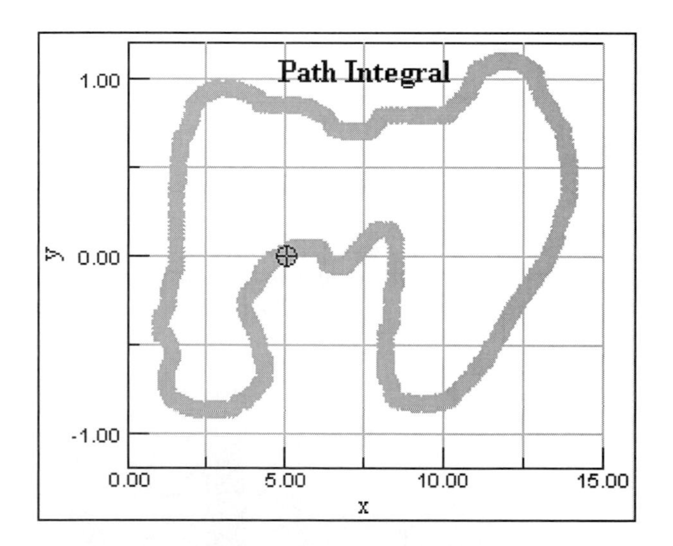

x	y	int
+5.02	-0.01	+33.06

Exercise 7.3.1: A cursor is shown in an x-y graph. The cursor can be dragged around the graph and its path is marked as it is moved. The data are sent to a *DataTable*, which shows x, y, and the value of the path integral, $\int \mathbf{F.dl}$. Data are shown for a cursor starting at $x = 5$ and $y = 0$.

Question 7.3.1: What is the value of **F**? Is it conservative?

Answer: dl/dl. *It is not conservative. To determine this, students must consider the best path to examine. The path shown is actually not very helpful. A more useful path is one with a component in only one direction at a time. For a path solely in the* x*-direction, one finds the value of the integral to be the distance traveled in the* x*-direction. The same thing occurs in the* y*-direction. Since this path integral depends only on the distance, not the displacement, this is not a conservative* **F**.

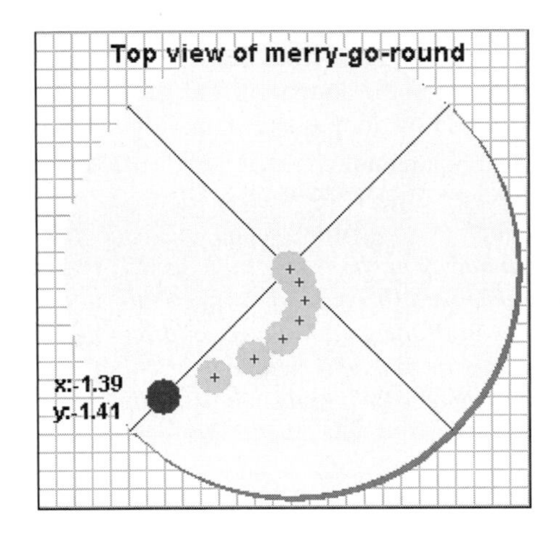

Exercise 7.3.2: A ball is shown on a rotating merry-go-round with a frictionless surface. The ball starts at the center of the merry-go-round and moves outward at a constant speed. Views from two reference frames are shown, one where the motion of the ball is shown as seen from an observer on the ground and the other (shown here) where the motion of the ball is shown as seen from an observer on the merry-go-round.

Question 7.3.2: How do you write the laws of the physics so that both frames are correct?

Answer: In order to get observers in the two reference frames to agree, we must invent a force that acts on objects in the accelerating frame. Here the accelerating frame is the rotating frame of the merry-go-round. An observer on the merry-go-round must invent a fictitious force to the right that causes the ball to accelerate to the right. Otherwise, the observer on the merry-go-round sees $\Sigma \mathbf{F} = 0$ *and an acceleration, which violates Newton's second law.*

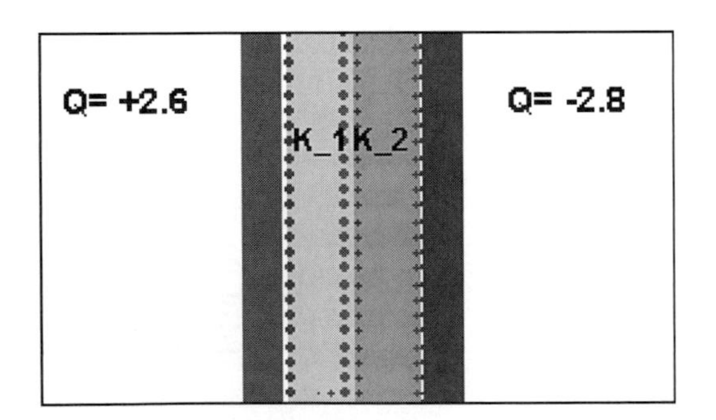

Exercise 7.3.3: Click-drag the dielectric blocks, dielectric constants $\kappa_1 = 2.4$ and κ_2, respectively, into the capacitor (charge given in μC and position given in centimeters). Observe how the electric field and the charge on the capacitor and the dielectric change when you move the dielectric. *Note: There is a 1 to 5% intrinsic error in the calculation of charge.*

Question 7.3.3: What is κ_2? You can measure the voltage and the electric field at any point by click-dragging.

Answer: 1.3. Use the electric field to determine the change in voltage across the plates. This orientation behaves like a capacitor in series. Besides a detailed calculation, what else can a student determine from looking at the animation? Well, students should be able to determine that $\kappa_2 < \kappa_1$. Students can make this determination by carefully looking at the bound charge on each dielectric. The larger the κ, the larger the bound charge on the surface of the dielectric. In fact, it appears from a qualitative observation that there is twice the bound charge on the κ_1 dielectric as compared to the κ_2 dielectric. Hence, κ_2 is approximately 0.5 κ_1, which is borne out by careful calculation.

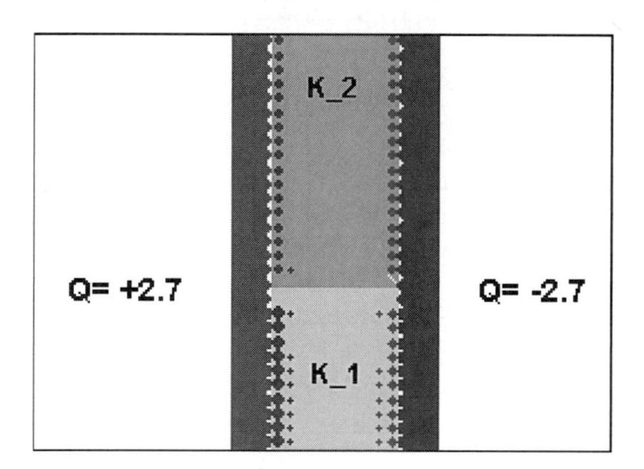

Exercise 7.3.4: Click-drag the dielectric blocks, dielectric constants $\kappa_1 = 4$ and κ_2 respectively, into the capacitor (charge given in μC and position given in centimeters). *Note: There is a 1 to 5% intrinsic error in the calculation of charge.*

Question 7.3.4: What is κ_2? You can measure the electric field at any point by click-dragging.

Answer: 2.5. Use the electric field to determine the change in voltage across the plates. This orientation behaves like a capacitor in parallel. Besides a detailed calculation, what else can a student determine from looking at the animation? Again, students should be able to determine that $\kappa_2 < \kappa_1$.

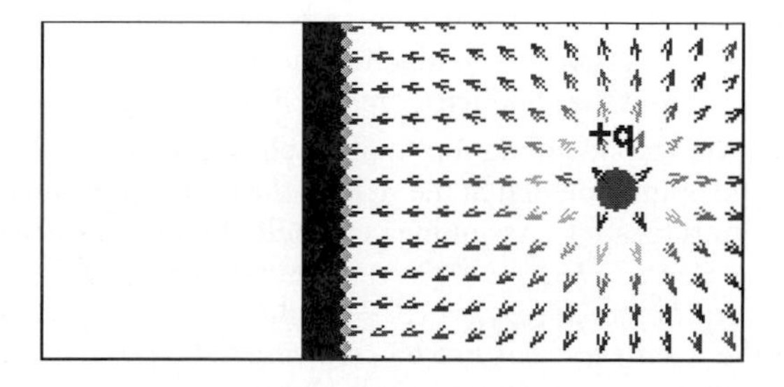

Exercise 7.3.5: The simulation shown represents a positive charge $(+q)$ near an infinite grounded conducting sheet. A second animation is shown (see the figure following Question 7.3.5) that depicts a similar situation without the conductor present. The rectangle is there for your reference only and does not carry any charge; nor does it affect the electric field.

Question 7.3.5: Use the second animation to place charge(s) where they belong to make the field in the second animation match the field in the first animation for $x > 0$. Assuming you could do this, where are the charge(s)?

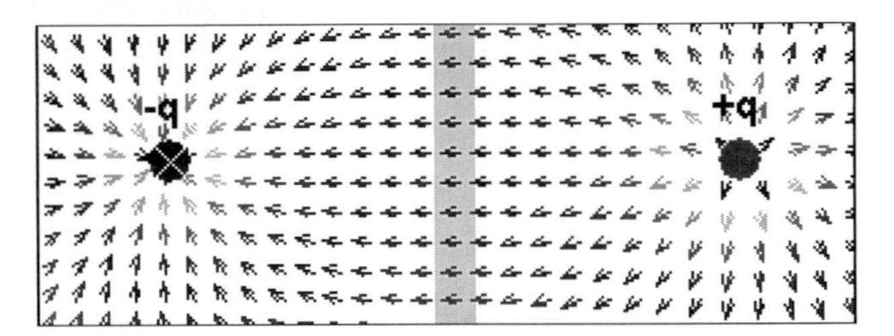

Answer: *A negative charge must be placed the same distance the original charge is from the plate, but on the other side, as shown in the preceding figure. This is a simple example of the method of images from electrostatics. Ask students to compare the regions of interest* x > 0 *and the region we do not care about,* x < 0*. What do they notice? As expected, placing another charge to the left of the rectangle exactly mimics the field in the region of interest. However, in the region* x < 0*, this is not the case. The electric field is indeed different there. Students must remember that since they were specifically interested in the region* x > 0*, placing a charge outside of this region is completely acceptable as long as its placement solves the boundary conditions.*

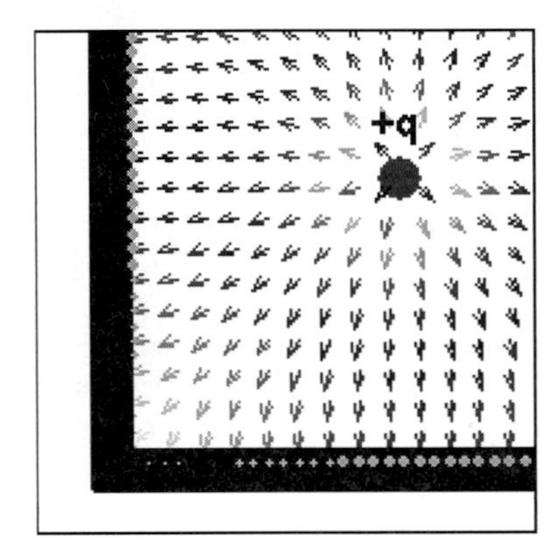

Exercise 7.3.6: The simulation shown represents a positive charge $(+q)$ at the position $(x, y) = (d, d)$ near two infinite grounded conducting sheets. The animation on the right depicts a similar situation without the conductor present. The rectangles are there for your reference only.

Question 7.3.6: Use the second animation to place charge(s) where they belong to make the field in the second animation match the field in the first animation for $x > 0, y > 0$ (the region where the $+q$ is located). Assuming you could do this, where are the charge(s)?

Answer: *There are three charges needed. One* +q *charge at* (x, y) = (−d, −d) *and two* −q *charges at* (x, y) = (d, −d) *and* (x, y) = (−d, d)*, respectively. The electric potential on the conductor will not be zero otherwise. This is another example of the method of images from electrostatics. Again, the electric field is indeed different outside of the region* *of interest. This is a good in-class warm-up for the out-of-class calculation of the electric potential and electric field of this configuration as the instructor knows the students' starting point for the calculation.*

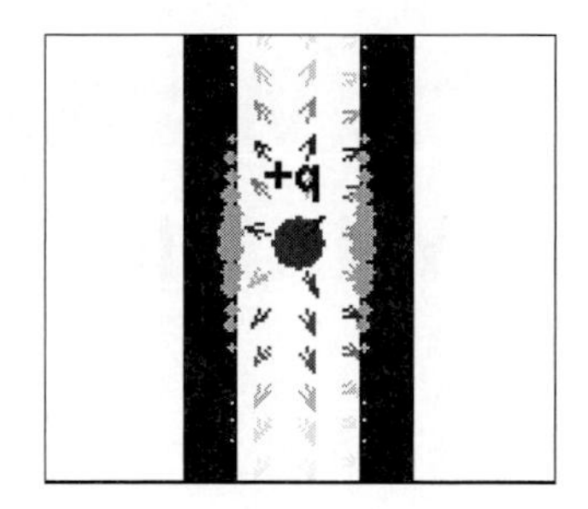

Exercise 7.3.7: The simulation shown represents a positive charge ($+q$) near two infinite grounded conducting sheets. The animation on the right depicts a similar situation without the conductor present. The rectangles are there for your reference only.

Question 7.3.7: Use the second animation to place charge(s) where they belong to make the field in the second animation match the field in the first animation for the middle region, $0.75 > x > -0.75$ (the region where the $+q$ is located). Assuming you could do this, where are the charge(s)?

Answer: *There are an infinite number of charges needed. Just in case your students get too enamored with the method of images, this problem will help point out the limitations in the method.*

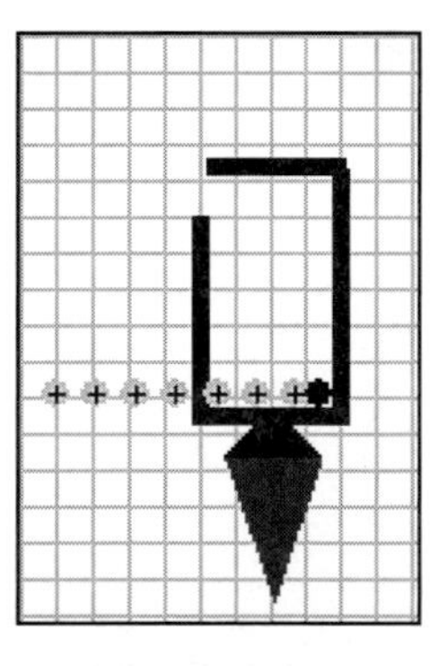

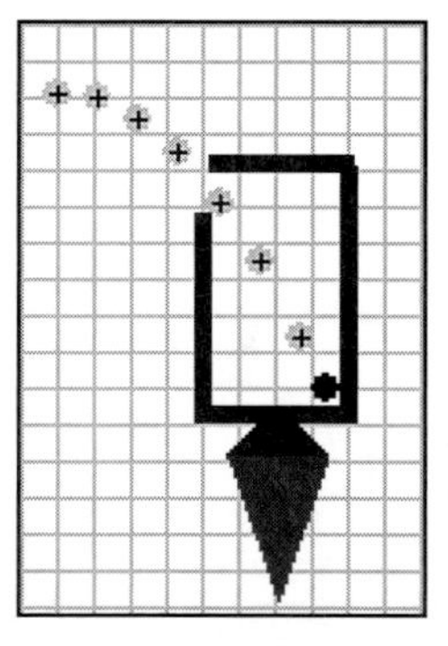

Exercise 7.3.8: The simulations shown here present a classic physics thought experiment, sometimes called a Gedanken experiment (from the German). An object is moving to the right in a straight line with constant speed. The animation on the left shows what a stationary observer would see. The animation on the right shows what an observer in the spaceship would see.

Question 7.3.8a: How would this motion appear to someone in an accelerating spaceship?

Answer: *The motion of objects as observed from the spaceship appears to be due to a force.*

Question 7.3.8b: How do you write the laws of the physics so that both frames are correct?

Answer: *In order to get observers in the two reference frames to agree, we must invent a force that acts on objects in the accelerating frame. Depending on the acceleration of the spaceship, we may call this fictitious force gravity. In fact, one cannot tell whether one is in a spaceship accelerating at 9.8 m/s*^{2} *far from the Earth, or in a nonaccelerating spaceship on the surface of the Earth (ignoring rotation of the Earth and other small effects). In fact, this equivalence between the two frames was Einstein's starting point for general relativity.*

CHAPTER 8

Mechanics, Waves, and Thermodynamics Problems

8.1 KINEMATICS

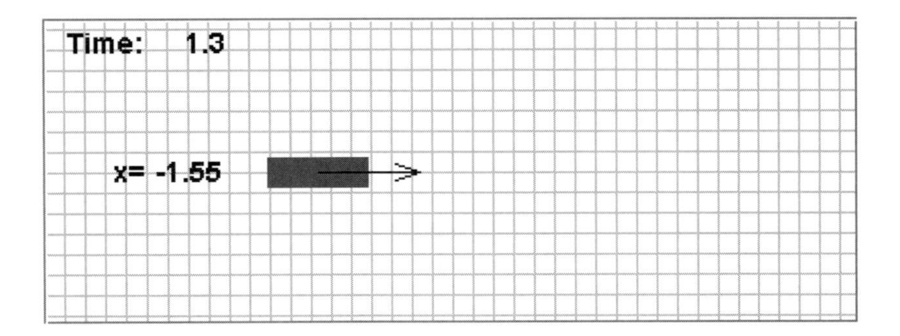

Problem 8.1.1: The rectangle starts at rest at $t = 0$ and moves to the right with increasing velocity. The acceleration of the rectangle is constant. The black arrow moves with the rectangle and has a constant length. The simulation runs from $t = 0$ to $t = 2$ and repeats. The student cannot control this applet in any way.

Question 8.1.1: Physicists use arrows to represent many things in diagrams. What vector quantity is being represented by the arrow in this simulation?

(a) displacement

(b) velocity

(c) acceleration

(d) speed

Answer: Acceleration. This problem can be given before acceleration is covered in class (i.e., as a WarmUp). Students can eliminate the three incorrect answers since they are clearly not constant.

Time: 1.3
x= -8.20
play pause <<step step>> reset

Problem 8.1.2: The animation begins with the puck at $x = -16$ at $t = 0$. It then moves to the right at a constant speed of 6 m/s until it bounces off of the wall at $t = 2$, after which it moves to the left with a constant speed of 4 m/s until the simulation stops at $t = 5$ s.

Question 8.1.2a: A hockey puck bounces off a wall and returns to its original starting point, where it is stopped as shown in the animation (position is shown in meters and time is in seconds). What is the average *speed* of the puck during the time it is in motion?

Answer: *4.8 m/s.*

Question 8.1.2b: A hockey puck bounces off a wall and returns to its original starting point, where it is stopped as shown in the animation. What is the average velocity of the puck during the time it is in motion?

Answer: *0.0 m/s.*

Question 8.1.2c: A hockey puck bounces off a wall and returns to its original starting point, where it is stopped as shown in the animation. What is the instantaneous velocity of the puck at $t =$ 3 second?

Answer: *−4.0 m/s.*

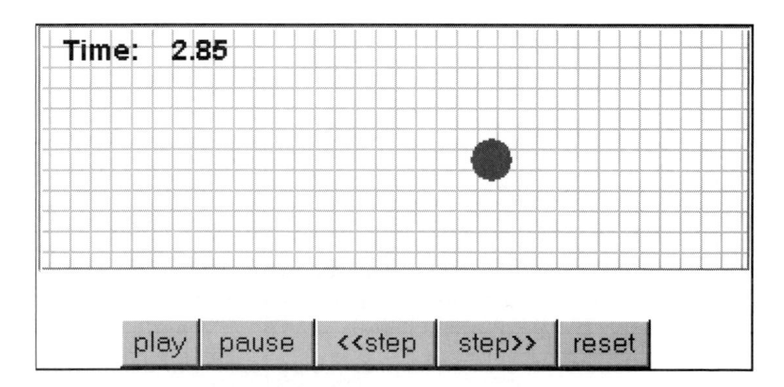

Problem 8.1.3: A ball moves across the screen with constant acceleration. Values near +/− 4.5 are effective. This problem is one of the most effective kinematics problems. The animation should start at time $t = 0$ with the trajectory set so that the ball is off the left- or right-hand side of the screen. That is, the ball should not be visible at $t = 0$ and move onto the grid at some later time. Many students simply cannot do a problem that does not give them an easy way to determine a value for the initial velocity, v_0.

Question 8.1.3a: What is the acceleration of the ball? You may click-drag the mouse inside the animation at any time to measure position.

Answer: Determined by the script. For example, the line in the script

```
document.Animator.addObject("circle","r = 10,x = 16*t,y = 18*t-4*t*t");
```

will display a ball with an acceleration of 8 units/time^2. We have observed that some students will mindlessly give an answer of 9.81 m/s^2 if the problem is recast to show vertical motion.

This problem is an effective tool for discussing measurement error since a straightforward application of $v = \Delta x/\Delta t$ *will likely give incorrect values due to the difficulty of measuring the center of the ball using the mouse. Even good students can disagree about the correct result. Students should be required to obtain an answer that is correct to better than 2%.*

Question 8.1.3b: What is the average velocity from time $t = 1$ to $t = 3$? What is the average acceleration? You may click-drag the mouse inside the animation at any time to measure position.

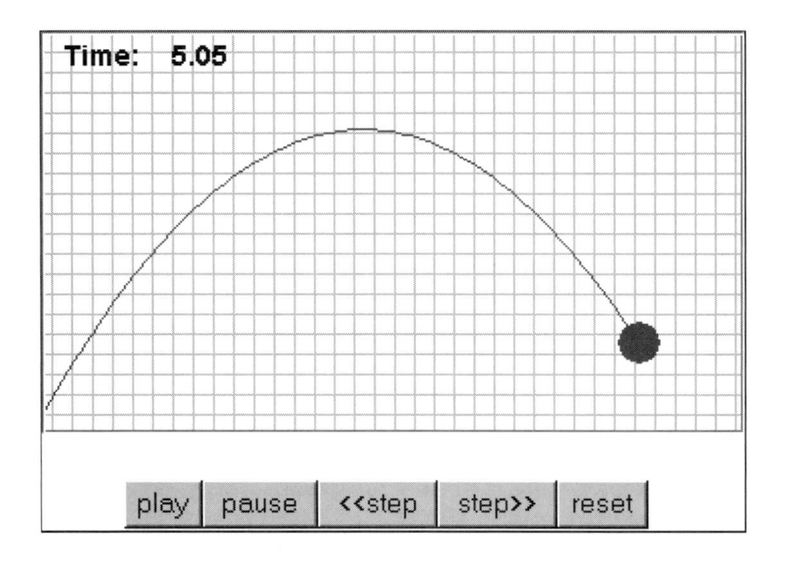

Answer: Determined by the script as in 8.1.3a. Different students may be assigned different time intervals in order to make the point that instantaneous and average acceleration have the same value if the acceleration is constant.

Problem 8.1.4: A ball moves across the screen in projectile motion leaving a visible trail to highlight its trajectory. Students may click-drag the mouse to measure coordinates. It is difficult to determine the velocity by measuring $\Delta\mathbf{r}/\Delta t$ using a single time step because of the one pixel limit imposed by the screen resolution.

Question 8.1.4: What is the minimum speed of the projectile? You may click-drag the mouse inside the animation at any time to measure position.

Answer: Since the acceleration of the ball is constant in the −y-direction, the projectile has its minimum speed at the maximum height when the y-component of its velocity is zero. The x-component of the velocity is constant and is easily measured by taking the distance traveled during a 5-second interval and dividing by time. This approach may not be obvious to students.

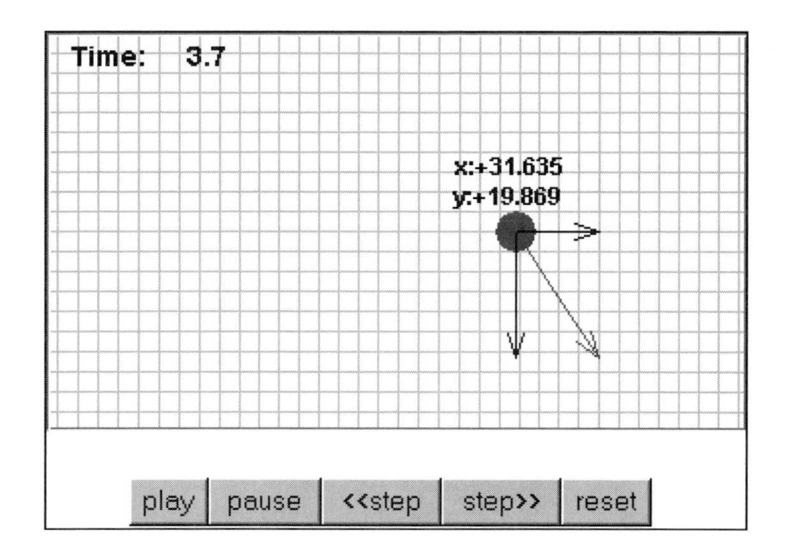

Problem 8.1.5: This animation presents a more elaborate version of Physlet Problem 8.1.4. The velocity components and particle position are shown as the animation progresses. This example demonstrates how it is possible to modify a script in order to add layers of complexity as students show mastery of basic concepts.

Question 8.1.5: What is the speed of the projectile at $t = 3.7$ s? What is the velocity of the particle? The acceleration is constant but not necessarily 9.8 m/s^2. The scale used to display velocity vectors is arbitrary. That is, do not assume that a vector of length five represents a velocity of 5 m/s.

Answer: Since x *and* y *are displayed to five significant figures, it is possible to obtain a good value of the velocity at* t = *3.7 s using the definition,* $\Delta\mathbf{r}/\Delta t$. *A number of students will attempt this problem by finding the initial velocity and the acceleration in order to use the constant acceleration kinematics equations given in the text. Here* $\mathbf{v} = (8.6\mathbf{i} + 12.8\mathbf{j})$ *m/s and* v = *15.4 m/s.*

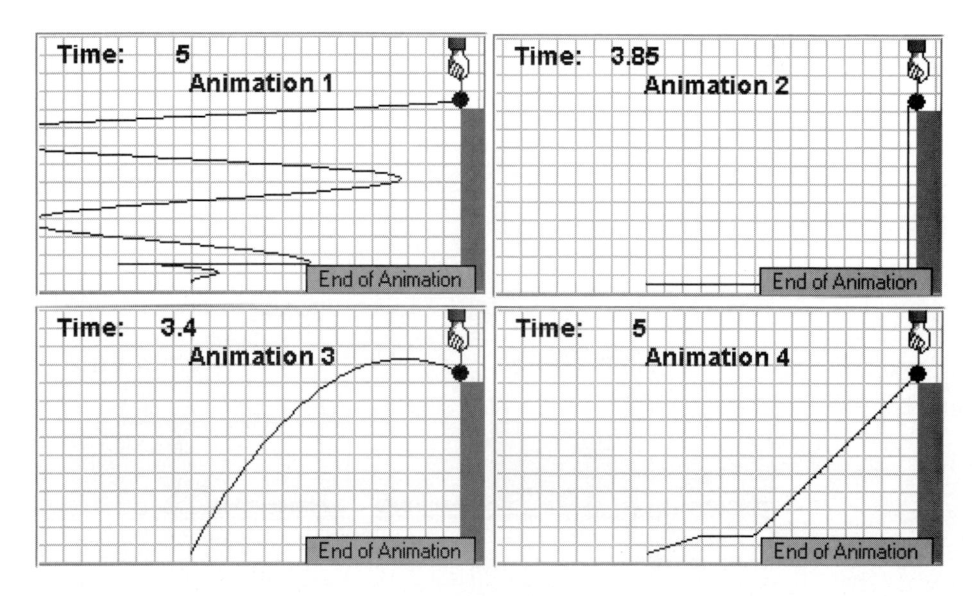

Problem 8.1.6: The animation depicts several ways a bowling ball can be lifted from rest onto a shelf (position is in meters and time is in seconds). Students may click-drag the mouse to measure coordinates. Students are asked to view all four animations depicted.

Question 8.1.6: Rank the paths by the displacement of the bowling ball during the animations (greatest first).

Answer: All tie. This example gets at the heart of the definition of displacement. While all of the bowling balls move through dramatically different paths, all of their starting and end points are identical; hence the displacements are all identical as well. This problem can be given as a WarmUp before class or as an in-class demonstration to see if students understand the concept of displacement.

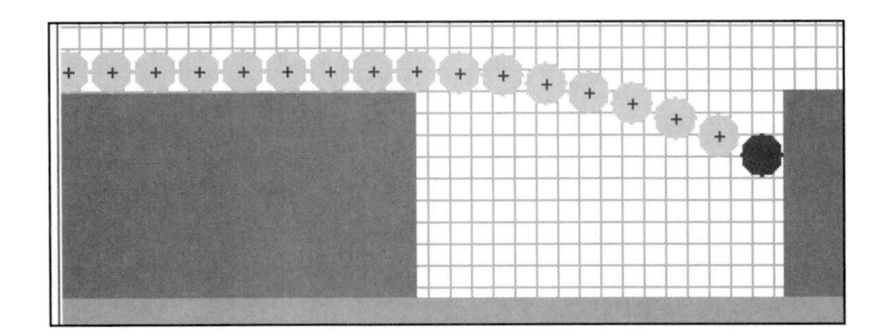

Problem 8.1.7: A ball rolls off a table at a constant speed. The ball hits another table when the animation ends at $t = 1$ second. Students may click-drag the mouse to measure coordinates. The ghost images are given to provide the path the ball follows.

Question 8.1.7: If the ball undergoes an elastic collision with the other table (v_y remains the same and v_x changes sign), where will the red ball land?

Answer: This kinematics question highlights the difference between a traditional book problem and a Physlet problem. It is completely up to the student to determine how to use the data that can be acquired from the animation. The most likely error in this calculation is for the student to measure the distance the ball will fall from the center of the ball to the floor instead of from the bottom of the ball to the floor. The answer depends on the separation between the tables and the initial velocity given to the ball; here the ball should land at $x = -0.3$ m.

8.2 NEWTON'S LAWS

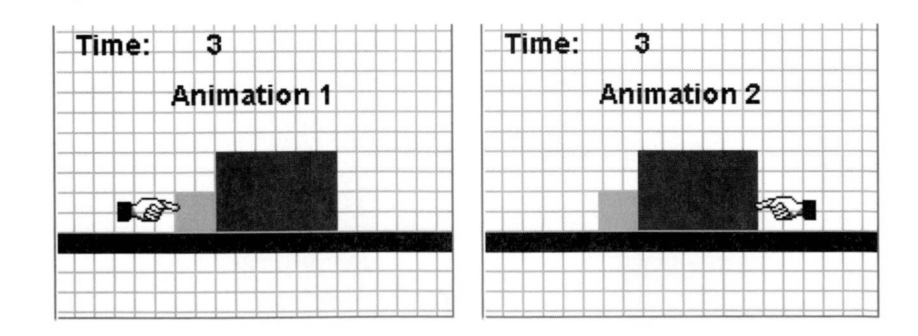

Problem 8.2.1: A small green 1.5-kg block receives a push and in the process pushes against a larger red 4.5-kg block as shown in Animation 1. Conversely, in Animation 2, the large red 4.5-kg block receives the same push, now from the opposite direction and in the process pushes against the small green 1.5-kg block as shown.

Question 8.2.1: Determine the force of the large block on the small block assuming that all surfaces are frictionless in both animations.

Answer: -11.7 *N/*-3.9 *N. This is a good problem to test whether students understand Newton's second and third laws and how to apply them for systems moving to the right and to the left. Here it is important for the student to measure the acceleration of the masses and relate this acceleration to the force the hand exerts on the system.*

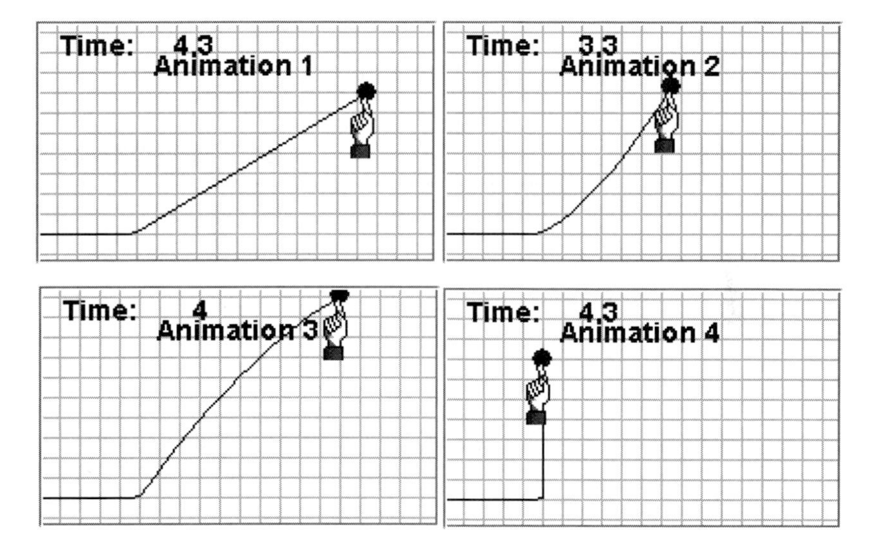

Problem 8.2.2: A satellite floats at constant speed when, at $t = 2$, its thrusters are suddenly engaged, producing a constant force perpendicular to its original motion (position is in meters and time is in seconds). Four animations with different paths are shown.

Question 8.2.2: Which animation correctly depicts the satellite's motion after the thrusters are first engaged?

Answer: Animation 2. This is an animated version of a question that appears on the FCI. As with the FCI question, this question verifies whether students understand the relationship between force and acceleration. Unlike the static figure in the FCI, seeing the animation helps many students visualize the relationship between force and acceleration.

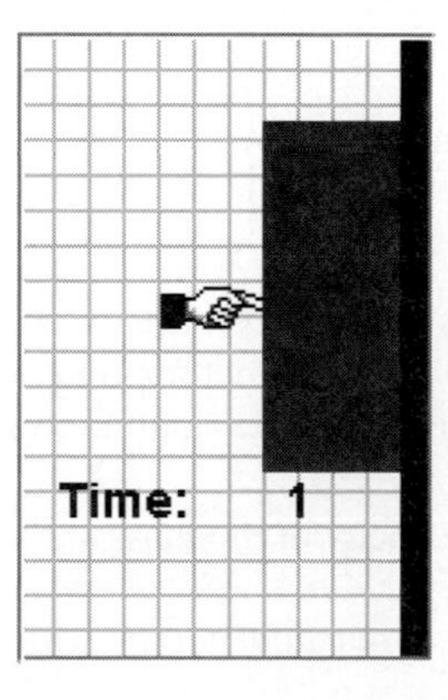

Problem 8.2.3: Consider a 2-kg physics textbook pressed against a wall as shown in the animation (position is in meters and time is in seconds). In this animation, the book is seen to slide down the wall.

Question 8.2.3: Given that $\mu_s = 0.5$ and that $\mu_k = 0.4$ between the wall and the textbook, determine the force of the hand that acts on the book.

Answer: *49 N, but it depends on the script. For this example, the book is moving at a constant speed, hence there is no acceleration and this is therefore also a statics problem. Students must realize that this is an important feature of the animation and make the determination of the force of the hand. This script can be modified easily so that the book is accelerating down the wall, which changes the force applied by the hand. An interesting way to give this type of problem is to pair the constant velocity problem with a constant acceleration problem to test students' understanding of Newton's laws and how an acceleration changes the result.*

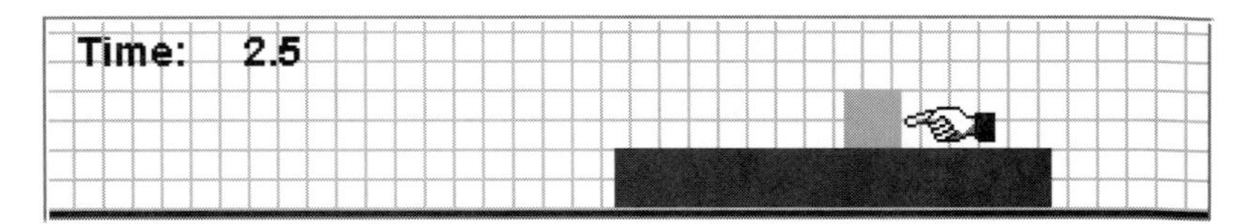

Problem 8.2.4: A small 10.0-kg block sits on a larger 100-kg block as shown (position is in meters and time is in seconds). The surface between the bottom block and the table is frictionless. At $t = 1$ second, the hand appears and the system accelerates to the left. The smaller block has a larger acceleration than the larger block.

Question 8.2.4: Given a push of $F = 50.0$ N on the top block, determine the coefficient of kinetic friction between the top and bottom block.

Answer: *0.33. Students must determine the acceleration of the larger block. This acceleration times the mass of the larger block is equal to the force of friction. In this problem, students are just as likely to determine the acceleration of the smaller block and use this variable to determine the coefficient of kinetic friction.*

8.3 WORK AND ENERGY

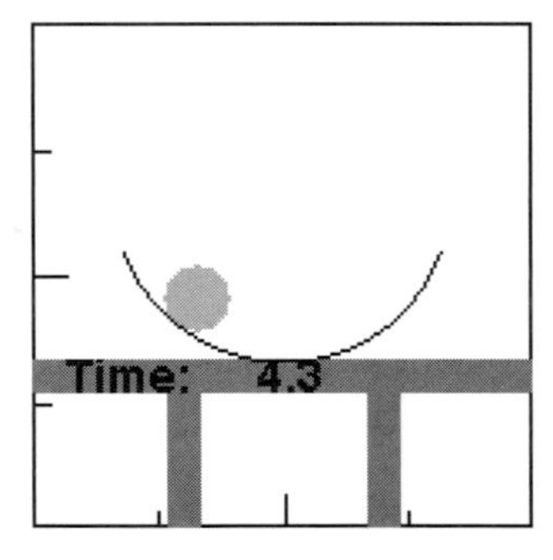

Problem 8.3.1: A mass of 2 kg is in a rather large bowl and moves as depicted in the animation (mouse click for position given in meters; time is given in seconds). The ball slides (frictionless surface) along the bowl surface, attaining the exact same height above the bottom of the bowl each time.

Question 8.3.1: Determine the velocity of the mass at the bottom of the bowl.

Answer: $v = 5.4$ *m/s. Students must determine the distance the ball falls and relate this to the change in potential energy, which equals the change in kinetic energy.*

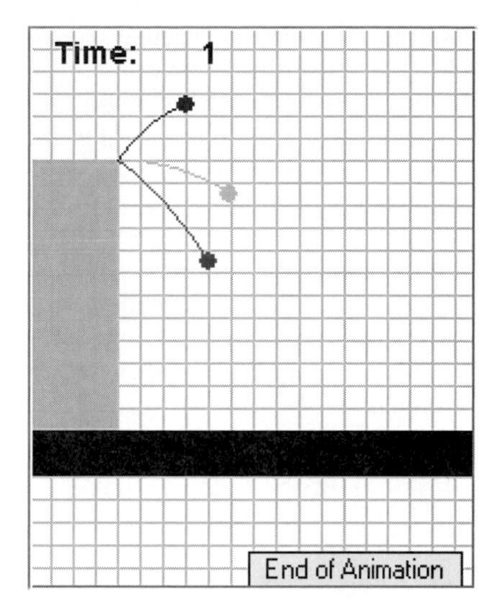

Problem 8.3.2: Three balls (a red ball launched at a negative angle, a green ball launched horizontally, and a blue ball launched with a positive launch angle) are thrown off the top of a building, all with the same speed but different launch angles (position is in meters and time is in seconds). The animation ends at the point depicted in the animation shown here.

Question 8.3.2: Rank the three balls according to which one hits the ground first and which one has the greatest speed upon impact with the ground.

Answer: Red, all tie. It is apparent from kinematics that the red ball will hit the ground first. However, it is not immediately clear that all three balls will land with the exact same speed. Using kinematics is a little complicated, so using the work-energy theorem is the way to go.

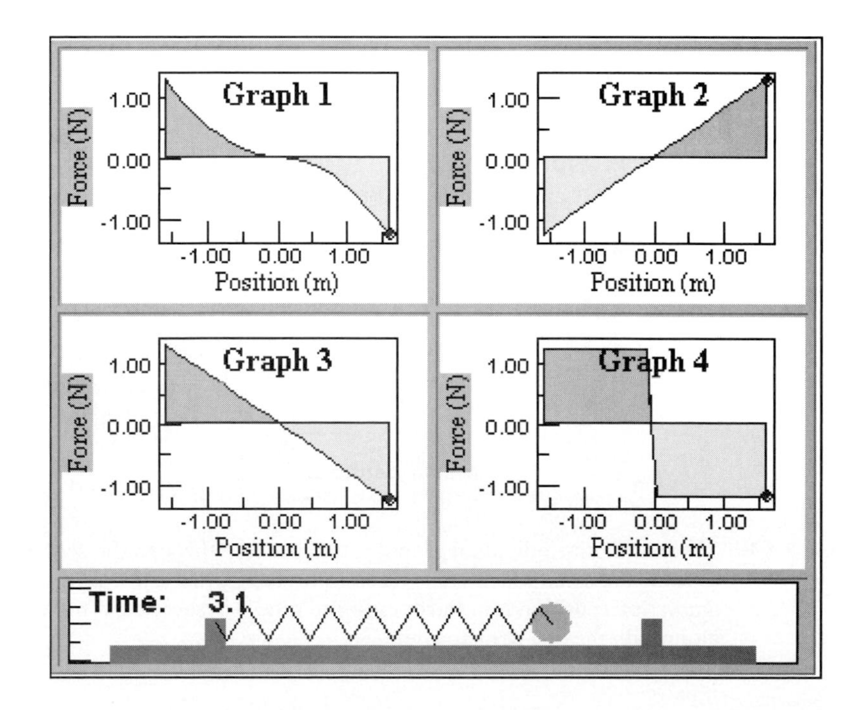

Problem 8.3.3: A ball on an air track is attached to a compressed spring (at $x = 0$ the spring is unstretched) as shown in the animation (position is in meters and time is in seconds). With the animation of the ball moving from maximum compression to maximum elongation, four graphs are drawn that show the area under a force versus position curve.

Question 8.3.3: Determine which area properly represents the work done by the spring during the animation (assume $v = 0$ at the beginning and end of the animation).

Answer: *Graph 3. Since this is a spring, the force should vary linearly with distance. This graph also shows why, for a spring, you cannot use the force dotted into the displacement, as the force varies. It is also apparent that at maximum extension of the spring, the velocity should be zero as the net work done by the spring is zero. This is also a very flexible script as it can be modified easily to show graphs for simple harmonic motion, momentum, and impulse problems.*

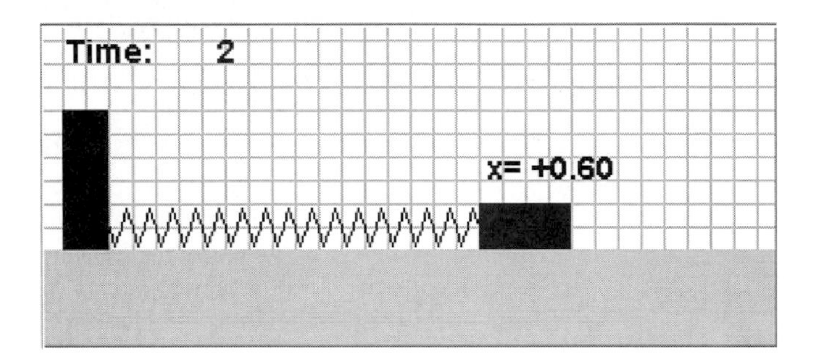

Problem 8.3.4: A 0.5-kg cart on an air track oscillates due to its attachment to a wall via a spring as shown in the animation (position in meters and time in seconds). The cart begins to be displaced from equilibrium and oscillates back and forth as shown in the animation.

Question 8.3.4: What is the spring constant of the spring?

Answer: k = *4.8 N/m. Again, use the work-energy theorem. Here, however, students must determine the equilibrium position to determine the speed of the block there. When this script is rewritten for simple harmonic motion problems, the period is used to determine the spring constant.*

8.4 GRAVITY

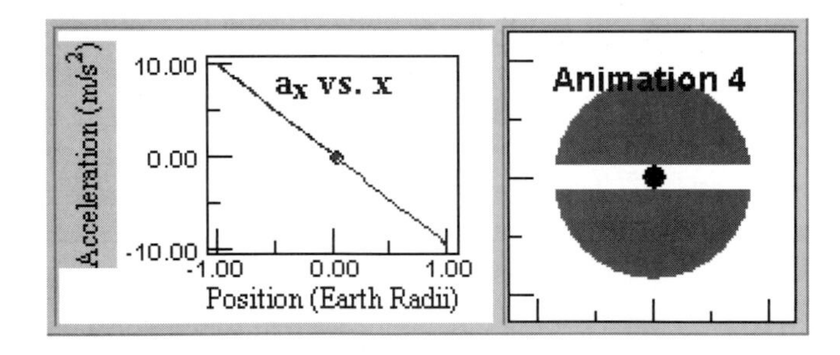

Problem 8.4.1: A very wealthy individual proposes to dig a hole through the center of the Earth and run a train (the small black circle) from one side of the Earth to the other as shown in the animation (position is in Earth radii and time is in seconds). Three other graphs are given along with the one depicted here.

Question 8.4.1: Which of the animations correctly depicts the motion of the train? Ignore frictional effects and assume the Earth is a uniform mass distribution.

Answer: *Animation 4, the animation shown. This is a standard in-text problem done in almost every introductory physics text. Here the standard student answers (9.8 m/s² on the left of center and −9.8 m/s² to the right of center, for example) look very unphysical when graphed.*

8.5 MOMENTUM

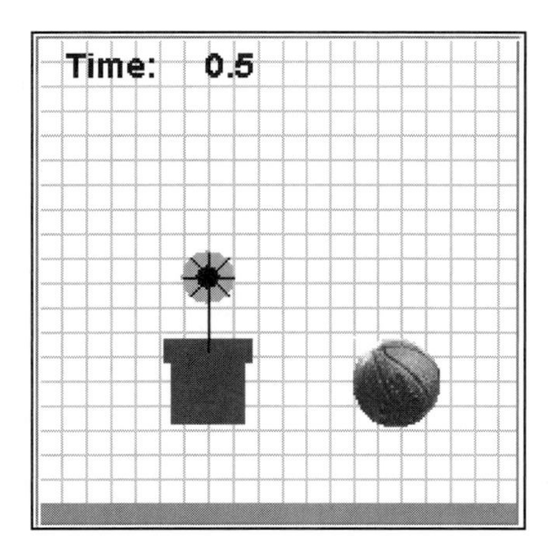

Problem 8.5.1: A ball and a flowerpot of identical mass fall from the same height and strike a table. The flowerpot hits the ground and stops there, while the basketball bounces back to its original height.

Question 8.5.1: Which undergoes the greater change in momentum after colliding with the table (position is in meters and time is in seconds), a flowerpot or a basketball of the exact same mass?

Answer: *The basketball. The change in momentum of the basketball is twice that of the flowerpot. As an in-class demo, this problem can be paused at the point of impact with the table (or at any point), and the instructor can ask the question a second time before resuming the animation.*

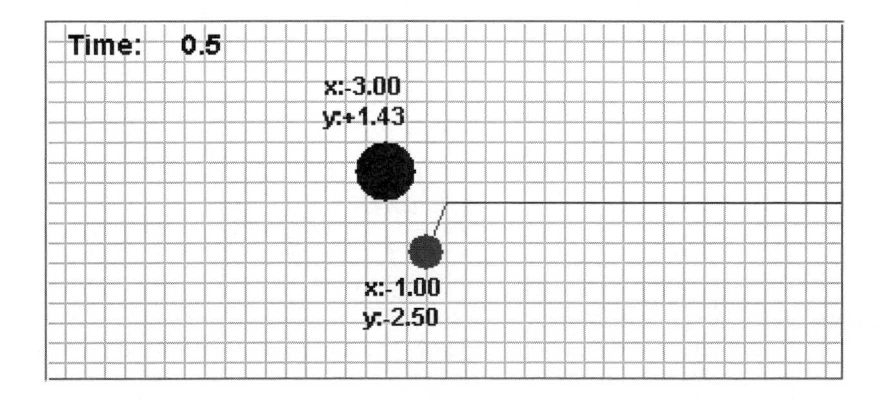

Problem 8.5.2: The small red mass moves at a constant speed to the left before striking the larger black mass as shown in the animation (position in meters and time in seconds). After the collision, the black mass and the red mass move at constant velocities. The trails are given to help the determination of angles.

Question 8.5.2: Determine the ratio of the red mass to the black mass: m/M.

Answer: m/M = *0.25. Students must determine the speed of the objects before and after the collision. Since we do not give a statement as to whether this is an elastic collision, students must rely on conservation of momentum. Hence, the angles from the* x*-axis must be determined. Students often assume perfectly elastic collisions unless they see the objects stick together, in which case they correctly determine an inelastic collision. Giving a partially elastic collision helps students with their problem-solving skills.*

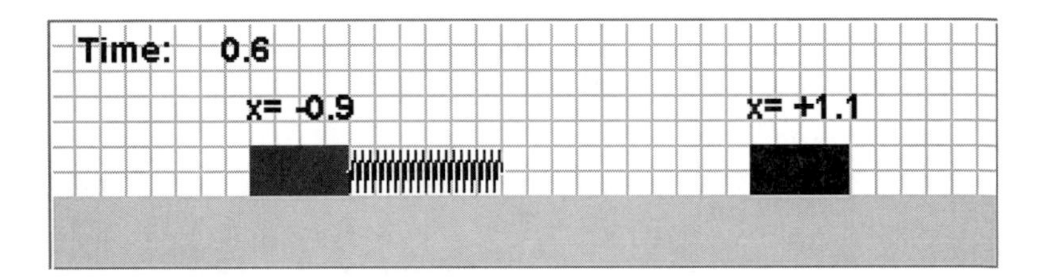

Problem 8.5.3: A spring is attached to a cart on an air track. With the spring compressed and locked, the cart is brought into contact with another cart (the carts have identical mass). When the spring is released, the carts travel as shown in the animation (position is in meters and time is in seconds).

Question 8.5.3: What is the velocity of the center of mass of the two carts after the spring is released?

Answer: 0. Since the net external force on the system is zero, the change in momentum of the center of mass is zero. Hence, when the spring is released the velocity of the center of mass is also zero.

8.6 ROTATIONAL DYNAMICS

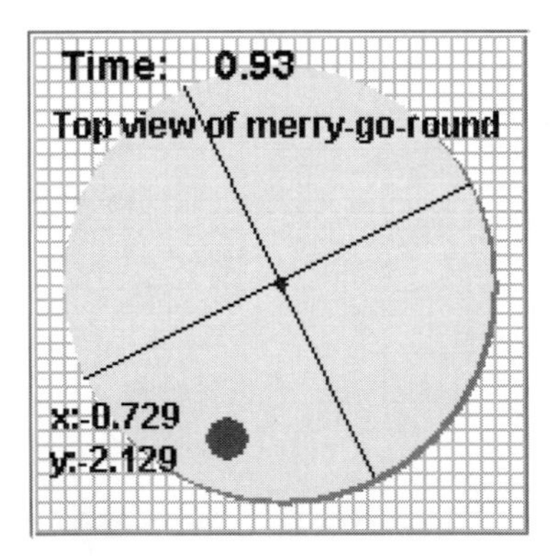

Problem 8.6.1: A child sits on a merry-go-round at the position marked by the red circle (position is in meters and time is in seconds). The animation shows a red circle; perhaps the child is a redhead, rotating around the edge of the merry-go-round at a constant angular speed.

Question 8.6.1: What is the child's angular displacement (in radians) after 0.44 seconds?

Answer: *2.1 radians. There are two main approaches to solving this problem. One is to measure the angle directly. This is a very imprecise approach. The preferred approach is to determine the angular frequency of the child and then use rotational kinematics for constant angular velocity. This script is again multipurpose as it can be used in questions of linear displacement and centripetal acceleration. The script can also be modified slightly to show angular accelerations.*

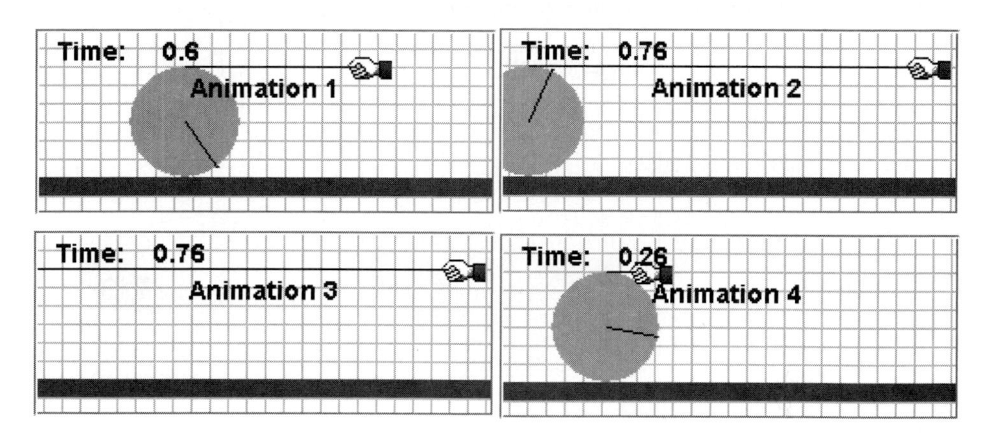

Problem 8.6.2: A wheel rolls without slipping while pulled by a string wrapped around its circumference as shown in the animation (position is given in meters and time is in seconds). Four animations are shown with varying rope and, therefore hand speeds. Animation 1 shows the hand moving at the same velocity as the translational velocity of the wheel. Animation 2 shows the hand moving at twice the translational velocity of the wheel. Animation 3 shows the hand moving at four times the translational velocity of the wheel. Finally, Animation 4 shows the hand moving at half the translational velocity of the wheel, which is why the string shortens.

Question 8.6.2: Which animation properly depicts the physical situation?

Answer: *Animation 2. This problem relies on the fact that the hub moves at the translational speed of the wheel, but the top of the wheel moves at twice the translational speed of the wheel.*

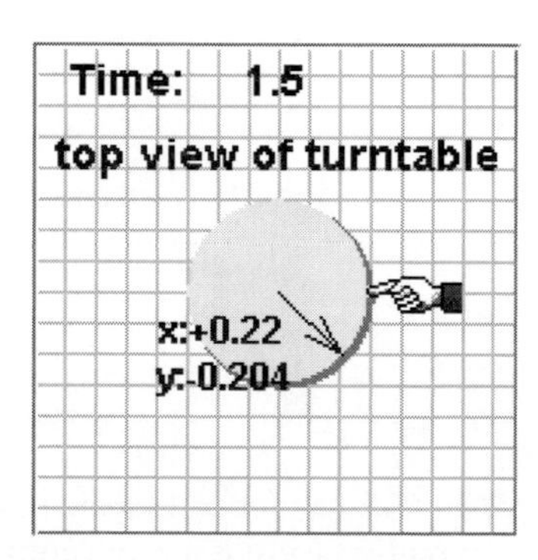

Problem 8.6.3: A turntable (a flat disk) of mass 5 kg is rotating at constant speed when your finger makes contact with the outer edge as shown in the animation (position of a point on the outer edge

of the turntable is shown in meters and time is shown in seconds). Friction between your finger and the turntable causes the turntable to stop.

Question 8.6.3: What is the average torque on the turntable by the frictional force?

Answer: -1.35 N · m. *Once students decide that the angular acceleration must be determined, they find the initial angular velocity. At this point students must relate the product of the angular acceleration and the moment of inertia to the torque. To continue this problem, questions can be asked regarding the force the hand applies and how this relates to the other definition of torque,* $\mathbf{r} \times \mathbf{F}$.

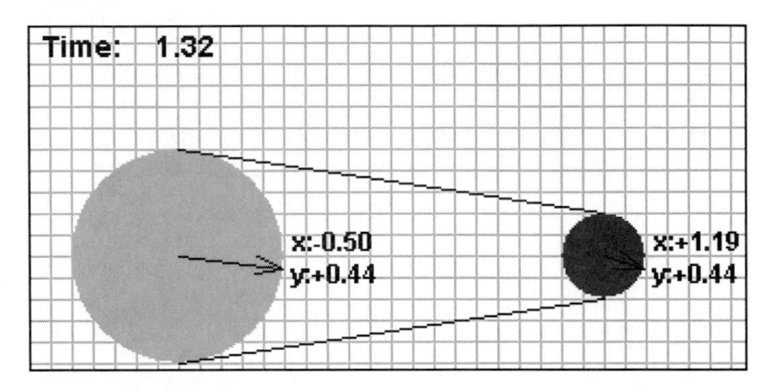

Problem 8.6.4: The animation depicts an idealized drivetrain for a bicycle. A large green disk (i.e., a flat cylinder) is used to rotate a small green disk of the same density and thickness via a massless chain that does not slip (position of a point on the edge of each disk is shown in meters and time is shown in seconds). In the animation, the larger green disk rotates at a slower angular speed than that of the smaller red disk.

Question 8.6.4: What is the ratio of the kinetic energy of the green disk to the kinetic energy of the red disk (K_{green}/K_{red})?

Answer: 6.3. *Since the densities are the same, the ratio of the masses of the two disks* (m_{green}/m_{red}) *is proportional to the square of the ratio of their radii. Students must then determine the angular speeds from the animation and then apply the relationship for the kinetic energy of rotation.*

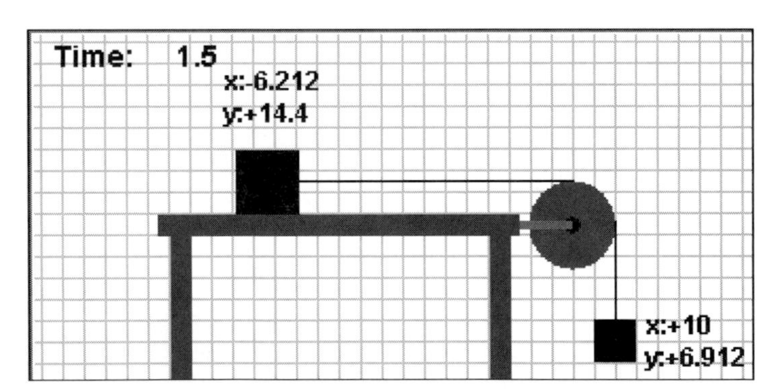

Problem 8.6.5: Two masses $M = 2.5$ kg and $m = 1$ kg are attached with a massless string over a pulley as shown in the animation (position is in centimeters and time is in seconds). The bearings in the pulley are frictionless and the string does not slip. As the animation progresses the two masses accelerate.

Question 8.6.5: What is the moment of inertia of the pulley?

Answer: $I = 0.27$ kg m^2. There are, of course, two approaches to this problem: forces/torques and energy. Using the force/torque method, students must determine the acceleration of the system. For the energy method, students must determine the velocity of the system when the hanging mass has moved a certain distance.

8.7 SIMPLE HARMONIC MOTION

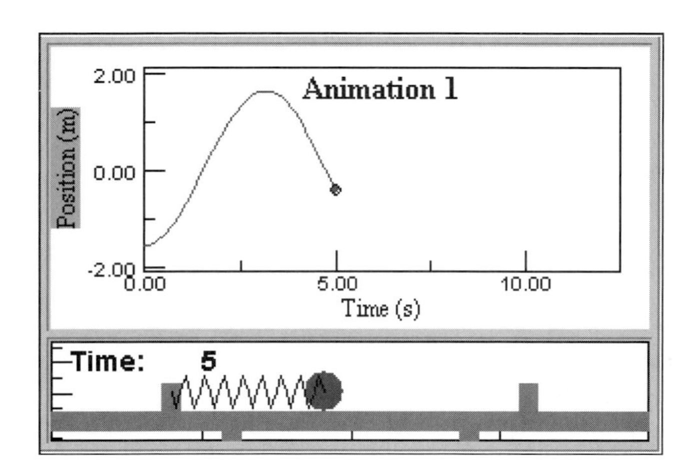

Problem 8.7.1: A ball on an air track is attached to a compressed spring as shown in the animation (position is in meters measured from equilibrium and time is in seconds). The ball moves from maximum compression at $t = 0$ to maximum elongation and back again while one of four possible position versus time graphs is shown.

Question 8.7.1: Determine which graph properly shows the position of the ball as a function of time.

Answer: Animation 1, the animation depicted here. Understanding the position versus time graph is the goal of this exercise. However, the script can be easily augmented to show velocity and acceleration versus time graphs.

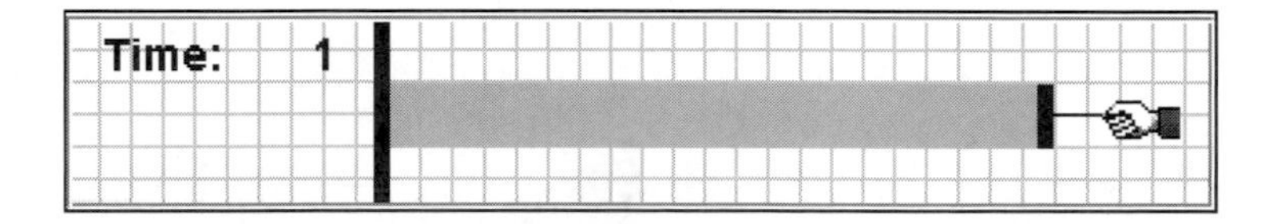

Problem 8.7.2: An aluminum bar (Young's modulus of 70×10^9 N/m^2) is stretched by a force as shown in the animation (position is in centimeters and time is in seconds). Assume the bar has a square cross section. The bar is stretched during the application of the force in the animation.

Question 8.7.2: What is the force required to elongate the aluminum?

Answer: 1.5×10^4 N. Students must apply the definition of the Young's modulus as the ratio of stress to strain.

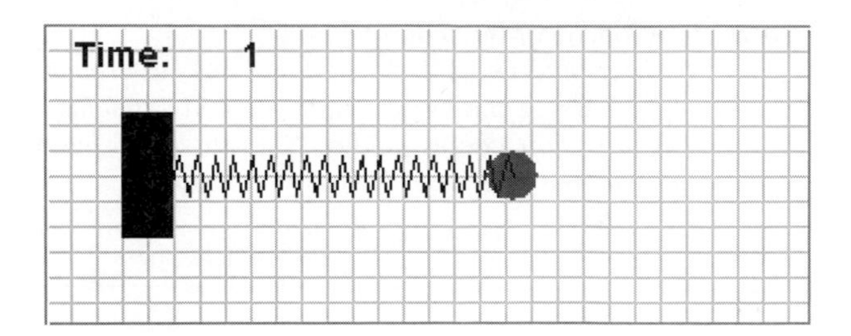

Problem 8.7.3: A 200-g mass is vibrating at the end of a spring as shown. Position is given (with a mouse click) in cm and time in seconds.

Question 8.7.3: What is the spring constant of the spring?

Answer: 3.5 N/m. Students must measure the period of the oscillation and then relate that period to the spring constant. Students will often measure the amplitude as well, even though it is a useless piece of information for this problem unless conservation of energy methods are used to find the spring constant.

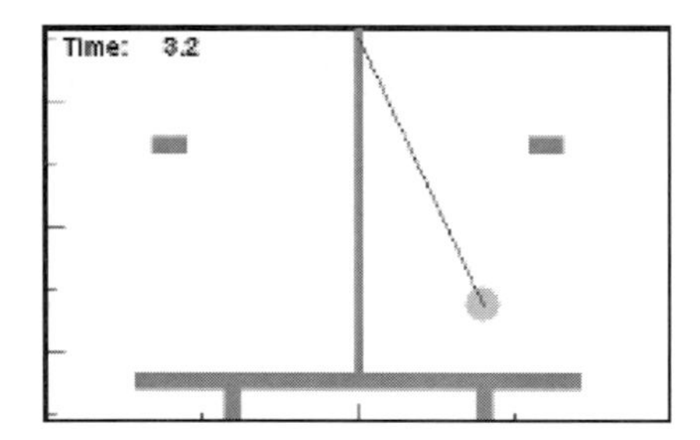

Problem 8.7.4: A pendulum is allowed to oscillate in an accelerating elevator as shown in the animation (position is in meters and time is in seconds).

Question 8.7.4: Determine the effective acceleration due to gravity by analyzing the motion.

Answer: 5.5 m/s². Period and pendulum length measurements are the only measurements necessary to determine the effective gravity in the elevator. Once the effective gravity is determined, the difference between g_{eff} and g denotes the acceleration of the elevator.

8.8 STATICS

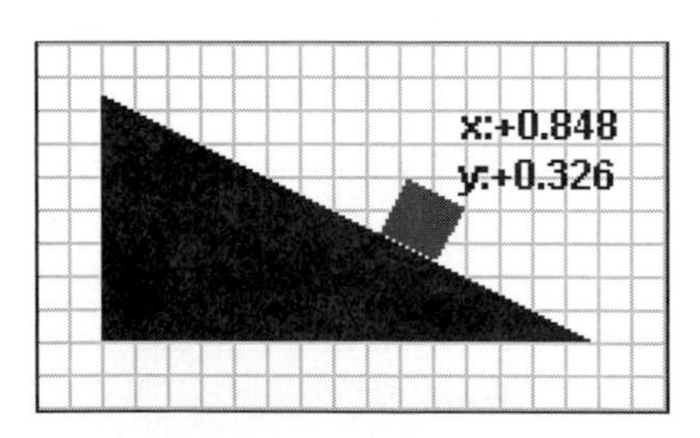

Problem 8.8.1: A 12-kg box slides down a rough ramp as shown in the animation (position is in meters and time is in seconds).

Question 8.8.1: From the animation determine the magnitude of the force of friction.

Answer: *52.6 N. Despite the motion seen in the animation, this is a statics problem. Since the block moves down the incline at a constant velocity, the sum of the forces on the block are zero. Therefore, the component of the weight down the incline is equal to the frictional force.*

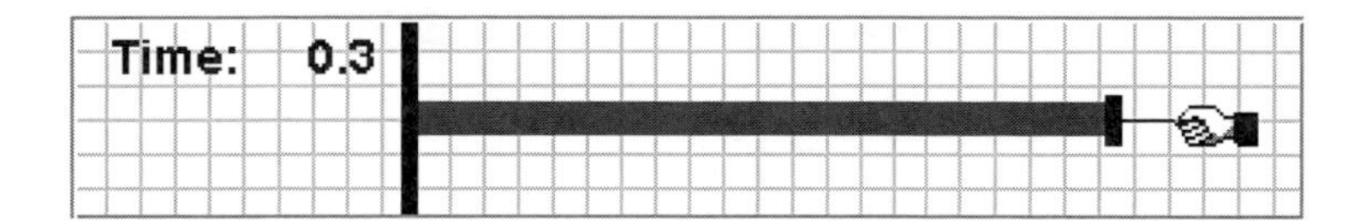

Problem 8.8.2: A metal bar is stretched by a force as shown in the animation (position is in centimeters and time is in seconds).

Question 8.8.2: What is the resultant strain in the material?

Answer: *0.05. The strain is the ratio of the change in length to the initial length of the bar. Students must therefore make position measurements related to the length of the rod and the change in the length of the rod.*

8.9 WAVES

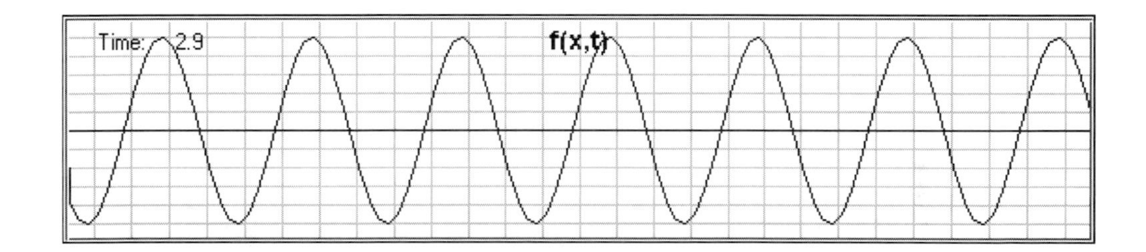

Problem 8.9.1: The animation shown here depicts a traveling wave moving to the left. Mouse-down in the animation to make position measurements.

Question 8.9.1a: Find the frequency of the wave shown in the animation (position is in centimeters and time is in seconds).

Question 8.9.1b: Find the wavelength of the wave shown in the animation (position is in centimeters and time is in seconds).

Question 8.9.1c: Find the velocity of the wave shown in the animation (position is in centimeters and time is in seconds).

Answers: *a = 0.5 cycles/second, b = 16 cm, c = −8.0 cm/second. For part a, students must use the relationship between the period and the frequency to determine the answer. Wavelength can be determined by a mouse-down in the animation, and the velocity can be determined by the relationship* $v = \lambda f$.

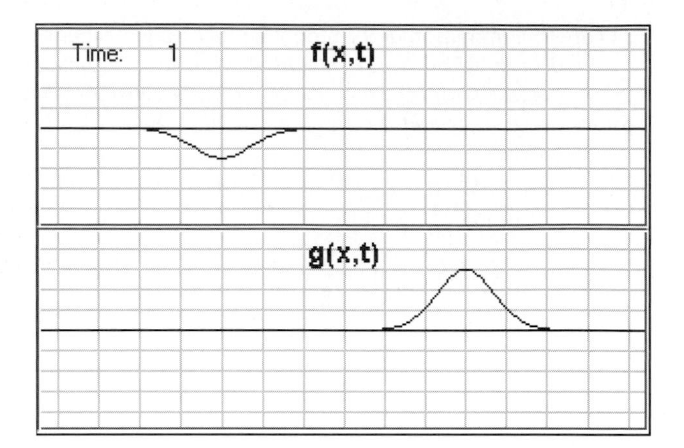

Problem 8.9.2: The preceding simulation shows disturbances on two identical strings (position is in centimeters and time is in seconds). As the animation proceeds, the top wave disturbance moves to the right and the bottom wave disturbance moves to the left.

Question 8.9.2: At $t = 2.5$ seconds, when the two waves overlap, what does the superposition of the two waves look like?

Answer: Their sum has a large peak, a depression, and then another large peak. Students must realize that these waves are of different heights and widths and will therefore not constructively interfere or destructively interfere. This script can be modified as an in-class demo by enabling the third panel, which adds the two waves together.

8.10 SOUND

Problem 8.10.1: The animation is a slow-motion representation of a cross section of a sound wave propagating in Lucite. A detector, the orange square, in the pipe measures the pressure (position given in meters and time given in seconds). The bar graph at the right measures the amplitude as determined by the detector. The instant depicted here shows the maximum amplitude of the sound wave.

Question 8.10.1: What is the speed of the sound wave?

Answer: 3500 m/s. The wavelength and frequency of the wave can be determined from the animation much like the preceding wave question.

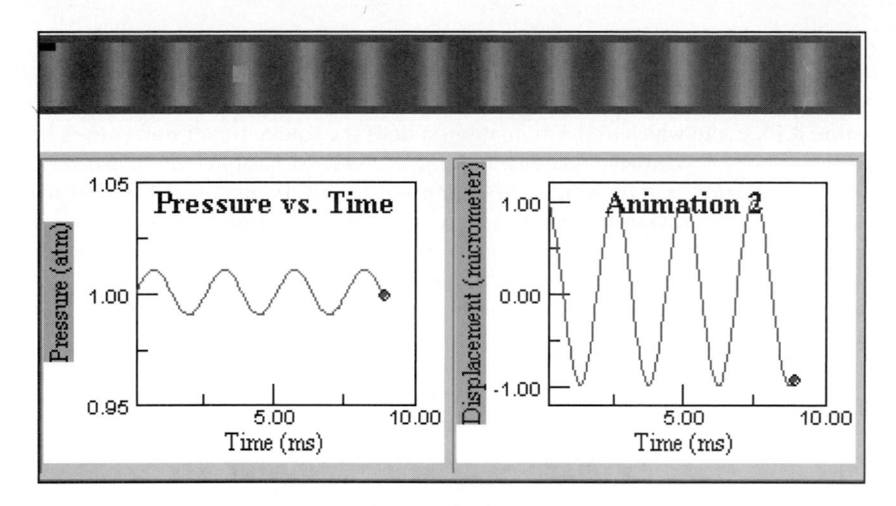

Problem 8.10.2: The animation represents a cross section of a sound wave propagating in a very long pipe. A detector, the orange square, placed in the pipe measures the pressure (position given in meters and time is given in milliseconds). There are four graphs with different depictions of pressure and displacement as a function of time.

Question 8.10.2: Which of the graphs properly represents the displacement of the air molecules in the pipe?

Answer: The animation shown here. Students must understand the relationship between the pressure variation and the displacement of the air molecules in a sound wave.

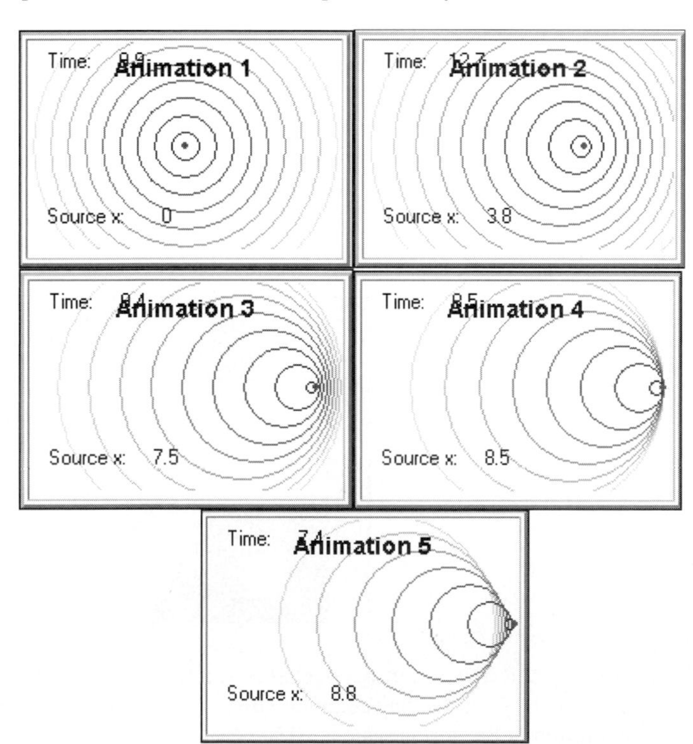

Problem 8.10.3: The animation represents a cross section of a three-dimensional sound wave propagating away from a moving source. Five animations are shown, each with a different source velocity.

Question 8.10.3: In which of the animation(s) does the source travel slower than the speed of sound?

Answer: Animations 1, 2, 3. Students must determine whether the source is moving slower than the wavefronts it produces. In Animation 1, it is clear from the animation and the wavefronts that the source is not moving. In Animations 2 and 3, it is clear from the animation that the part of the wavefronts that leads the source is moving away from the source. In the final two animations, it is clear that the leading wavefronts do not move away from the source and, therefore, the source moves at or faster than the speed of sound.

8.11 FLUIDS

Problem 8.11.1: Blood flows from left to right in an artery with a partial blockage in the animation (position is in centimeters and times is in seconds). A blood platelet is shown moving through the artery. In Animation 1, the platelet moves at the same speed during the animation. In Animations 2 and 3, the platelet moves through the blockage slower and faster, respectively, before it returns to its original speed in the unclogged region on the right.

Question 8.11.1: Which animation properly represents the motion of the platelet as it moves through and past the blockage?

Answer: Animation 3. This problem relies solely on the continuity of the fluid. Since the volume of the fluid flowing from the left through the blockage must be the same as through the blockage, the velocity of the fluid in the blockage must be faster. At this point a subsequent question can be asked using Bernoulli's equation: Compare the pressure of the unblocked artery to that of the blocked artery. Bernoulli's equation tells us that the pressure in the blocked region is reduced, which can lead to a further closing of the artery.

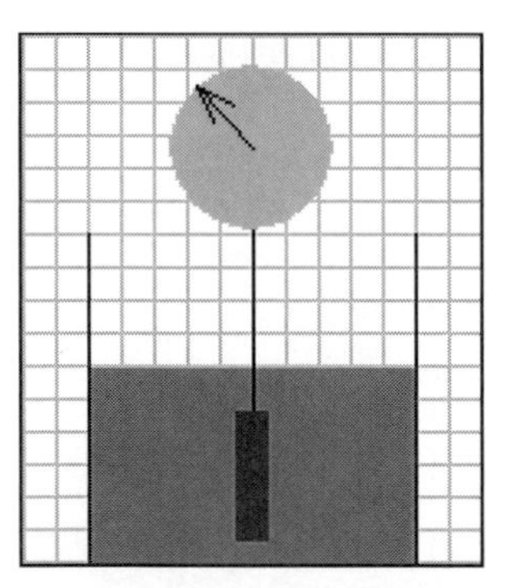

Problem 8.11.2: An object is lowered into a bucket of water as shown in the animation. As the object is lowered, the scale reading changes. The initial reading on the spring scale is 19 N. One full revolution of the spring scale is 10 N.

Question 8.11.2: Find the density of the object being immersed in the water bucket.

Answer: *1.9 g/cm^3. As the object is dropped into the water, the buoyant force changes until the object is completely submerged. At this point, the buoyant force is $\rho_w V_0$, the sum of the forces on the object is zero since the buoyant force plus the force of the scale equals the weight of the object. Given the readings of the scale, the use of the volume of the object is not necessary. Besides not being necessary, the use of the volume is impossible as it is not given nor can it be determined from the animation. Students will often make up a thickness for the object, then measure the length and width to "determine" the volume of the object.*

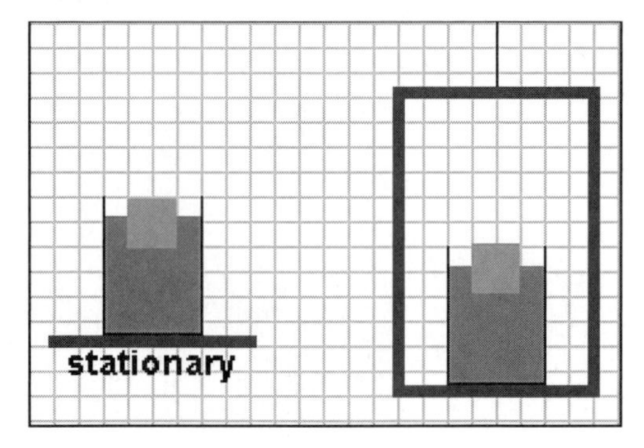

Problem 8.11.3: A block of wood in a bucket of water is placed in an elevator that moves as shown in the animation (position is in meters and time is in seconds). There are three animations to choose from: one where the block is more submerged than the stationary case, one where the block is less submerged than the stationary case, and the animation shown. The picture on the left represents the orientation of the wood when the elevator is stationary.

Question 8.11.3: Which animation correctly depicts the new orientation of the wood while the elevator moves as shown?

Answer: *Animation 1, the animation shown here. In this animation, the elevator is accelerating upward at a constant rate. Students will usually acknowledge that this will affect the apparent weight of the block. In equilibrium, however, the buoyant force (the "weight" of the fluid displaced) is equal to the "weight" of the object. Hence, in this problem since both the buoyant force and the "weight" of the object change in the same way, the object floats as it would normally.*

8.12 THERMODYNAMICS

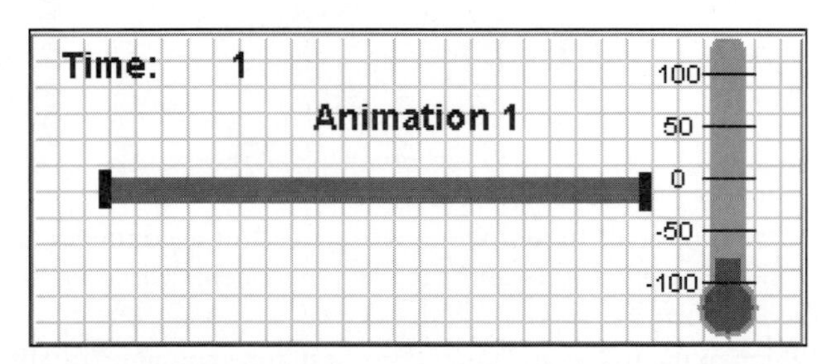

Problem 8.12.1: Metal bars are heated as shown in the animation (position is in centimeters and time is in seconds). Four metal bars undergo the exact same change in length. The bars are the same size at different temperatures. A thermometer (scale given in Celsius) shows a temperature increase. The change in temperature for each bar is identical even though the starting temperatures are not the same.

Question 8.12.1: Are all of the metal bars made of the same material?

Answer: *Yes. The bars undergo the exact same change in length per length for the exact same change in temperature. Therefore, the bars are made out of the same material. This script can be modified so that the bars undergo different length changes for the same temperature change.*

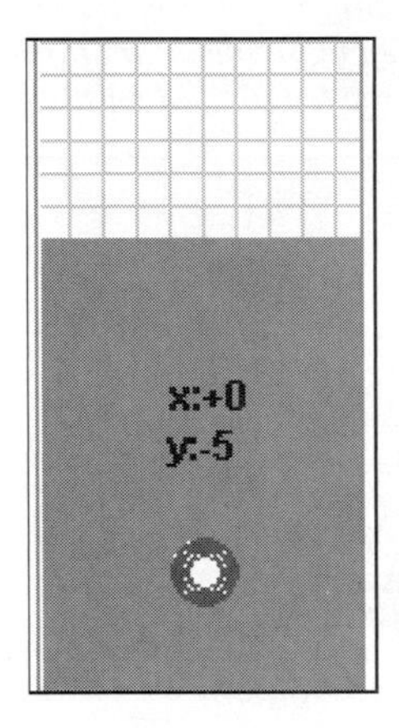

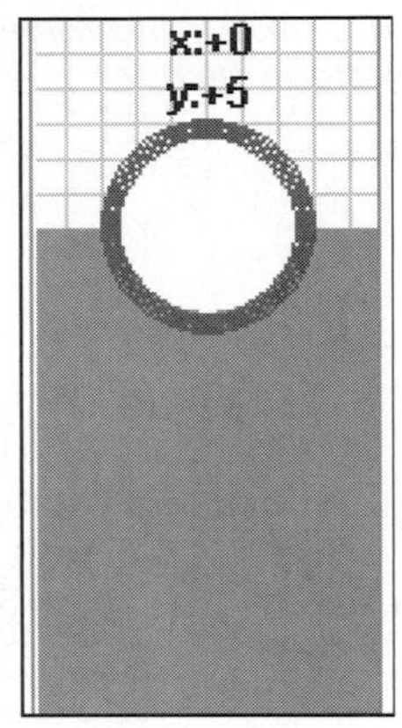

Problem 8.12.2: A spherical air bubble rises in a large pool of water as shown in the animation (position is in meters and time is in seconds). Mouse-down in the animation to determine position. The figures show the beginning and end of the animation.

Question 8.12.2: What is the ratio of the final to the initial temperature of the air in the balloon?

Answer: *13.6. Students must use the ideal gas law and determine the pressure from the sum of atmospheric pressure and* ρgh *of the water. Since this is an air bubble under water, the number of molecules in it remains constant.*

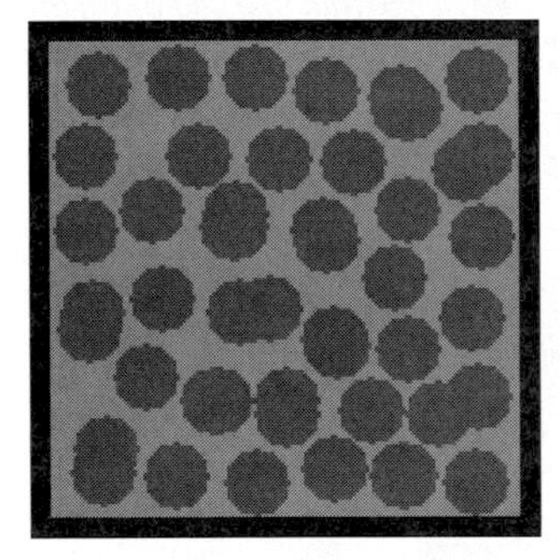

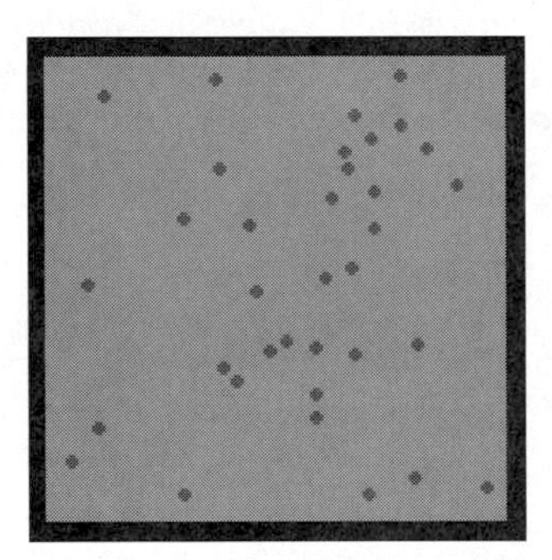

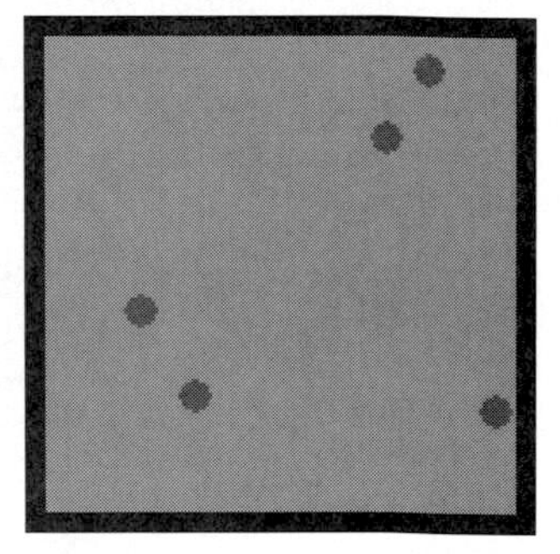

Problem 8.12.3: Numerous small hard spheres are placed in a container as shown in the animation (position in centimeters and time is in seconds). Three animations are shown with the same volume.

Question 8.12.3: Which of the following animations best approximates (qualitatively) an ideal gas?

Answer: *Animation 2. In Animation 1, the volume of the molecules takes up most of the volume in the box and therefore this is not an ideal gas. Animation 3 is not an ideal gas because there are too few molecules in the box. Animation 2 is the ideal gas since the molecules do not take up much of the box, do not often collide with each other, and there are many molecules in the box.*

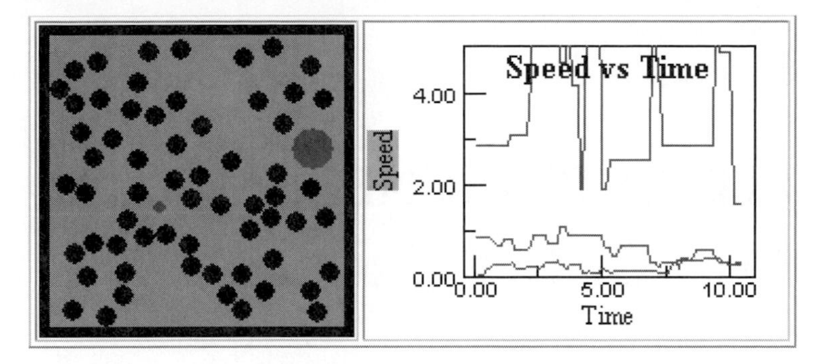

Problem 8.12.4: Numerous small hard spheres representing molecules are placed in a container as shown in the animation (position in centimeters and time in seconds). Here the relative masses of the molecules are RED 80 : GREEN 20 : BLUE/BLACK 1.

Question 8.12.4: Which particle type has the greatest average kinetic energy?

Answer: *All tie. The graph shows that the small blue sphere has the fastest speed, the green middle-sized sphere has a slightly faster speed, and the large red sphere has the slowest speed. Despite this, the equipartition theorem says that the average kinetic energies of these spheres (molecules) should be the same.*

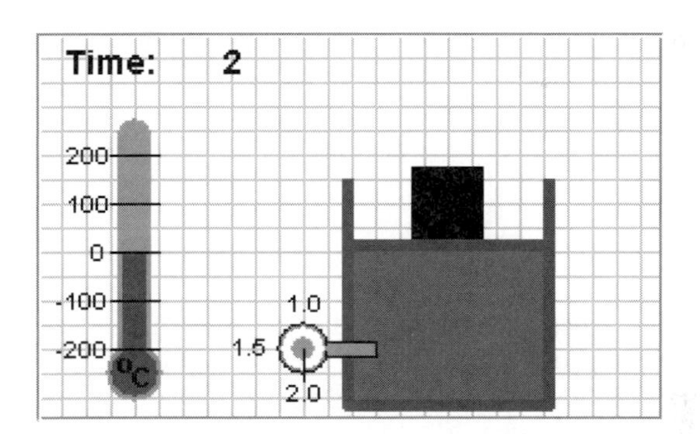

Problem 8.12.5: A mass is placed on a container (the dimension you cannot see has a length of 28 cm) filled with a gas as shown in the animation (pressure is given in atmospheres, position is in centimeters, and time is in seconds). During the animation, the pressure gauge continues to read 2 atm and the volume in the container decreases. The thermometer shows the temperature of the gas in the cylinder as the animation proceeds.

Question 8.12.5: What is the work done by the gas?

Answer: *−181.5 J. Since the pressure remains constant, the work is* P ΔV *and since the volume change is negative the work done by the gas is negative (while the work done on the gas is positive). The area is given, and the difference in height can be determined by a mouse-down to determine position.*

CHAPTER 9

ELECTROMAGNETISM AND OPTICS PROBLEMS

9.1 ELECTROSTATICS

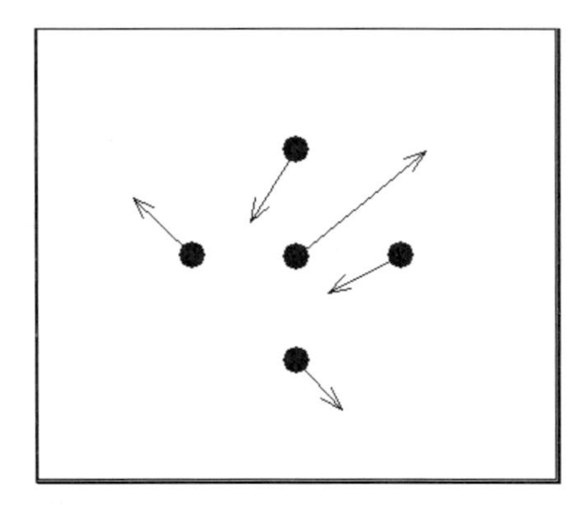

Problem 9.1.1: Five objects are shown on the screen along with vectors that are proportional to the net force on each object. Vectors are redrawn if any object is click-dragged to a new position. Notice that the inverse square law allows one to determine the nature of the force between any two charges (i.e., attractive or repulsive) if they are brought close together.

Question 9.1.1: How many charges have like signs? You can click-drag on any charge to change its position.

Answer: Three charges have one sign; two charges have the opposite sign. This is observed by selecting an arbitrary charge as a test charge and dragging it close to each of the remaining charges in order to observe the direction of the force.

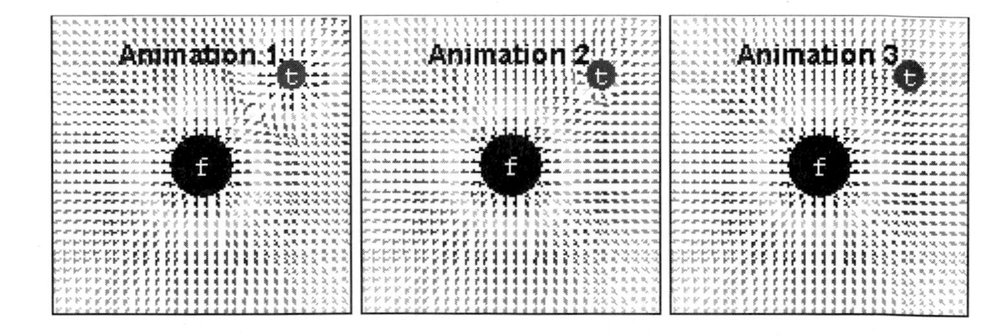

Problem 9.1.2: One fixed charge and one "test" charge is shown in each animation. The fixed charge is labeled by an *f* and the test charge is labeled by a *t*. You can click-drag on the "test" charge to change its position. There are three animations depicted.

Question 9.1.2: In which animation is the "test" charge a true test charge?

Answer: Animation 3. As students move the "test" charge in each animation, the field changes or remains the same. The animation where the test charge leaves the original field unaffected is the correct animation.

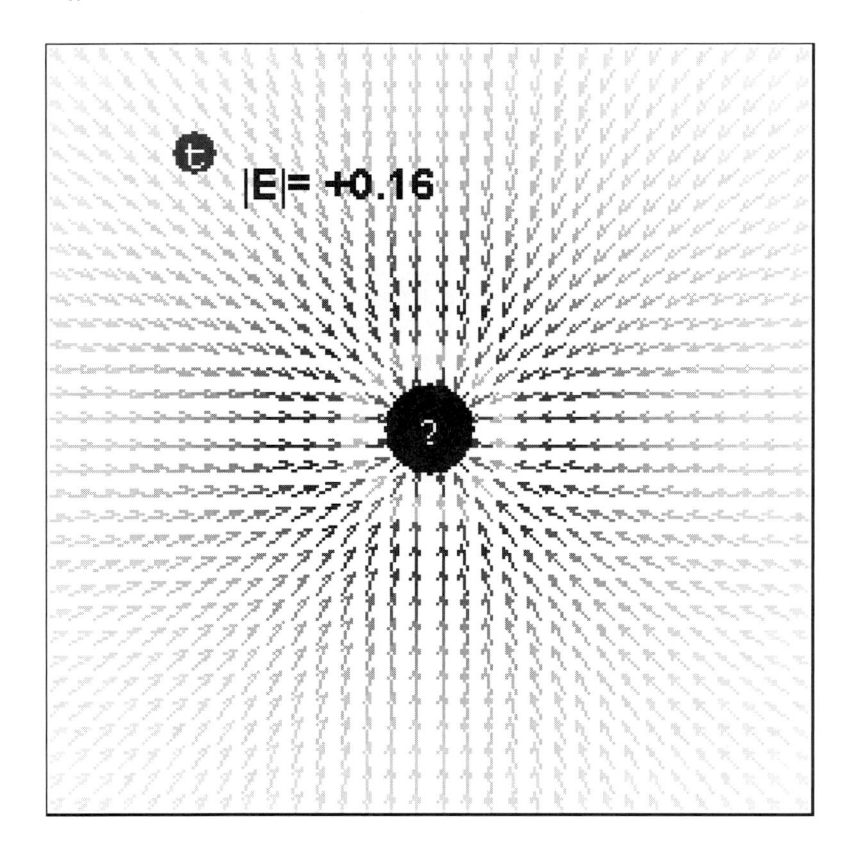

Problem 9.1.3: A fixed particle with an unknown charge uniformly distributed about its volume is shown. The vectors shown point in the direction of the electric field and the color of the vectors represent the field's magnitude. Distance is measured in m and the field is given in N/C. The small test charge is dragable to any position in the animation.

Question 9.1.3a: Determine the charge on the particle in Coulombs by dragging around the test charge.

Answer: Depends on the script, here 0.39 nC. Since the magnitude of the electric field is given in N/C, measurements can be made to determine the field at a given radius from the center of the charge distribution. Since the field is radial, the student can use the fact that $E = kq/r^2$.

Question 9.1.3b: Determine the number of excess electrons that are on the sphere by dragging the test charge.

Answer: Depends on the script, here 2.4×10^9 *electrons. Given a correct answer in 9.1.3a, this question is a conversion based on the number of electrons in a Coulomb.*

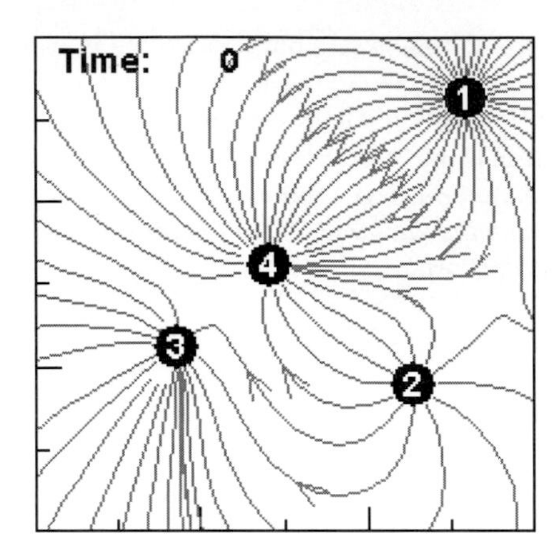

Problem 9.1.4: Four charged particles are shown with their electric field lines. These lines are redrawn every time the charges are dragged to a new position. You can click-drag any of these particles. If you overlap two charges, their charge values will add.

Question 9.1.4: What observation can you make concerning the group of charges?

Answer: Depends on the script. Students should drag around the charges to see how the field changes. In fact, here the total charge is zero.

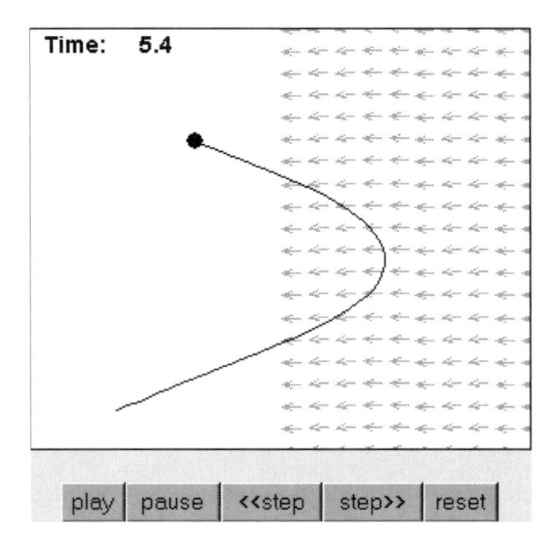

Problem 9.1.5: A black ball moves at a constant speed until it encounters a constant vector field that changes its direction. Time is shown in the left-hand corner of the simulation. Position can be measured by click-dragging the mouse.

Question 9.1.5: A 1-mg particle with a charge of 2 μC is fired into an unknown electric field as shown. Find the magnitude of the electric field. Click-drag to measure the position in meters. Time is measured in seconds.

Answer: Depends on the script. This simulation has an electric field of 1.5 N/C. The trajectory of the particle to the left of the field is that of a particle with a constant velocity. In the electric field, the particle accelerates to the left. Students must consider the motion in the x-direction to find this acceleration, then relate m***a*** *to* q***E***.

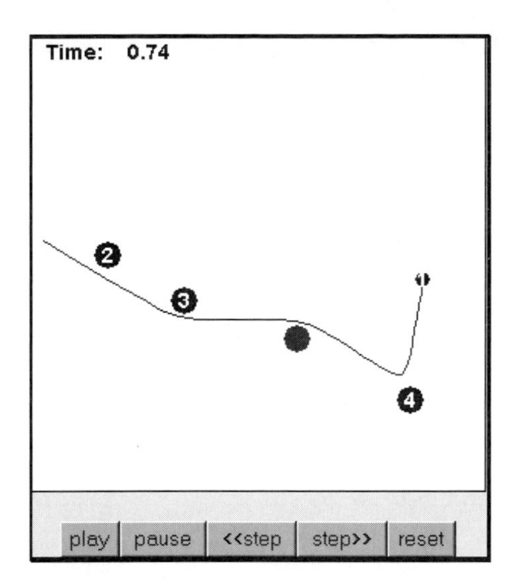

Problem 9.1.6: A blue charge moves under the action of the electric field created by four fixed charges. Its trajectory is drawn as a blue line as it moves.

Question 9.1.6: A blue charge is fired past 4 fixed charges as shown in the simulation. The red charge (the unmarked charge) is positive. Determine the signs of the unknown charges. You may consider neutral to be a possible answer.

Answer: The moving charge is negative since it is attracted toward the positive charge. Charge 2 is neutral since it does not deflect the moving charge. Charge 3 is positive since it attracts the moving charge. Charge 4 is negative since it repels the moving charge.

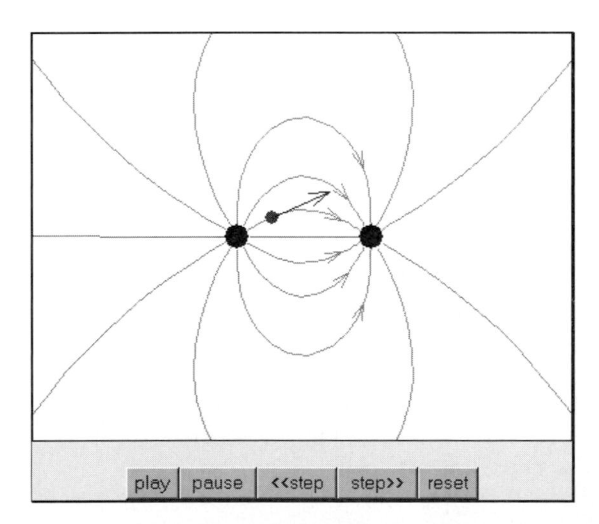

Problem 9.1.7: Two fixed charges are shown as black circles along with a dragable test charge drawn in red. The dark blue arrow attached to the test charge represents the force on the test charge. The light blue lines represent the field lines. When the "play" button is pressed, the test charge accelerates due to the electric field. The test charge marks its trajectory with a red line as it moves under the influence of the fixed charges.

Question 9.1.7a: Click-drag the test charge and describe the relationship, if any, between the dark blue force vector and the light blue electric field lines.

Answer: *The force vector is always tangent to the electric field lines. The force is larger wherever there are many field lines and smaller wherever there are few field lines.*

Question 9.1.7b: Place the test charge on a field line and start the animation. Does the trajectory of the test charge follow the electric field lines? Explain.

Answer: *No, the trajectory does not follow the field lines. The field lines show the direction of the force on the positive test charge.*

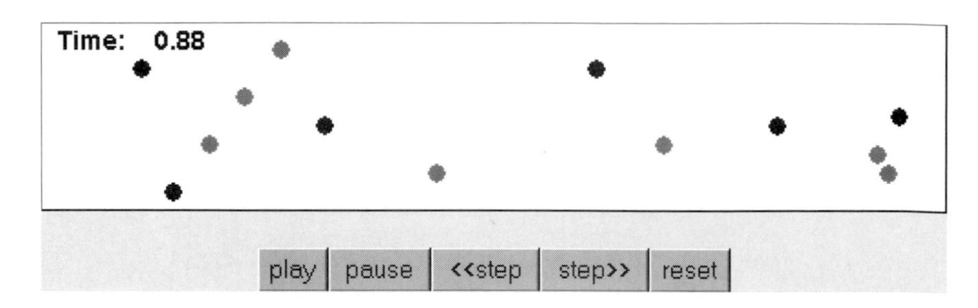

Problem 9.1.8: Blue and green spheres move to the right and left, respectively. Each sphere represents a charge and the color signifies the sign of the charge.

Question 9.1.8: A drift tube is shown here. The green dots represent atomic clusters of charge 1 nC and the blue ones clusters of charge -2 nC. What is the total current to the right?

Answer: *Count the number of positive charges entering the drift tube from the left,* N_1*, and the number of negative charges entering from the right,* N_2*, during a 1 second interval. The total current can be calculated using*

$$I = Q_1N_1 + Q_2N_2$$

Students will often make a sign mistake since negatively charged particles traveling to the left produce a positive current to the right.

9.2 GAUSS'S LAW

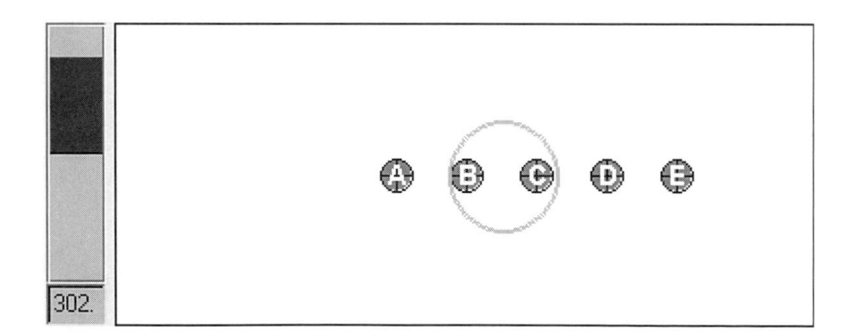

Problem 9.2.1: The animation shows several line charges pointing out of the page and a cylindrical electric flux detector. The bar graph displays the electric flux passing through the cylindrical flux detector. Drag the surface and observe flux readings. The Gaussian surface cannot enclose a portion of a line charge but may enclose two line charges.

Question 9.2.1: Which pair(s) of line charges is (are) identical in size but opposite in magnitude?

Answer: A and B and also D and E. By dragging around the Gaussian surface, students can determine the opposite line charges by enclosing zero net charge. When zero net charge is enclosed by the Gaussian surface, the flux is zero.

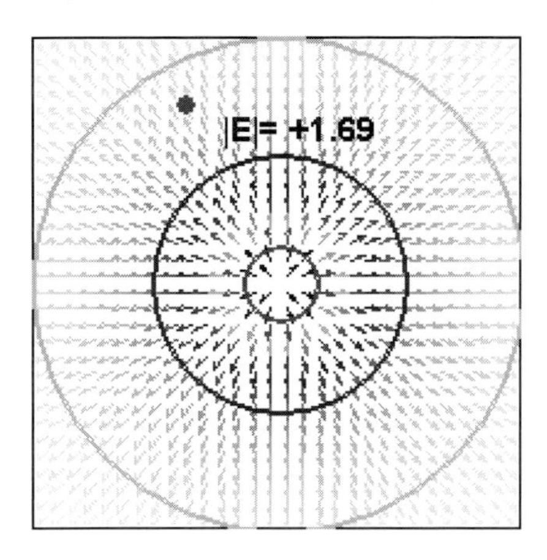

Problem 9.2.2: An unknown charge is located on the screen along with three spherical shells (red[inner], blue[middle], and green[outer]). You can only see the part of the sphere that is in the plane of the page. A test charge is also shown that measures the electric field (in N/C) at that point (position is given in meters).

Question 9.2.2: Calculate the flux through each spherical shell. You can click-drag on the test charge to change its position.

Answer: All tie at 352 Nm^2/C. There are several ways to come to this conclusion. First, a novice problem solver will determine the magnitude of the electric field and multiply by the surface area of the sphere at all three surfaces. A more expert approach would be to realize that the flux is related to the charge enclosed and that the field vectors do not indicate any other charges besides the one at the origin. Hence, only one measurement is required.

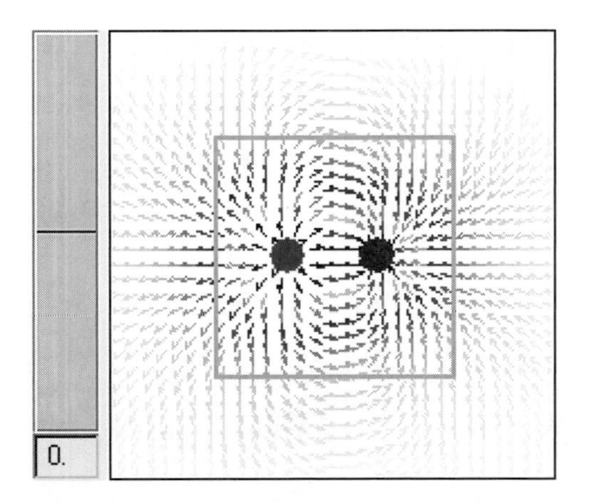

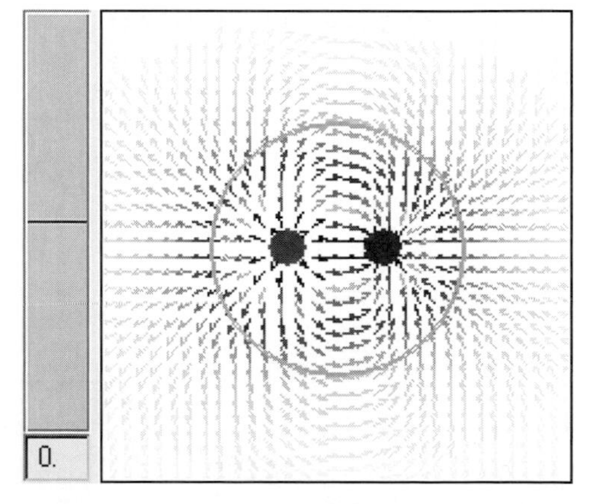

Problem 9.2.3: An electric dipole is shown surrounded by two electric flux detectors. The bar graph displays the electric flux passing through a cubical flux detector and a spherical flux detector. Observe the flux reading.

Question 9.2.3: Calculate the electric field at the surface of each detector using Gauss's law. The detectors are fixed in place.

Answer: The electric field cannot be determined from the information given. No symmetry. While Gauss's law is always true, sometimes it is not a useful calculational tool for determining the electric field. This is just such a case. Since field information is given in terms of field vectors, students can determine that the electric field is not constant (magnitude and direction) on each Gaussian surface.

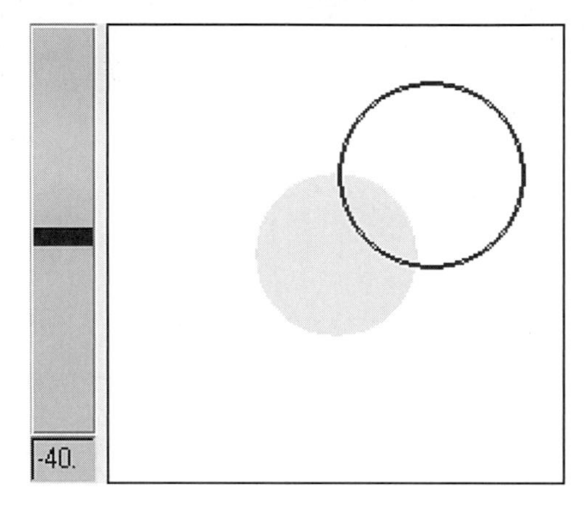

Problem 9.2.4: An unknown charge distribution is shown in the animation with a gray sphere for reference. The bar graph displays the electric flux passing through a cubical flux detector and several spherical flux detectors (the largest is shown). The resolution of the detector is 1 mC.

Question 9.2.4: Drag each surface and observe flux readings. Describe the charge distribution.

Answer: There is a uniform positive charge distribution at a radius smaller than the radius of the gray sphere and a very small negative charge distribution at a radius greater than the radius of the gray sphere. Using the various Gaussian surfaces allows the student to determine the charge enclosed. However, students using too coarse a detector will miss the small negative charge distribution outside of the gray sphere.

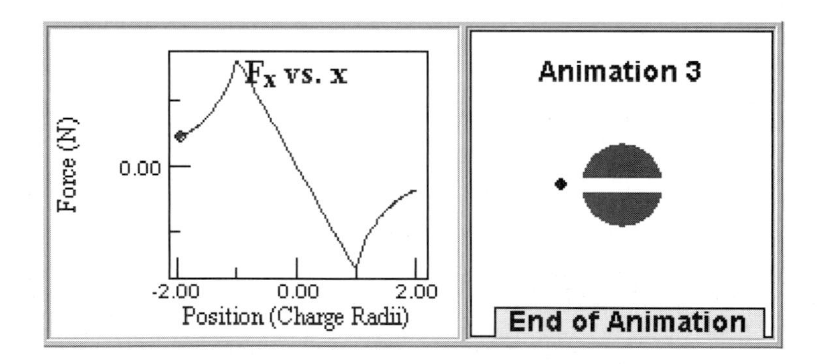

Problem 9.2.5: A test charge starts at rest and is attracted to a fixed negative charge distribution that has a small tunnel that allows the test charge to pass through as shown in the animation (position is in charge radii and time is in seconds).

Question 9.2.5: If the charge is uniform, which of the animations correctly depicts the motion of the test charge?

Answer: Animation 3, the animation shown. This problem is very similar to Problem 8.4.1 except we are now also interested in the force (and therefore the electric field) outside of the uniform charge distribution. Going out from zero, the field increases linearly until the surface, then falls as $1/r^2$.

9.3 ELECTRIC POTENTIALS

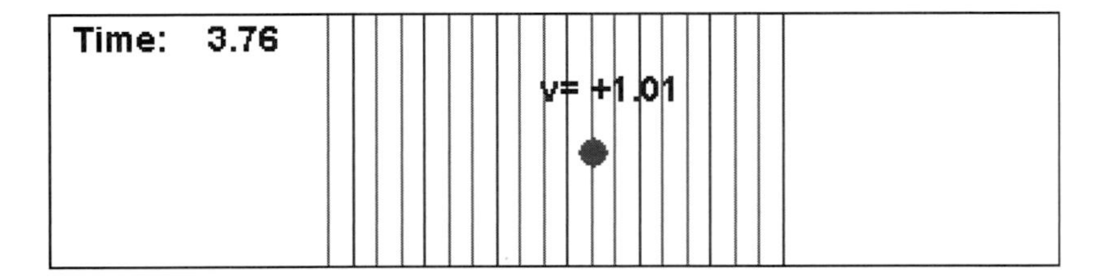

Problem 9.3.1: In this animation the potential changes from 0 V to 1 V as shown by the equipotential lines. The particle's velocity is displayed in meters/second. Click-drag to place the 1 μC test charge anywhere in the animation before you press play. You can also use the mouse to measure the potential at a point.

Question 9.3.1: What is the mass of the particle?

Answer: 1 gram. In order to determine the mass of the particle it must be placed in the region with the electric field, which is the region where the electric potential changes. Students can then determine the change in kinetic energy and relate it to the particle's change in potential energy.

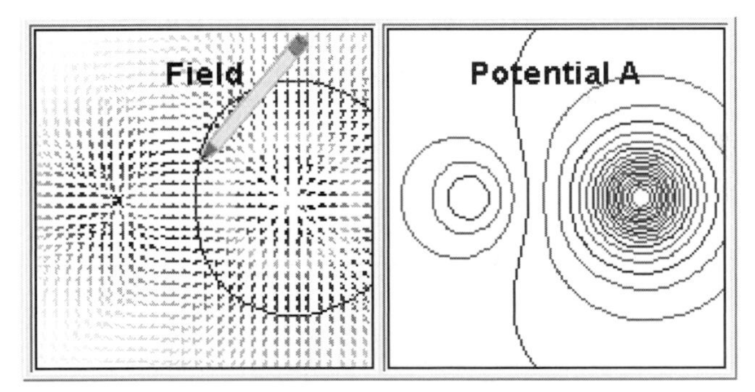

Problem 9.3.2: The panel on the left displays an electric field plot. The arrows in the field plot represent the direction and the colors represent the magnitude of the electric field. The panel on the right displays one of four possible sets of contour lines.

Question 9.3.2: Draw the equipotential lines for this field by dragging the pencil (at its tip) after clicking the "draw on" button. After you have drawn your lines, determine which potential plot best corresponds to your potential plot.

Answer: Potential A, the potential shown. Here the student must understand that the equipotential surfaces must be perpendicular to the electric field lines (here the electric field vectors). Besides drawing the lines with the pencil, the option of double-clicking to automatically place a contour (equipotential) line is also given.

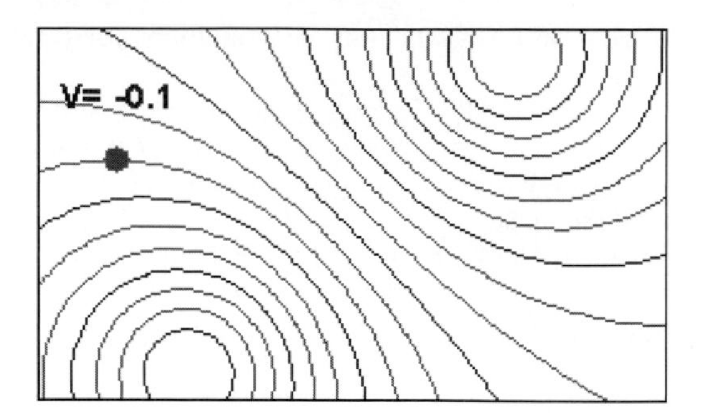

Problem 9.3.3: An equipotential plot is shown here. The electric potential measured in Volts is shown next to the electron.

Question 9.3.3: How much work must an external force do in order to move an electron (the red object) from $[x, y] = [1.1, -1.5]$ to $[x, y] = [-1.6, 1.2]$?

Answer: 0 Volts. The starting point and the end point are on the same equipotential surface. Therefore, the amount of work done by the external force is zero because the object moves perpendicular to the electric field lines and hence the electric force does no work on the particle.

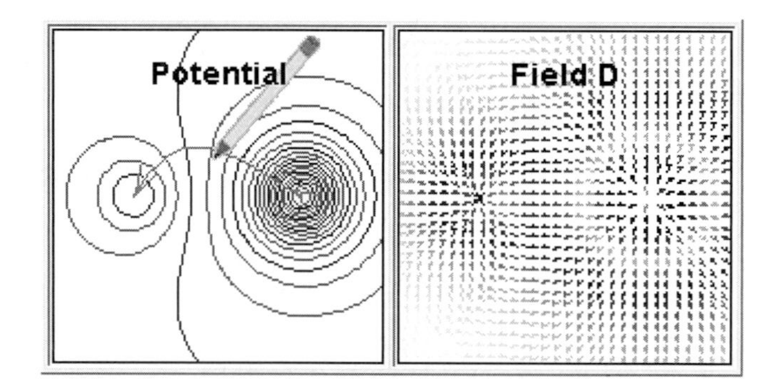

Problem 9.3.4: The panel on the left displays an equipotential plot. The contours represent points at the exact same potential. The panel on the right displays one of four possible sets of electric field vectors.

Question 9.3.4: Draw the electric field lines for this potential by dragging the pencil (at its tip) after clicking the "draw on" button. After you have drawn your lines, determine which field best corresponds to your potential plot.

Answer: Field D, the field shown. Note that this problem is precisely the inverse of 9.6.2! Here the student must again understand that the equipotential surfaces must be perpendicular to the electric field lines (here the electric field vectors). Besides drawing the electric field lines with the pencil, the option of double-clicking to automatically place a field line is also given.

9.4 CAPACITORS

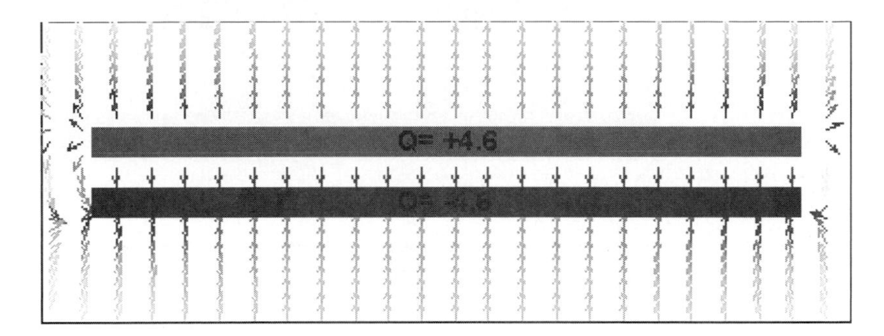

Problem 9.4.1: Four simulations represent parallel plate capacitors. You can click-drag and observe the electric potential at that point. The capacitors vary in width, length, and separation. The charge on each plate is also given with a 1 to 5% intrinsic error in the calculation of charge.

Question 9.4.1: For which of the four capacitors would you expect the effect of fringing to be minimal (i.e., the capacitance is $C = \epsilon_0 A/d$?

Answer: The third and fourth capacitors. As with the animation shown here, in order to have the capacitance given as C = ϵ_0A/d, *the electric field must be (mostly) uniform between the plates. Students can click-drag in the animation to measure the electric field there.*

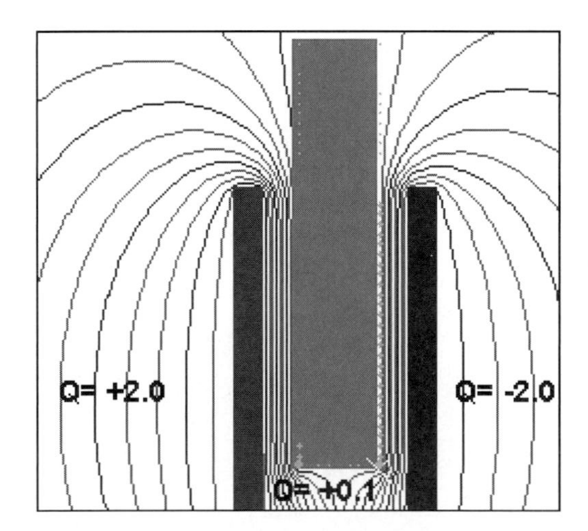

Problem 9.4.2: A capacitor made up of two plates is shown with a moveable conductor that can be placed in between the plates. The capacitance of a parallel plate capacitor can be altered by dragging (drag from the middle) a conducting block in between the two plates (charge given in μC and position given in centimeters). The light red and light blue circles represent the position of the charge on the conductor, and the equipotential surfaces are shown.

Question 9.4.2: What is the ratio of the new capacitance to the old capacitance?

Answer: 1.7. The capacitor now acts like two identical capacitors that add in series. The separation between the "plates" also changes.

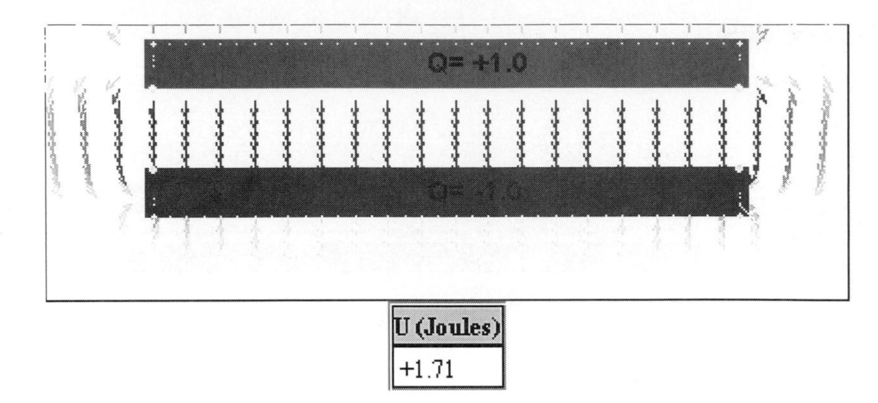

Problem 9.4.3: The capacitor plates depicted here are maintained at a constant charge (charge given in μC and position given in centimeters). The light-colored spheres are representative of the charge on the plates and the electric field vectors are shown. The *DataTable* beneath the capacitor gives the energy of the system in Joules.

Question 9.4.3: What is the capacitance of the system when the plates are 0.5 cm apart?

Answer: *0.45 pF. Students must use first drag the bottom plate into place 4 cm from the top plate. As this is done, the energy stored in the system changes, as shown by the energy DataTable. Then one can use the relationship among the capacitance, charge, and energy stored in the capacitor.*

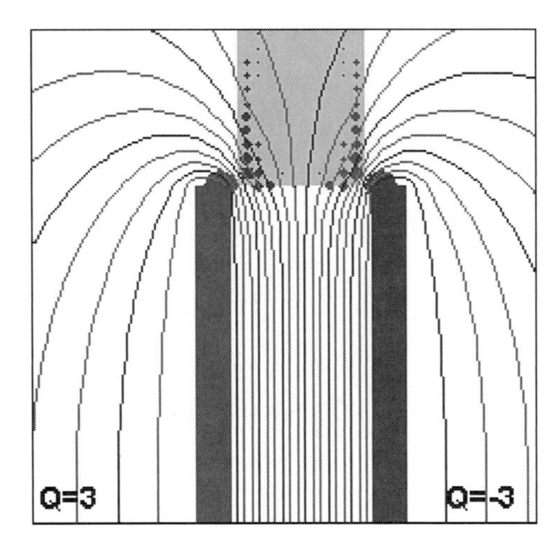

Problem 9.4.4: Click-drag the dielectric block into the capacitor (charge given in μC and position given in centimeters). Observe how the electric field and the charge on the capacitor and the dielectric change when you move the dielectric. The light red and light blue circles represent the position of the charge on the dielectric. You can measure the voltage and the electric field at any point by click-dragging.

Question 9.4.4: What is the dielectric constant of the slab?

Answer: *2.4. When the dielectric is moved in place between the capacitor plates, the electric field (and therefore the electric potential) decreases proportional to the dielectric constant.*

9.5 CIRCUITS

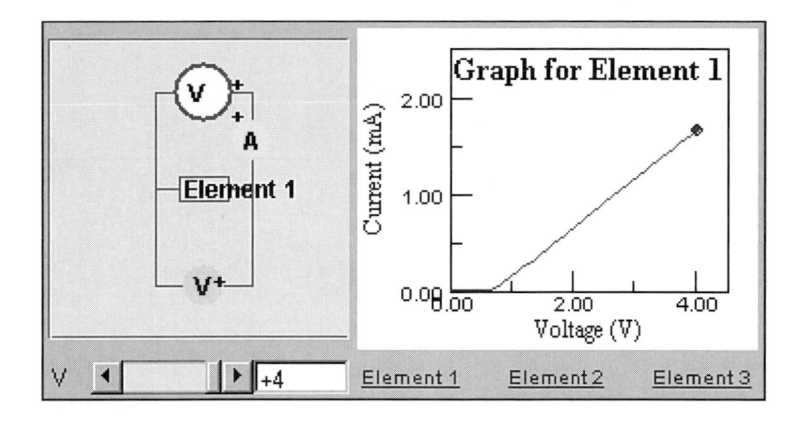

Problem 9.5.1: There are three unknown circuit elements connected to a voltage source as shown. The voltage is controlled by the slider and given in Volts. You can measure the voltage and current by placing the mouse cursor over either meter or by looking at the graph. You may drag the slider to the right to change the voltage shown in the graph.

Question 9.5.1: Determine which elements obey Ohm's law.

Answer: Circuit element 2 only. The graph shown does not show a circuit element that obeys Ohm's law because the voltage does not vary linearly with the current.

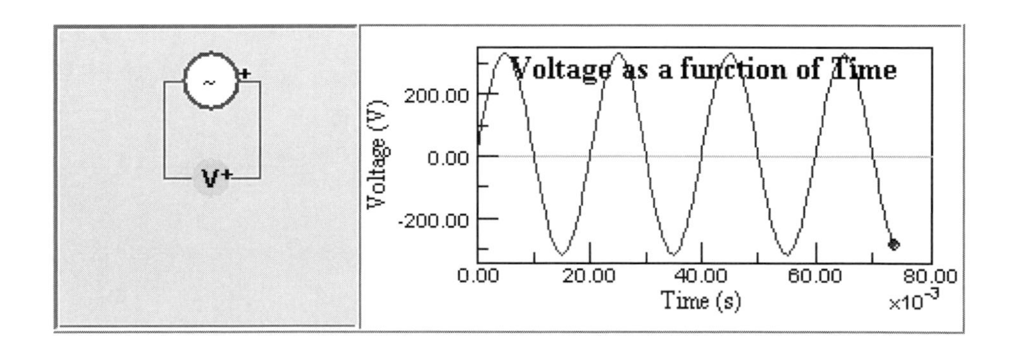

Problem 9.5.2: An oscilloscope is connected to a sine ac supply voltage as shown above. By clicking in the graph window, you can read the voltage values or you can consider the time evolution by looking at the graph.

Question 9.5.2: Use the graph to determine the maximum voltage V_{max} and the effective voltage V_{rms}.

Answer: 325 V is the maximum voltage and 230 V is the rms voltage. The maximum voltage can be read off the graph and the rms value is 0.707 of the maximum value.

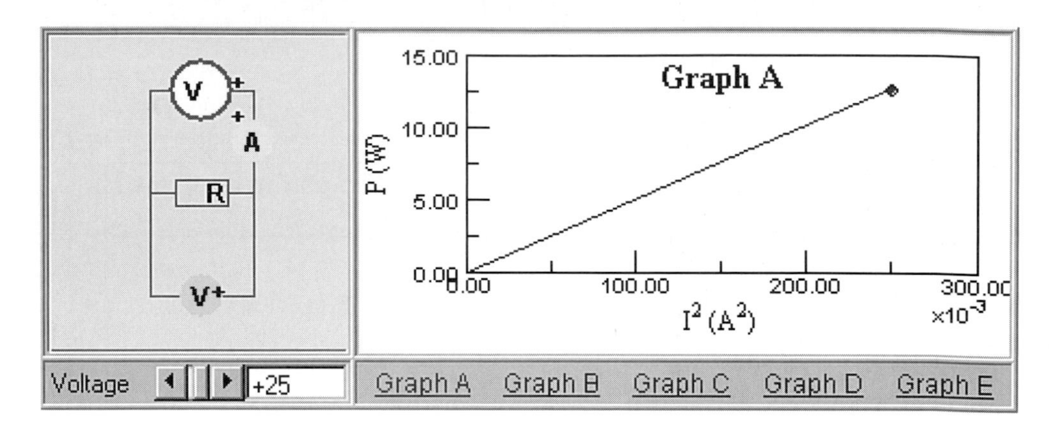

Problem 9.5.3: There is a resistor in a circuit connected to a voltage source as shown. The voltage is controlled by the slider and given in Volts. You can measure the voltage and current by placing the mouse cursor over either meter or by looking at the graphs. Graphs show various combinations of current, resistance, and power.

Question 9.5.3: Determine which graph(s) correctly represent the variables shown.

Answer: Graphs A, D, and E. The power dissipated by the resistor is I^2R.

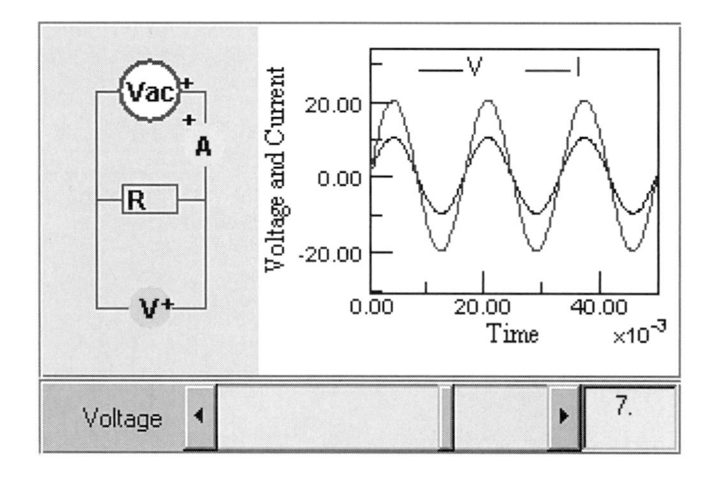

Problem 9.5.4: An unknown resistor is connected to an ac voltage source as shown. The rms source voltage is controlled by the slider. You can click-drag inside the graph to measure coordinates and place the mouse on a circuit component to obtain readings.

Question 9.5.4: Find the unknown resistance if current is measured in milliamps (mA).

Answer: 500 Ω. Since Ohm's law can be applied either to rms values or to instantaneous values, choose the maximum values of the current and voltage and relate them to the resistance.

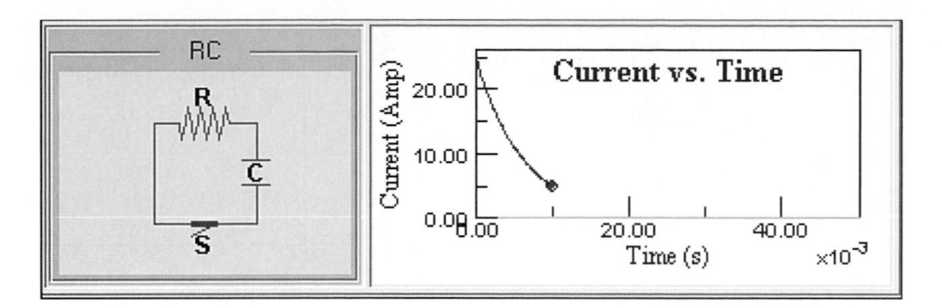

Problem 9.5.5: The circuit shown here is an RC circuit with a capacitance of 0.30 μF. The graph displays current versus time. You may read values from the graph clicking the mouse anywhere on the graph.

Question 9.5.5: Find the value of the resistor.

Answer: 20 kΩ. Students can click on the graph to determine when the current is 0.367 of its maximum value. Once this time is determined—it is the RC time constant—it can be related to the resistance.

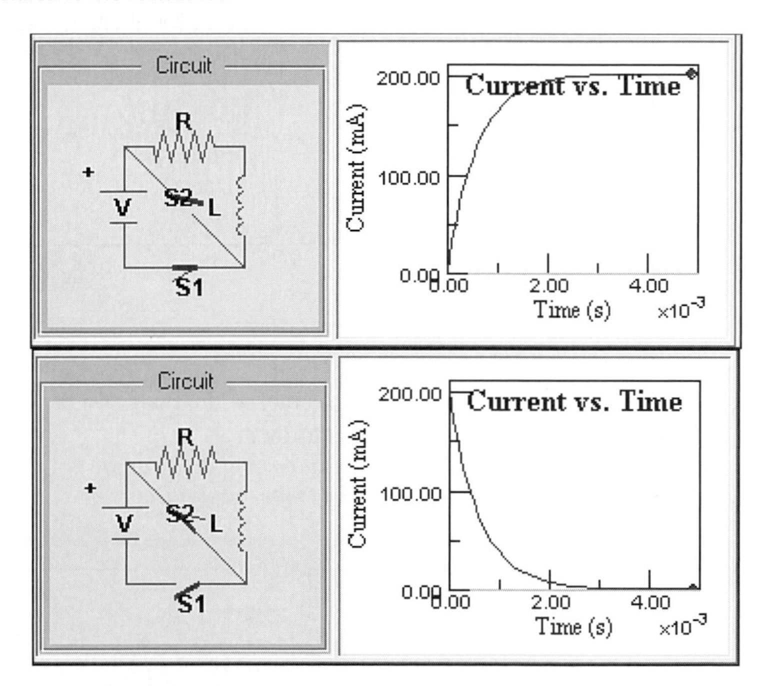

Problem 9.5.6: The circuit with a battery, an inductor, and a 20-Ω resistor can have a battery connected or disconnected depending on which switch is open. The battery is initially connected (switch 1 is closed and switch 2 is open) and then the battery is disconnected. When the battery is disconnected, the system evolves in time. You may look at the current from the graph after switch 2 is closed and switch 1 is open.

Question 9.5.6: What is the inductance of the inductor?

Answer: 12 mH. It is up to the user to determine which circuit to use in determining the time constant. For charging, the time that corresponds to 0.632 of the maximum current is the time constant. For the discharge, it is 0.367 of the maximum current.

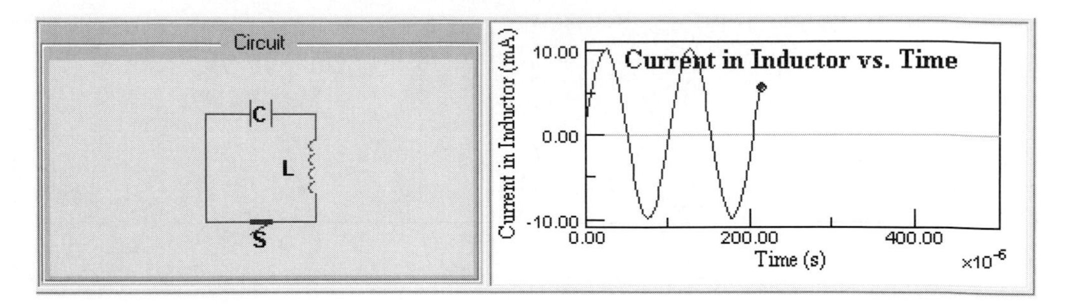

Problem 9.5.7: The circuit shown with an inductor and an initially charged 2-μF capacitor evolves in time. You may determine the current from the graph.

Question 9.5.7: What is the inductance of the inductor?

Answer: *0.126 mH. In an LC circuit, the frequency of the oscillation is related to the inductance and the capacitance in the circuit.*

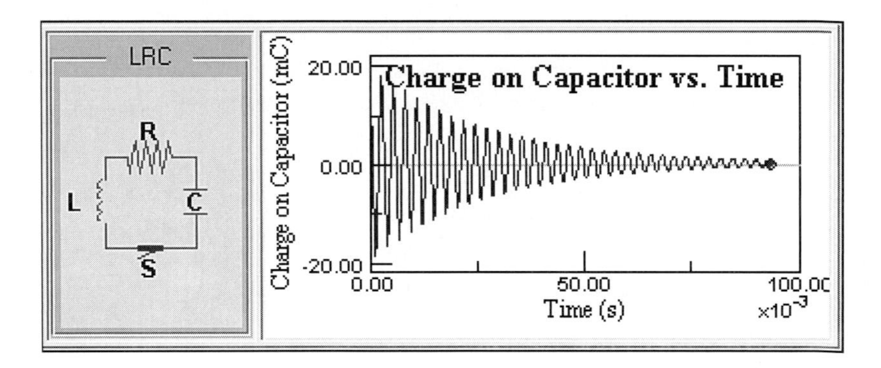

Problem 9.5.8: The circuit shown with an inductor, a 3.0-Ω resistor, and an initially charged 4.8-μF capacitor evolves in time. You may look at the charge on the capacitor from the graph.

Question 9.5.8: What is the inductance of the inductor?

Answer: *40 mH. Use the exponential decay of the graph to determine the time constant; then relate the time constant to the inductance.*

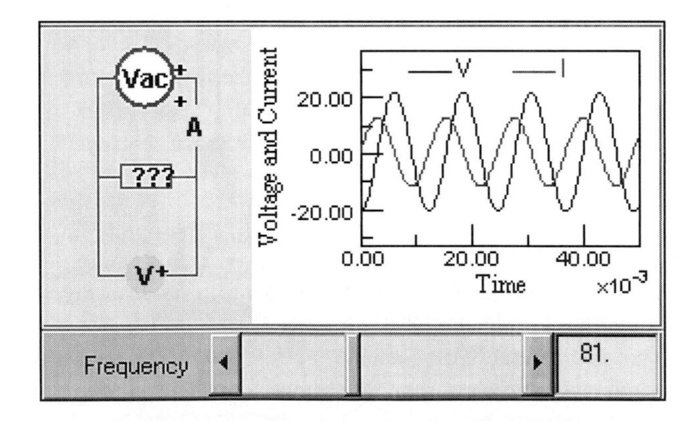

Problem 9.5.9: An unknown circuit element is connected to an ac voltage source as shown. A graph of the voltage and current is also shown.

Question 9.5.9: What is the element in the circuit, a resistor, an inductor, or a capacitor?

Answer: A capacitor. By looking at the graph, one can determine that the current leads the voltage by 90° independent of the frequency chosen in the slider.

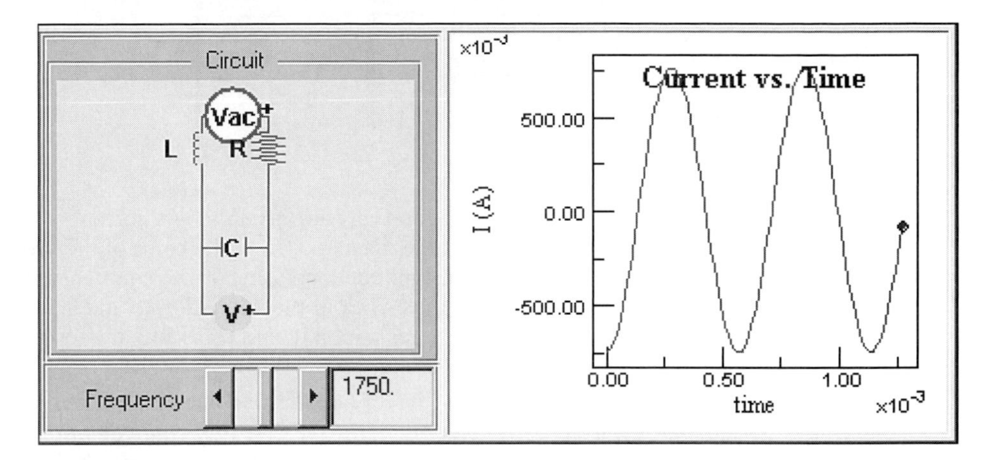

Problem 9.5.10: An LRC circuit is shown on the left, where the frequency of the ac power source is controlled by a slider. The graph on the right shows the current as a function of time.

Question 9.5.10: Find the resonant frequency of the LRC circuit.

Answer: 1000 Hz. An LRC circuit, given L, R, and C, has a resonant frequency, a frequency where the current flowing through the circuit is a maximum. The current changes as the frequency changes. Students must find the resonant frequency by finding the maximum current through the circuit.

9.6 MAGNETIC FIELDS

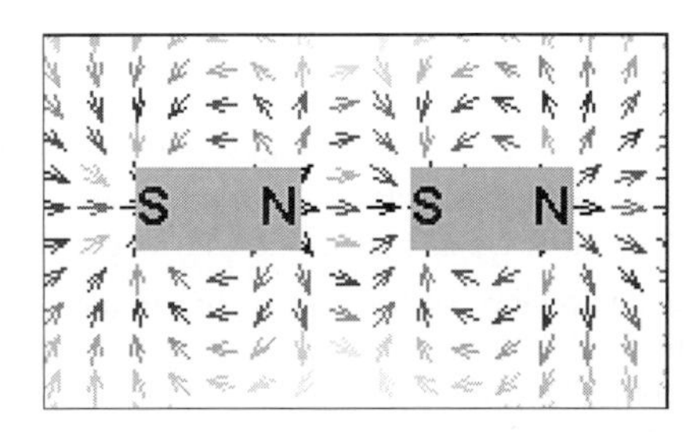

Problem 9.6.1: The animation represents two typical bar magnets each with a North and South pole. The arrows represent the direction of the magnetic field. The color of the arrows represents the magnitude of the field, with magnitude increasing as the color changes from blue to green to red to black. You may drag either magnet and double-click anywhere inside the animation to add a magnetic field line, and mouse-down to read the magnitude of the magnetic field at that point. There are two animations depicted, one with a mislabeled magnet.

Question 9.6.1: Which animation correctly depicts a properly labeled magnet?

Answer: Animation A, the animation shown. The correct animation shows the magnetic field vectors coming out of the North poles and heading into the South poles.

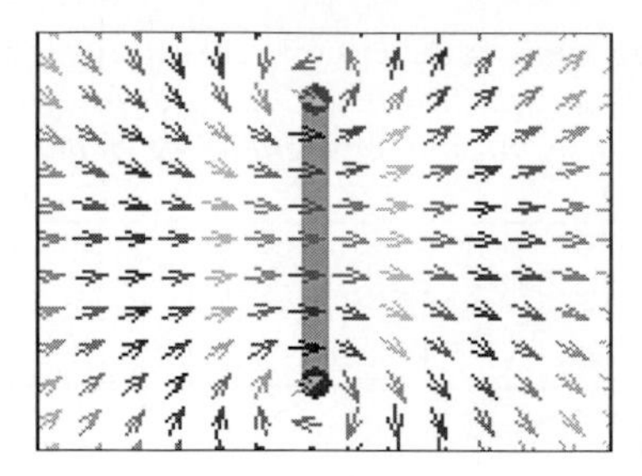

Problem 9.6.2: A cross section of a circular wire loop carrying an unknown current is shown here. The arrows represent the direction of the magnetic field. The color of the arrows represents the magnitude of the field with magnitude increasing as the color changes from blue to green to red to black. You can double-click in the animation to add magnetic field lines, click-drag the center of the loop to reposition it, and click-drag the top or bottom of the loop to change its size.

Question 9.6.2: Does current flow out of the red (the top) end or the blue (the bottom) end?

Answer: The current comes out of the red (the top) end and into the blue (the bottom) end. Use the right-hand rule on either cross section and compare the curl of your fingers with the magnetic field vectors.

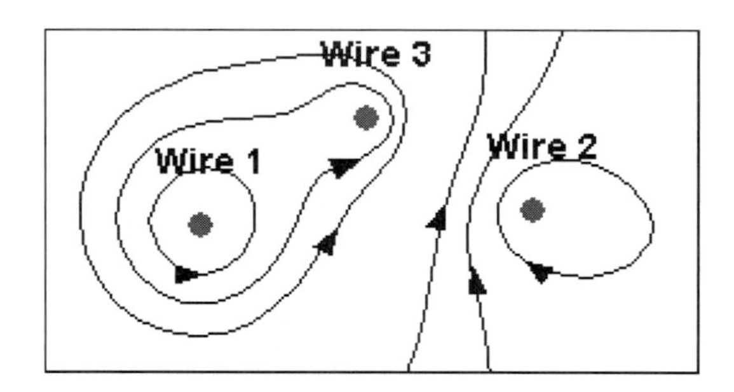

Problem 9.6.3: A cross section of three wires carrying unknown currents is shown here. You can double-click anywhere inside the animation to draw a magnetic field line. You can also click-drag the wires, but this will erase any field line that you have drawn.

Question 9.6.3: Which wires are carrying current out of the plane of the simulation, that is, out of the computer monitor?

Answer: Wire 1 and Wire 3. Again, students must use the right-hand rule to compare the magnetic field lines on either cross section of wire to the curl of their fingers and the direction that their thumb points.

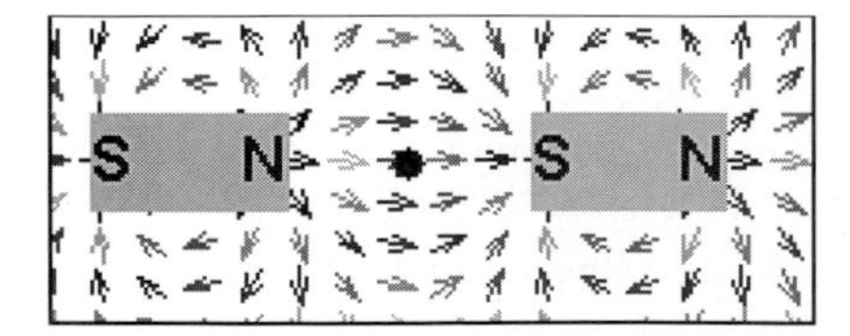

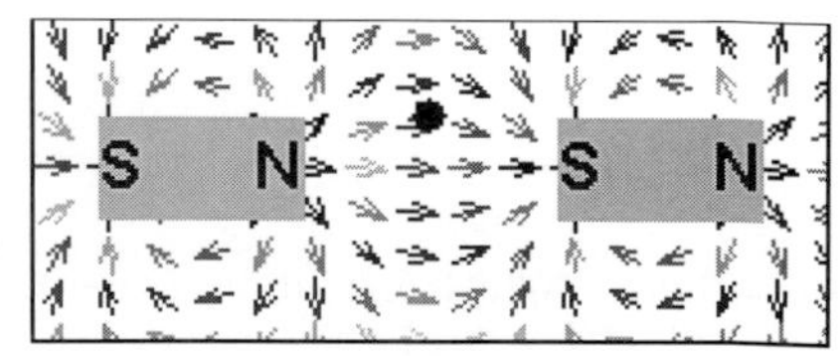

Problem 9.6.4: The animations shown here represent two typical bar magnets each with a North and South pole. The arrows represent the direction of the magnetic field. A wire is placed between the magnets and a current that comes out of the page can be turned on. The figure on the left represents the position of the wire when the current is off, and the figure on the right depicts one of four possible positions of the wire after the current is turned on.

Question 9.6.4: Which animation properly represents the deflection of the wire when the current is on?

Answer: Animation 4, the animation shown. Students must use the right-hand rule to relate the direction of the current and the direction of the magnetic field to the direction of the force experienced by the wire.

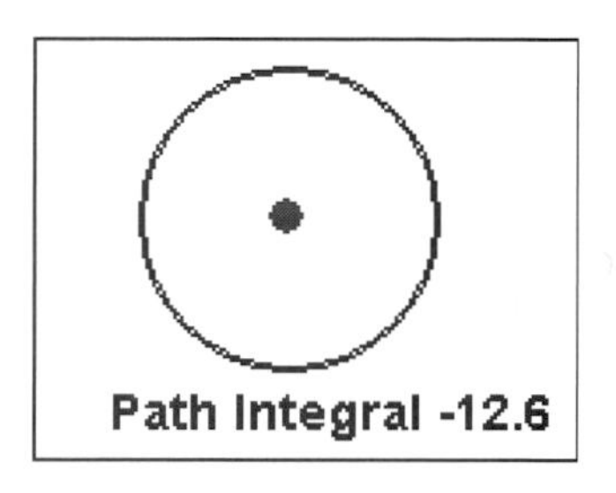

Problem 9.6.5: A wire with current flowing through it perpendicular to the page is shown (position is given in meters and the integral is given in 10^{-10} Tesla-meters). Also shown is a choice of two detectors (a square one and a circular one) that displays the integral of *B* "dot" *dl*. Choose a detector and observe readings. You may drag a detector from its original position if you wish.

Question 9.6.5: Use your reading(s) and Ampere's law to determine the current flowing through the wire.

Answer: I = *−1.0 mA (− means into page). You cannot use the square detector, but you can use the circular detector shown here. Students must then relate the value of the path integral to the current enclosed by that path (Ampere's law).*

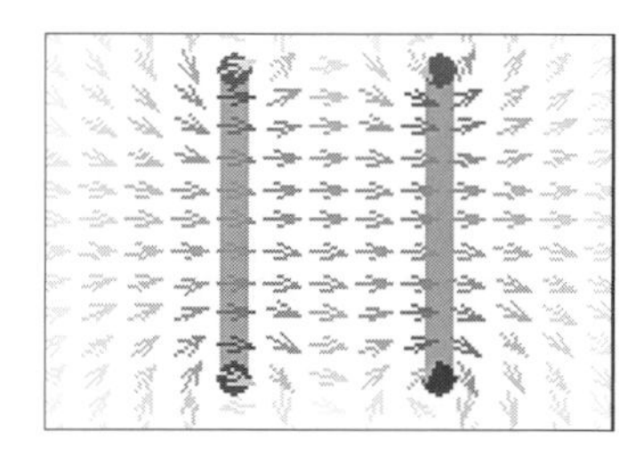

Problem 9.6.6: A cross section of a two circular wire loops carrying the exact same current is shown here (position given in centimeters and magnetic field given in milli-Tesla). You can click-drag to read the magnitude of the magnetic field.

Question 9.6.6: What is the current in each loop?

Answer: 6.0 A. Use the law of Biot-Savart and the magnetic field near a loop of wire to determine the current by adding up the contribution to the magnetic field from each loop. But where to do this? The easiest method is to choose the midpoint between the two coils on the x*-axis.*

9.7 FARADAY'S LAW

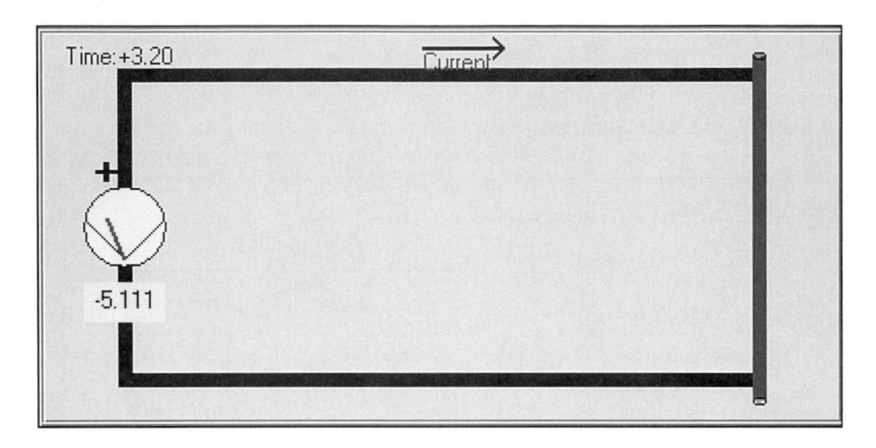

Problem 9.7.1: The animation shows a top view of four wires and a galvanometer in the lab. Current flowing into the + terminal (i.e., counterclockwise) will deflect the meter to the right. During the time interval $t = 2$ s to $t = 8$ s a magnet (not shown) is slowly pushed completely through the rectangle from the bottom toward the top. (You can also think of the magnet as being pushed out of the computer monitor toward the user.) The animation starts at $t = 0$ and stops at $t = 10$ seconds.

Question 9.7.1: Observe the meter reading during this simulation. Which pole was inserted first? North or South?

Answer: *The North pole of the magnet. A clockwise current implies an induced magnetic field into the page. From Lenz's law, the magnetic field that creates this current must be increasing out of the page.*

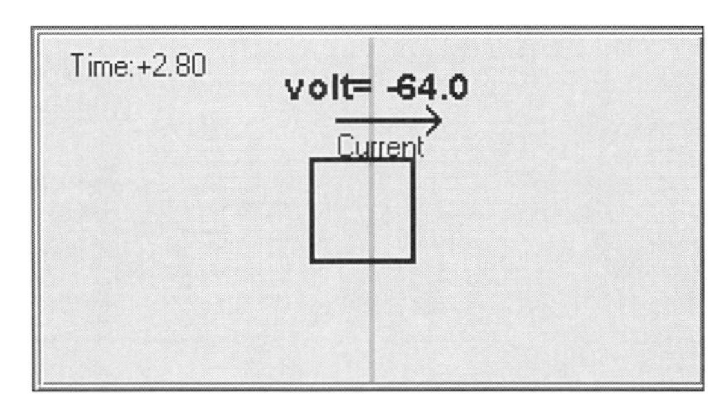

Problem 9.7.2: A loop of wire travels from the right to the left through an inhomogeneous magnetic field. (The green line is at $x = 0$ is for your reference.) The induced emf in the loop is shown in Volts and the direction of the induced current is also shown in the animation.

Question 9.7.2: Describe the magnetic field perpendicular to the computer screen.

Answer: $B_z(x < 0) = constant$, $B_z(x > 0) = constant$, *and* $B_z(x < 0) < B_z(x > 0)$. *From Lenz's law, we know that the magnetic field must be increasing out of the page across the green line. Note that some students will assume that the magnetic field on the left of the green line is zero and the magnetic field on the right is positive (out of the page). However, this is not the case; the magnetic field on the right of the line must be greater than on the left of the line.*

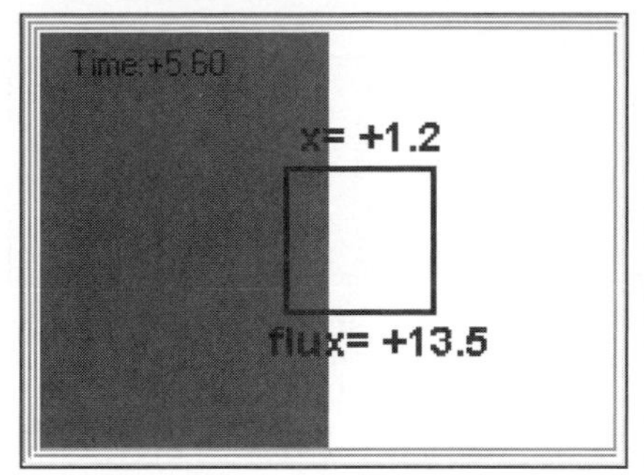

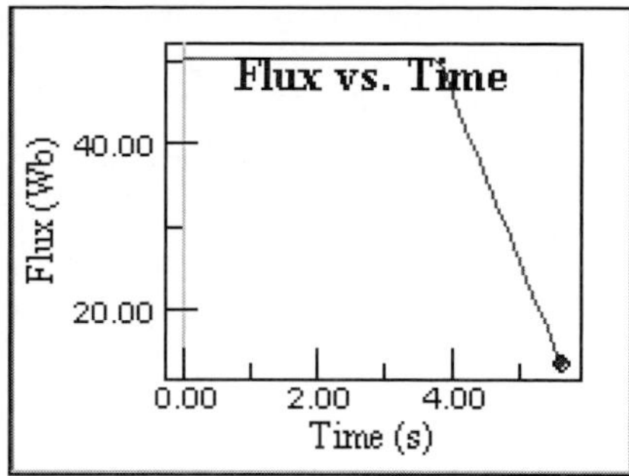

Problem 9.7.3: A loop of wire moves to the right through a magnetic field (a positive magnetic field is out of the screen) that abruptly goes to zero at $x = 0$.

Question 9.7.3: Given the flux graph shown here, what are the magnitude and direction of the *emf* as the loop of wire undergoes the transition at $x = 0$?

Answer: 19.2 V, counterclockwise. The change in flux is negative and, therefore, Lenz's law states that the current is counterclockwise. Since the change in flux per unit time is constant, students can calculate $\Delta\Phi_B/\Delta t$ *from the graph.*

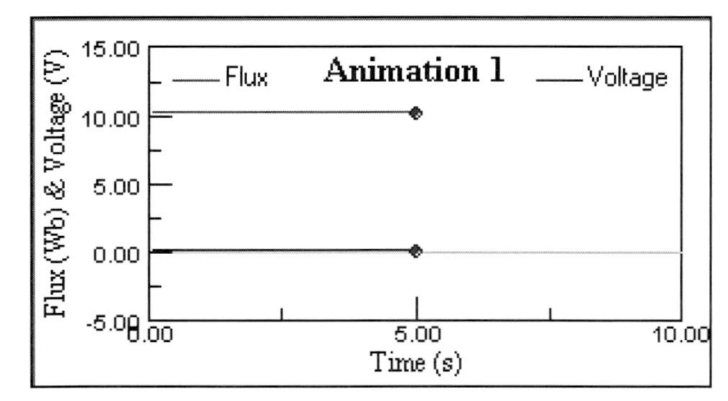

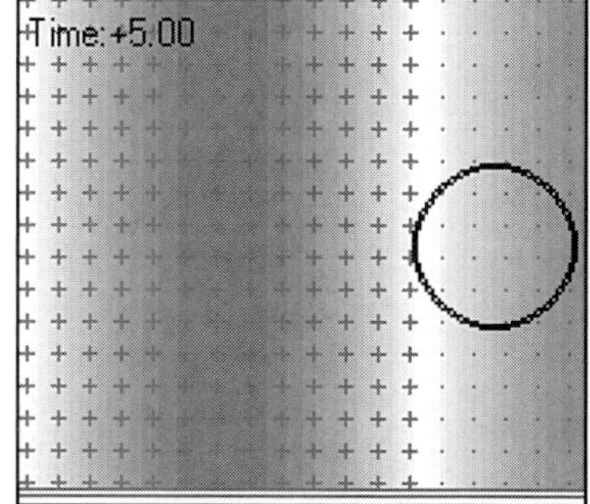

Problem 9.7.4: The animation shown here represents a moving wire loop in a changing magnetic field (red indicates a field out of the screen and blue indicates a field into the screen). The graph shows both the flux through the detector and the induced emf in the wire. There are four animations with different combinations of flux and emf values.

Question 9.7.4: Which animation correctly represents both the voltage and flux through the loop as a function of time?

Answer: Animation 1, the animation shown. As the flux remains constant in time, there is no induced emf in the wire. Even though the magnetic field varies in time, the loop itself moves to the right along with the magnetic field, and therefore the flux remains constant and no emf is induced.

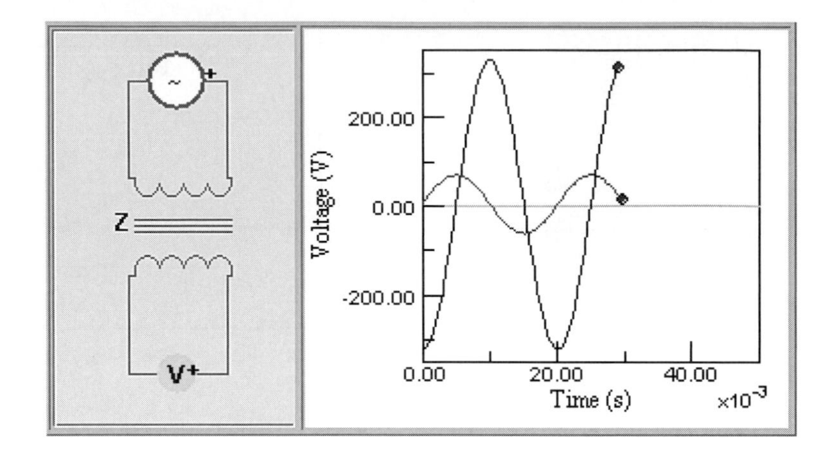

Problem 9.7.5: A transformer is connected to a wall socket as shown here. The graph shows the primary voltage V_p (black) and the secondary voltage V_s (red) as a function of time.

Question 9.7.5: How many turns N_s does the secondary coil contain when the primary coil contains $N_p = 200$ turns?

Answer: *40 turns. Despite the change in phase across the transformer, the ratio of voltages is related to the ratio of turns.*

9.8 ELECTROMAGNETIC WAVES

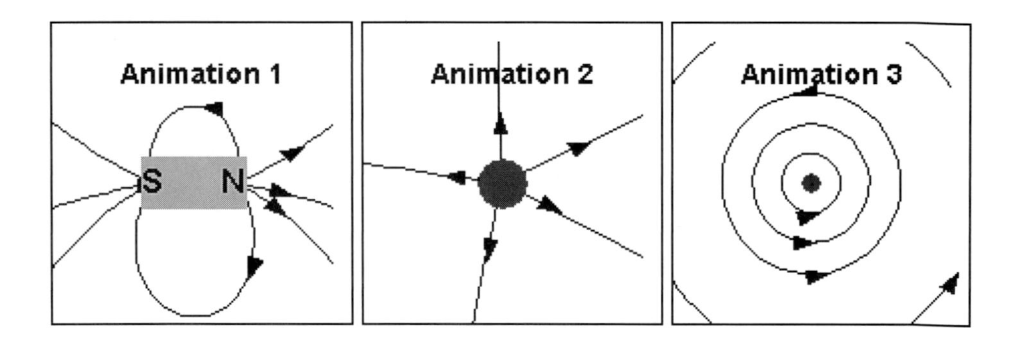

Problem 9.8.1: The animations shown here represent objects and their magnetic fields. There are three animations given.

Question 9.8.1: Which animation(s) correctly represents physical objects?

Answer: *Animations 1 and 3. Due to the fact that there are no magnetic monopoles, the second animation is not physical in classical electromagnetism (theoretical physicists, however, speculate that there must be at least one magnetic monopole in the universe). The other two animations show magnetic field lines that close on themselves.*

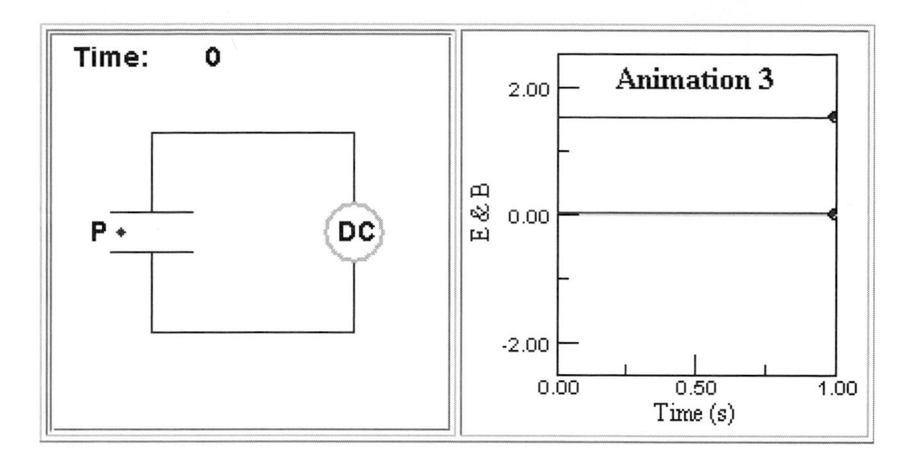

Problem 9.8.2: A dc circuit with a capacitor is shown fully charged. On the right are four graphs representing the electric field and magnetic field at point *P* between the capacitor plates.

Question 9.8.2: Which graph correctly represents the electric field and magnetic field between the plates as a function of time?

Answer: *Animation 3, the animation shown. The electric field is constant, and so the magnetic field is zero. It is a changing electric field that creates a magnetic field.*

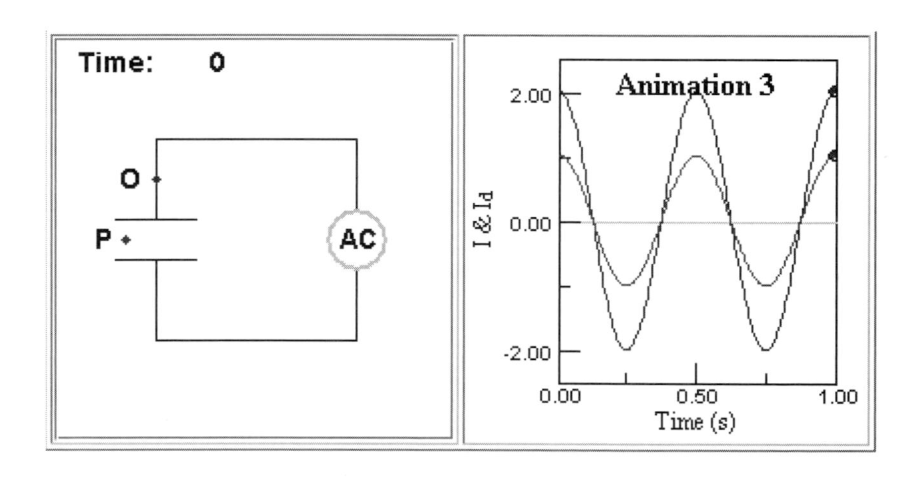

Problem 9.8.3: An ac circuit with a capacitor is shown. On the right are four graphs representing the current at point *O* and the displacement current at point *P*.

Question 9.8.3: Which animation correctly represents both the current at point *O* and the displacement current at point *P*?

Answer: *Animation 2; the displacement current must equal the current (not shown). As the current changes, so must the displacement current.*

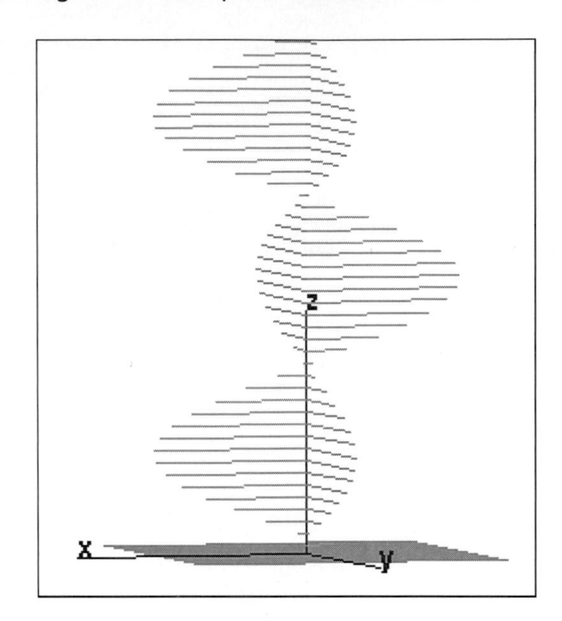

Problem 9.8.4: The animation shown here represents a traveling electromagnetic wave in the z-direction. The electric and magnetic field are represented with the magnetic field not drawn to scale. (This static picture does not properly represent the problem; please look at the CD.)

Question 9.8.4: Which wave is the electric field and which one is the magnetic field?

Answer: The red wave is the electric field and the blue wave is the magnetic field. This must be the case from the right-hand rule. The cross product of the electric field with the magnetic field points in the direction of propagation of the electromagnetic wave.

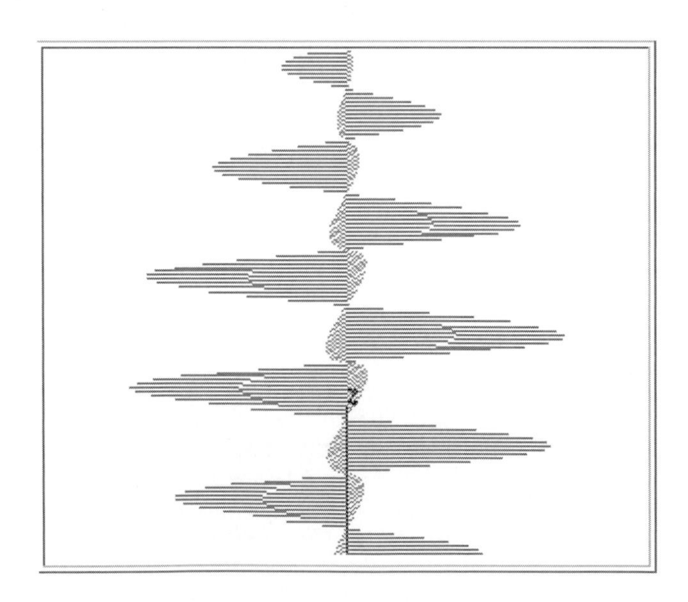

Problem 9.8.5: The animation shown here represents a traveling electromagnetic wave in the z-direction. The electric and magnetic field are represented with the magnetic field not drawn to scale. (This static picture does not properly represent the problem; please look at the CD.)

Question 9.8.5: Is the wave an example of an AM or FM signal?

Answer: *AM. The amplitude of the wave changes; therefore it is amplitude modulation.*

9.9 OPTICS

Problem 9.9.1: A beam of parallel rays propagating to the right is refracted upon entering and leaving another medium. Both the position of the source and the angle of the source can be adjusted by click-dragging the circular hotspots. Angles may also be measured by click-dragging away from the hotspots.

Question 9.9.1a: A light source emits a beam as shown in the simulation. What is the index of refraction of the substance? You can click-drag both the position and the ray angle for this source.

Answer: $n = 2.4$.

Question 9.9.1b: Click-drag the light source into the region of higher index of refraction. Adjust the angle of the source in order to demonstrate total internal reflection. What is the angle of total internal reflection?

Answer: $\theta = 24$ *degrees.*

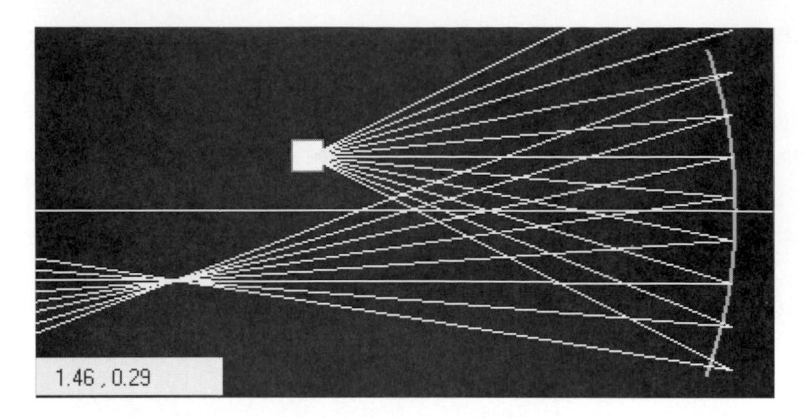

Problem 9.9.2: A point source emits a cone of light that eventually strikes a mirror. The position of the source may be adjusted using the mouse.

Question 9.9.2: A point source is located to the left of a mirror. You can drag this point source to any position. Find the focal length of the mirror.

Answer: f = *1.3 m. This problem is easy if the source is dragged onto the optic axis and its position adjusted so that the incident and the reflected rays overlap. This occurs when the source is at the radius of curvature of the mirror.*

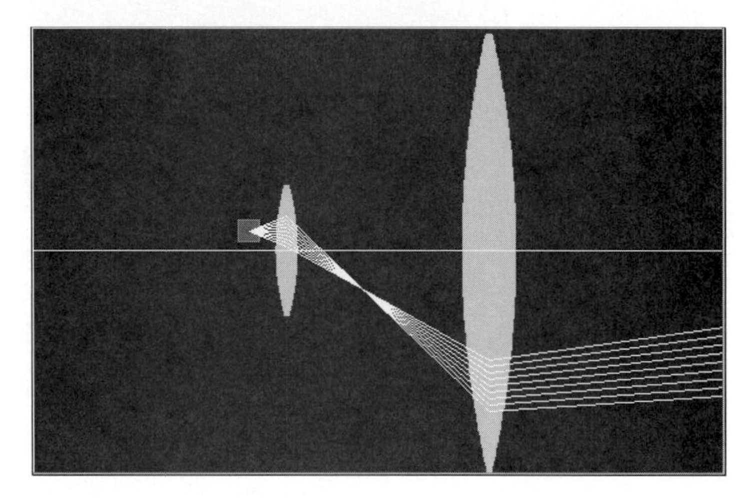

Problem 9.9.3: A point source emits a cone of light before it passes through two lenses. The position of the source may be adjusted using the mouse.

Question 9.9.3: Two lenses, an eyepiece and an objective, are used to make a microscope. Where should the object be placed for optimal viewing by a relaxed eye? You may focus the microscope by click-dragging the object into position. Position is measured in cm.

Answer: The microscope is in focus when the light rays leaving the second lens are parallel. This corresponds to the object being at infinity, the optimal distance for viewing by a relaxed eye. Many students find this result confusing since they assume that objects at infinity must appear very small.

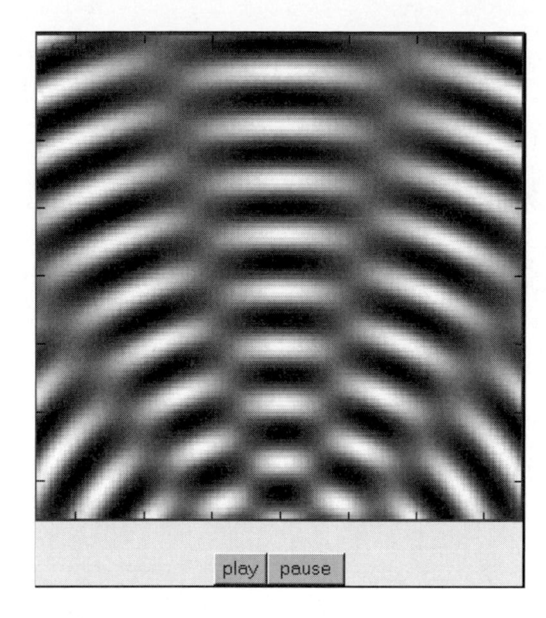

Problem 9.9.4: A ripple tank wave pattern is created and displayed whenever the play button is pressed. The mouse can be used to make coordinate measurements.

Question 9.9.4: A double slit is hidden somewhere below the animation. What is the slit separation? You can measure (x, y) coordinates by click-dragging inside the animation. Assume all measurements are given nanometers (nm).

Answer: 2 nm.

CHAPTER 10

MODERN PHYSICS PROBLEMS

10.1 SPECIAL RELATIVITY

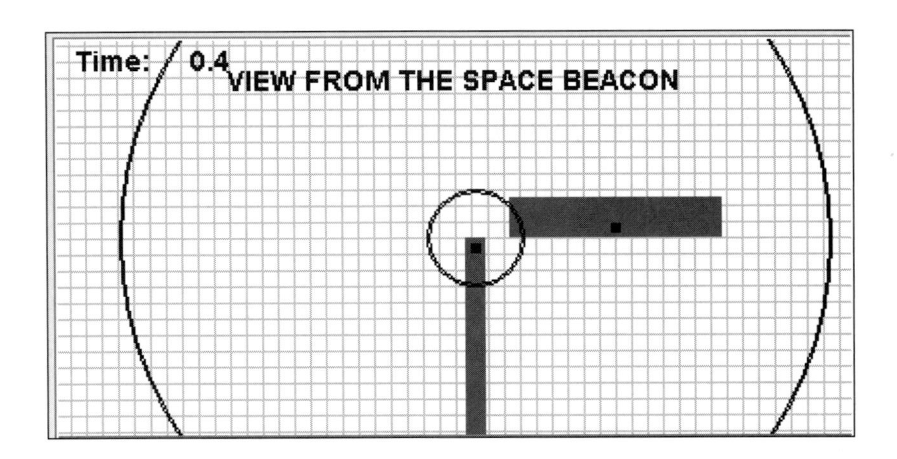

Problem 10.1.1: A moving red rectangle simulates a space ship. A space beacon, the vertical green rectangle, emits a light flash when each end passes by.

Question 10.1.1a: A spaceship (red) flies close to a space beacon (green) at 70% of the speed of light. The beacon emits light flashes as shown in the animation. The time shown in the upper left-hand corner is the time as measured in the reference frame of the beacon. Click-drag to measure the size of objects in meters. What is the time difference between these light flashes as seen by the crew inside the space ship?

Answer: 0.11 s.

Question 10.1.1b: What is the length of the spaceship as measured by the crew of the spaceship?

Answer: 150 m.

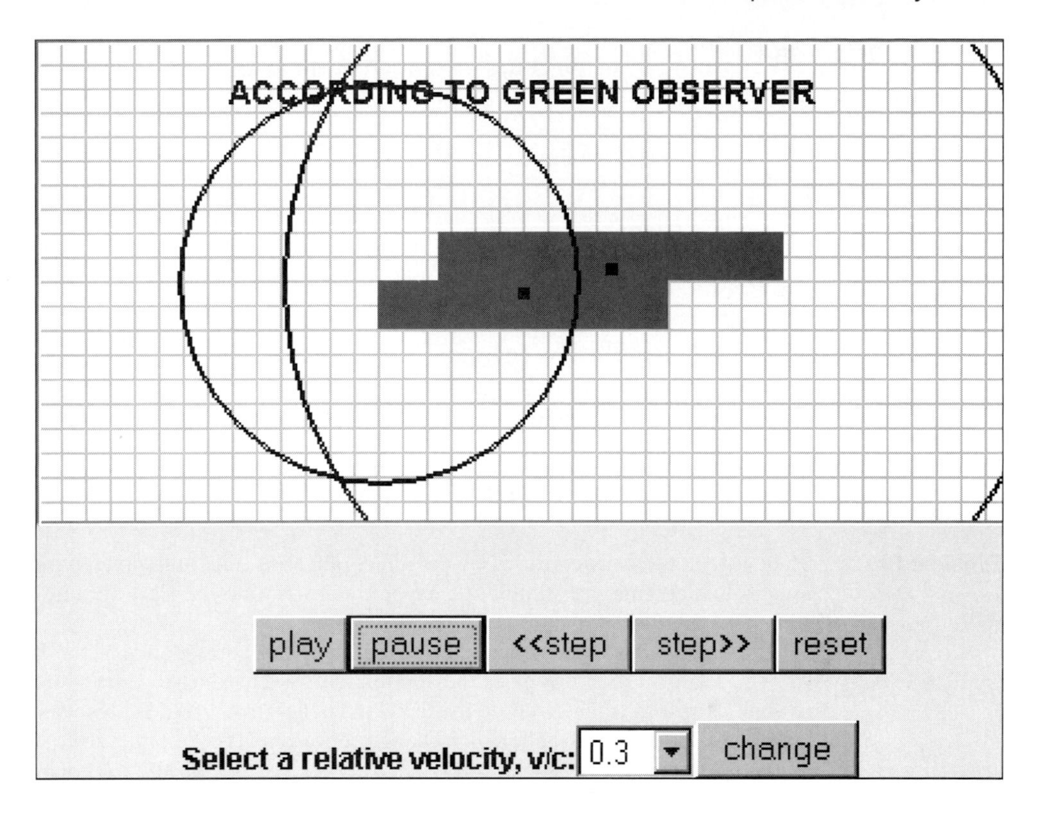

Problem 10.1.2: A moving red rectangle simulates a space ship passing directly over another space ship, a green rectangle. A light flash occurs when the left sides coincide and again when the right sides coincide.

Question 10.1.2: Two measuring sticks pass close by each other with relative speeds comparable to the speed of light. A light flash occurs when the right ends of the sticks coincide, event 1, and again when the left ends coincide, event 2. Which of the following is true for all relative velocities? You can view the events in either the green reference frame or the red reference frame.

(a) There is a unique speed when the events are simultaneous in both reference frames.

(b) Event 1 always occurs before event 2.

(c) At a given speed, both the red and green observers will always agree on which event occurred first.

(d) None of the above.

Answer: *None of the above.*

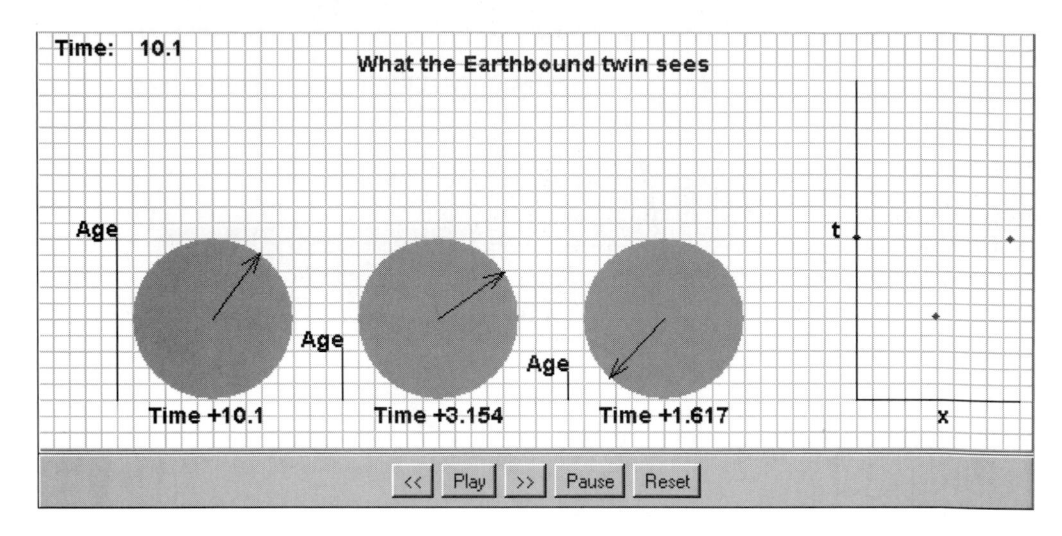

Problem 10.1.3: Students can be easily confused by the effect of transit time in the twin paradox. Unlike Minkowski diagrams, this stimulation explicitly shows what the Earth-bound twin sees by adding and subtracting the transit time to the display.

The diagram has three moving dots. The black dot corresponds to the Earth-bound twin. The leading green dot gives the simultaneous (according to Earth) position of the traveling twin. The trailing red dot, which will change to blue, indicates what the Earth-bound twin actually sees. (The color change represents the Doppler shift as the Earth-bound twin looks at his brother.) It lags the other red dot on the outbound leg, as the image of the traveling twin received by the Earth-bound twin is late due to the light travel time. On the return leg, this dot turns blue and rapidly catches up to the green dot.

The clock on the left shows proper time for the Earth-bound observer. It moves at a constant rate, with one full rotation corresponding to one grid spacing on the time axis of the diagram.

The middle clock corresponds to the leading green dot. This is the time in the twin's frame of reference and is time dilated according to special theory of relativity.

The clock on the right shows what the Earth-bound observer would actually see through a telescope focused on his traveling twin's clock. This time takes into account both the relativistic Doppler effect and the light travel time. Note that it is slow up until almost the very end, when it speeds up as the blue-shifted signals from the traveling twin begin to arrive back at Earth.

To the left of each clock is a time bar, indicating the total number of rotations of the clock (i.e., the age). The left-most bar shows the age of the Earth-bound twin, the middle bar gives the simultaneous (according to Earth) age of the traveling twin, and the right-most bar gives the age of the traveling twin as viewed by the Earth-bound twin through his telescope.

Question 10.1.3a: How fast is the moving twin traveling? Assume time is measured in years and distance is measured in light years.

Answer: 0.95 c.

Question 10.1.3b: What is the age difference between the twins when they reunite?

Answer: 13.8 yr.

Question 10.1.3c: How can you convert the readings from the third clock to the second clock?

Answer: If the earthbound twin were to subtract the light travel time from these readings, he would obtain the values shown on the second clock.

Question 10.1.3d: Some dots follow world lines in the diagram, others do not. Which dots follow world lines?

Answer: The red turning blue dot does not follow a world line since its position and time are the apparent coordinates as seen from the Earth.

10.2 HYDROGENIC WAVEFUNCTIONS

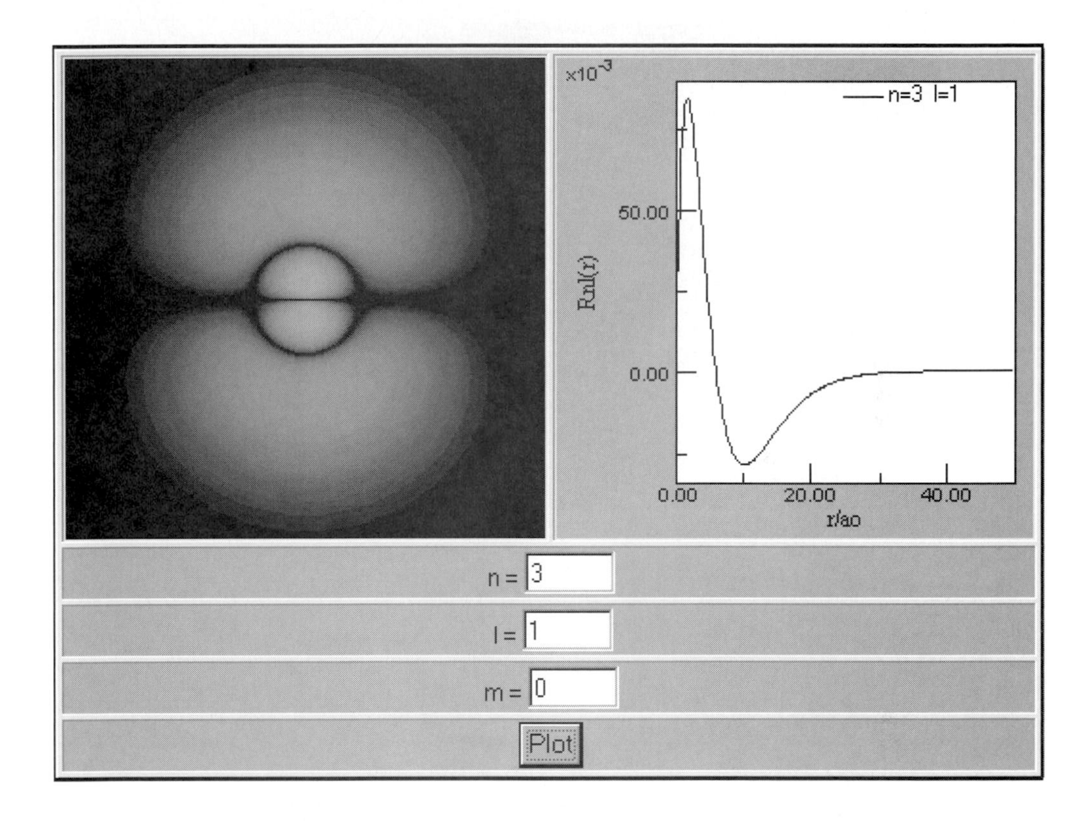

Problem 10.2.1: The two Physlets show a density plot of the hydrogenic wavefunction and the solution to the radial equation. The word *density* refers to a method for plotting 3-D information on a two-dimensional screen. Here it has nothing to do with the probability density in quantum mechanics. The radial solutions used here are the associated Laguerre polynomials scaled with the Bohr radius, a_0, set equal to one.

Question 10.2.1a: Make multiple plots of the wavefunctions for $n = 3$. How is the behavior of the radial wavefunction for $l = 0$ different than for $l = 1$ or 2? Does the radial wavefunction depend on m? Try this for a few other values of the principal quantum number and see if your conclusion holds.

Answer: The radial wavefunction does not depend on m.

Question 10.2.1b: For $n = 3$, how many times does the radial wave function cross zero (change signs) for each possible value of l? Try this for a few other values for the principal quantum number and see if your conclusion holds.

Answer: The number of zero crossing, not counting the zero at origin, is $n - 1 - l$.

Question 10.2.1c: For a given principal quantum number, there is a maximum value for l. The graph of the radial wavefunction in this case should have only one maximum value. Obtain a general formula relating the radius for this maximum value for all n.

Answer: Radius of the hydrogenic wavefunction increases as the square of the radius, r^2. *The exact relationship for the maximum of radial wavefunction is* $a_0 \cdot n \cdot l$, *where* a_0 *is the Bohr radius. Since this Physlet uses units with* $a_0 = 1$, *the maximum will occur at* $n \cdot l$.

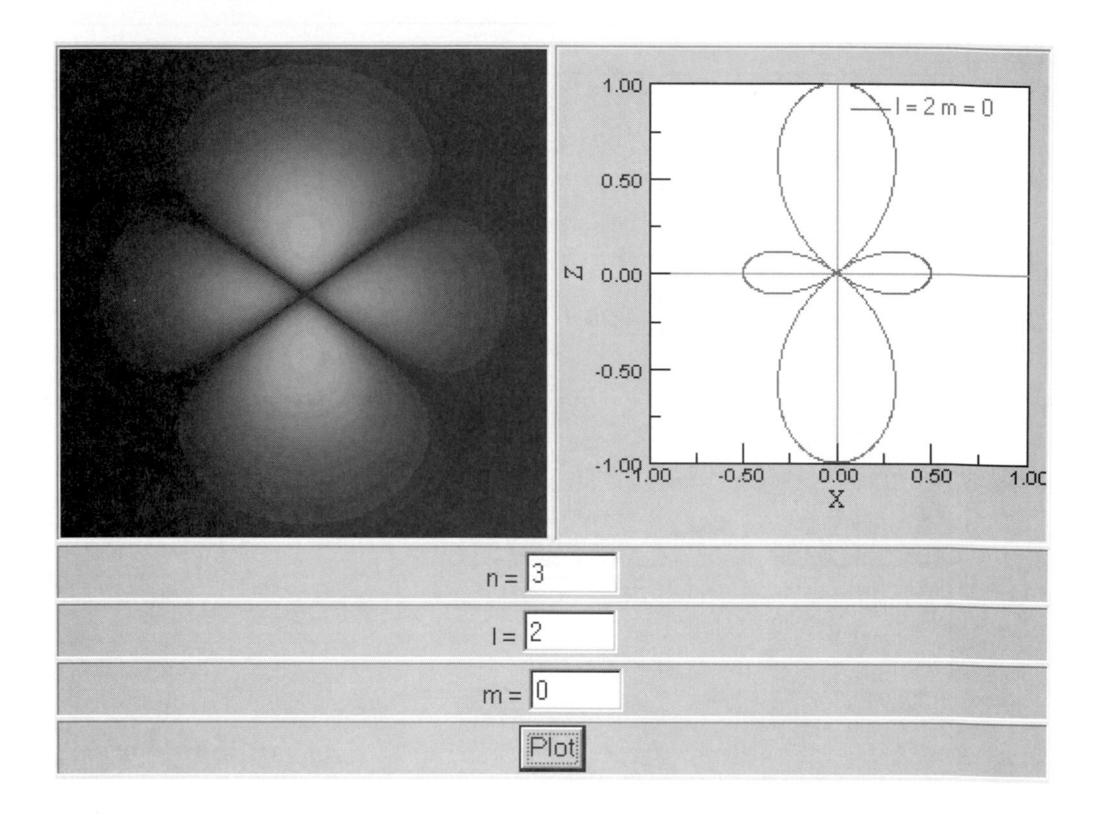

Problem 10.2.2: The two Physlets show a density plot of the hydrogenic wavefunction and the solution to the angular, that is, polar, equation. The word *density* refers to a method for plotting 3-D information on a two-dimensional screen. Here it has nothing to do with the probability density in quantum mechanics. The polar solutions used here are the unnormalized associated Legendre polynomials, $P_{lm}(q, f)$. Note that the x- and z-coordinates range from -1 to $+1$.

Question 10.2.2a: Make multiple plots of the wavefunctions. On which quantum numbers does the angular wavefunction depend? Be systematic. Change one number at a time.

Answer: The angular wavefunction depends on the l *and* m *quantum numbers.*

Question 10.2.1b: For any given values of l and m_l, does the angular wavefunction change when m is changed to $-m$? Does the total wavefunction change? Explain.

Answer: The phase of the wavefunction changes when m *is changed to* $-$m.

Question 10.2.1c: Notice the dependence of the number of lobes on l and m_l. Obtain a general formula for this dependence.

Answer: The number of lobes is $2(l - m + 1)$.

Question 10.2.1d: For $l = 1$ and $m_l = 0$, determine the angles for which the wavefunction is a maximum and a minimum. Explain your results in terms of the formula for the wavefunction for this state.

Answer: When l = *1 and* m = *0, the wavefunction is a maximum at* $\theta = 0$ *and* π *and a minimum at* $\theta = \pi/2$. *The spherical harmonic for this state is a cosine function.*

10.3 SQUARE WELLS AND THE SCHRÖDINGER EQUATION

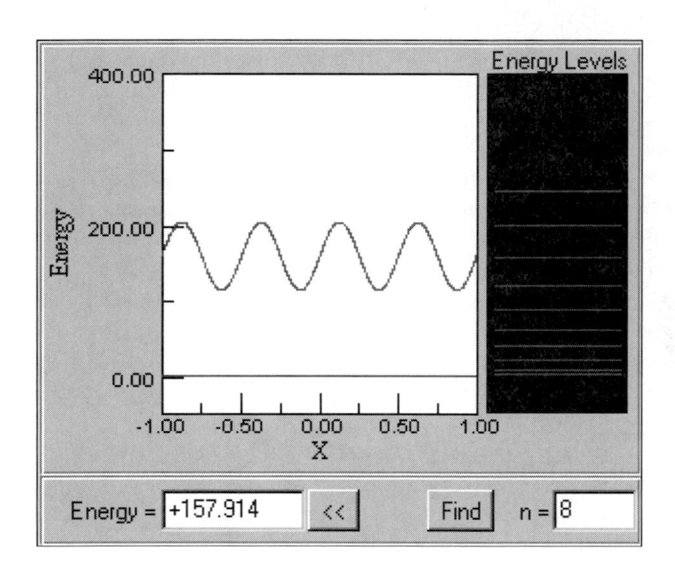

Problem 10.3.1: This Physlet shows the solution to Schrödinger's equation for a particle inside an infinite square well. It is solved using the "shooting method" in which an initial guess for the energy is made. After each iteration, the result is compared to the known boundary value and the energy is refined until an acceptable tolerance level is reached. In this problem the trial solution to the wavefunction is calculated from left to right. The boundary conditions for this problem are that

$$\Psi(-a) = \Psi(+a) = 0$$

where a is half the width of the well.

There are several important features to realize when presenting this exercise to students:

1. Students can be confused by the plotting of the wavefunction with the vertical axis shifting to the energy eigenvalue. To them this may mean that the wavefunction didn't go to zero at the boundary. Most books do this in order to compare the different wave functions. See Chapter 18, Section 18.3, for an alternate plotting method using data connections that does not shift the wavefunction.
2. The walls of the graph are hard (i.e., the potential at the walls is always infinite). This will have important implications in other exercises, where new potentials are defined within the walls of the graph.

3. It will always be possible to get a mathematical solution to the differential equation, but the important question for a physicist is "Does the solution have physical meaning?" Solutions will have physical meaning if they satisfy the boundary conditions.
4. Use the mouse:
   - Left-click in the graph for graph coordinates.
   - Right-click in the graph to take a snapshot of the current graph.
   - Left-click-drag the mouse inside the energy level spectrum to change energy levels and wavefunction of the particle.

Question 10.3.1a: What are the energy levels for the first four states? What functional dependence of the energy level on the quantum number do your results indicate? What is the mass of the particle in the well shown here if the horizontal axis is in meters and the energy is in joules?

Answer: The first four levels are 2.467, 9.878, 22.207, and 39.478. The ratios of these energies to the ground state energy are 1 : 4 : 9 : 16. These ratios indicate that there is an n^2 *dependence between the energy and the principal quantum number. If the units used in the simulation are MKS, then the mass of the particle would be* $5.57\ 10^{-69}$.

Question 10.3.1b: In order to keep from having to deal with very small and very large numbers, computer simulations often set some combination of constants equation to 1. Using your answers to the first question and the theoretical values for the energy levels of an electron in the well, determine this scaling combination and the units of energy used in the Schrödinger equation.

Answer: The combination of units $\hbar^2/2\,m$ *has been set equal to one.*

Question 10.3.1c: Note where $x = 0$ is located in comparison to the cell wall. Some texts place the $x = 0$ point at a cell wall. Does this difference affect the energy levels and/or the wavefunctions? Explain.

Answer: The choice of origin does not affect the energy levels. It also does not affect physics of the wavefunctions, only the way that the wavefunctions are expressed in terms of sinusoidal functions.

Question 10.3.1d: The "parity" of a wavefunction is defined to be

$$\textit{even} \quad \text{if} \quad \Psi(x) = \Psi(-x)$$

$$\textit{odd} \quad \text{if} \quad \Psi(x) - = \Psi(-x)$$

What is the parity of each of the wavefunctions for the first six energy levels? What general conclusion can you draw regarding the quantum number and the parity for an arbitrary energy level?

Answer: The ground state, $n = 1$*, has even parity and the first excited state,* $n = 2$*, has odd parity. This switching between even and odd parity will continue as the quantum number is increased.*

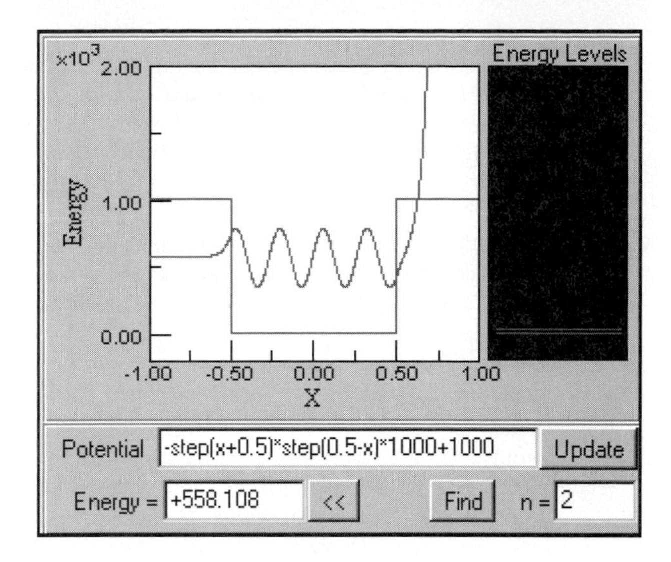

Problem 10.3.2: This exercise is designed to study the finite square well and to show how the shooting method can be used to determine eigenvalues. Although the finite square-well potential problem is more realistic then the infinite well, it is difficult to solve because it yields transcendental equations. With the finite potential, it is possible for the particle to be bound or unbound. A bound level is one whose energy is less than the well depth. See Problem 10.3.1 for additional hints.

Question 10.3.2a: Explain how you can observe that an energy value is acceptable as you right-click-drag.

Answer: Right-click and drag in the energy spectrum and you should observe the right side of the wavefunction flipping from negative to positive. This is a sign that you have passed through a physical solution to the Schrödinger equation [i.e., that the boundary condition at $\Psi(x = \pm 1) = 0$]. The correct boundary condition is assumed at the left side of the graph and the wavefunction is calculated in the direction of increasing x.

Question 10.3.2b: By changing the principal quantum number, determine the first four bound state energies for this well. How do the energies corresponding to the same quantum number compare for finite and infinite potential wells of the same width?

Answer: The bound state energies are 8.72, 34.8, 78.3, and 139. These energies are smaller than an infinite square well of the same width because the wavefunction is slightly broader due to its penetration into the classically forbidden region.

Question 10.3.2c: Determine the number of bound states for this well. A solution of Schrödinger's equation for this problem indicates that the total number of bound states is the next largest integer above the product of the width divided by pi (π) and the square root of the depth. (Note the scale of the vertical axis in the graph.) Does this hold true for your results? Identify in the equation for the potential the parameter that determines the width of the well and make it half as wide. Does the number of bound states still equal the predicted value?

Answer: The finite well obeys the formula since it has 11 bound states. Since the numerical algorithm must assume that the finite well is contained inside an infinite square well, it may he hard to tell if the eleventh state is bound or unbound. You can argue that the eleventh state is bound because the wavefunction is exponentially decaying in the wall region. The twelfth state is bounced by the infinite well but not by the finite well.

Reducing the width of the well by a factor of 2 results in six bound states. This result is still in agreement with the formula.

Question 10.3.2d: Decrease the depth of the well to 200 units while keeping the bottom of the well at 0 units and the width at 1.0 units. What happens to the wavefunction at the boundary of the finite well? What happens to the energy levels? Does the formula for the number of bound states still hold?

Answer: The wavefunction extends further into the wall region and the wavefunction energies decrease. There are five bound states. This result is still in agreement with the formula.

Question 10.3.2e: Do your conclusions regarding parity for wavefunctions in the infinite square well (see Problem 10.3.1d) still hold for the finite square well?

Answer: The parity is determined by the symmetry of the potential. Since the symmetry has not changed, the conclusions about parity are still valid.

PART THREE
REFERENCE

CHAPTER 11
RESOURCES

The primary reference for Physlets is the Davidson College WebPhysics server. New Physlets are being developed, documentation and examples are being improved, and the occasional bug is being squashed. However, the Web is not always available or convenient. This book provides documentation in two additional formats.

The CD enclosed in the back of this book contains a snapshot of the Physlets and the documentation that existed on our server shortly before publication. Since much of the documentation is machine generated, it may be overwhelming. But it is accurate in the sense that it was generated from the raw Java source code. Every method that is accessible from browser-based JavaScript is listed. Every inheritance is documented.

We have based this book on four Physlets with wide applicability. The next few chapters provide additional technical details, programming tips, and comments for these Physlets. The first two Physlets, *Animator* and *EField*, are usually used to model phenomena and to generate data. The next two, *DataGraph* and *DataTable*, are designed to present data after they have been processed by a data connection. Extensive examples are available on the CD.

The final chapter, Chapter 18, briefly introduces a number of additional Physlets that implement the scripting model described in this book. The script examples for these Physlets are not examples of pedagogy; they are proof-of-concept scripts that we have used to check the basic functionality of each Physlet.

You can also find a few examples in Part Two of the book that use older version 3—or below—Physlets. These older Physlets are not supported and documentation is not included. Problems that use these Physlets are included as examples of pedagogy. Some documentation for older Physlets is available on our Web site, but these Physlets are likely to change as they are written in the coming months.

11.1 AVAILABLE PHYSLETS

As mentioned in Chapter 5, the Java language underwent a significant change as it evolved from versions 1.0 to 1.1. For example, the mechanism whereby user actions, such as mouse drags and button clicks, were processed has changed. These changes required a significant amount of recoding. The evolution from Java version 1.1 to 1.2 added functionality but, fortunately, did not significantly change the core language.[1] We are confident that applets written for Java 1.1 will run on future Java platforms and have adopted this version for new Physlets. Expect to see most existing Java 1.0 Physlets converted to Java 1.1 in the near future.

1.1.1 Java 1.1

- *Animator* was originally designed for the animation of geometric shapes and images along predefined trajectories. Recent versions allow objects to move in response to particle interactions as well as external forces.
- *BField* displays the magnetic field from straight wires or coils using color-coded vectors. External fields can also be specified.
- *BlackBody* displays the radiation spectrum at different temperatures. The temperature can be changed by dragging the peak of the blackbody curve, by entering the value into a text field, and by scripting.
- *Circuits* contains a collection of applets designed to model common dc and ac configurations.
- *DataGraph* plots data sets and functions. It was designed to record data from other Physlets using real-time interapplet communication.
- *DataTable* displays columns of numbers. It too was designed to record data from other Physlets using real-time interapplet communication.
- *EField* plots fields produced by a potential function and fixed point charges. Test charges that move under the action of the field can be included in the simulation.
- *Eigenvalue* calculates the energy spectrum of the Schrödinger equation using the shooting method and displays the wavefunction. Dragging the mouse inside the energy level diagram shows how only certain energy eigenvalues are able to satisfy the boundary conditions.
- *EMWave* shows a three-dimensional display of a traveling wave. The wave can be scripted to model linear as well as circularly polarized light.
- *Faraday* presents the classic problem of a wire sliding on top of a U-shaped conductor embedded in a magnetic field. A galvanometer display shows the current flow as the wire is moved.
- *Hydrogenic* is a package that consists of three applets designed to display radial wavefunctions, angular wavefunctions, and probability density plots.

---

[1]Java version 1.2 was recently renamed the Java 2 platform by Sun Microsystems. This change was purely for marketing.

- *Molecular* uses a hard disk model to teach kinetic theory.
- *Optics* models an optics bench with lenses, mirrors, apertures, and sources.
- *Poisson* solves the boundary value problem for conductors and simple charge-density distributions using the relaxation method. Contour and field plot options are provided.

11.1.2 Java 1.0

- *Doppler* demonstrates the classical and relativistic Doppler effect. It was the first Physlet written.
- *Lorentz* allows the user to show the electric field from a moving point charge. The magnetic field can also be displayed.
- *Minkowski* presents overlapping space-time diagrams for two different reference frames. Events can be attached to either frame, thereby demonstrating length contraction, time dilation, and other relativistic effects.
- *QMTime* solves the time evolution of the Schrödinger equation in one dimension. Rather than displaying separate graphs for real and imaginary components of the wavefunction, the amplitude is presented and color-coded to show the phase.
- *Retard* displays the electric field from a point charge including the time retardation. Common radiation fields such as synchrotron radiation and free-electron laser wigglers are preprogrammed.
- *RippleTank* models a water-filled tank containing one or more point-source exciters.
- *Superposition* creates a multipanel display containing two time-dependent waves and their sum.

11.2 JAVADOC OUTPUT

The Physlet documentation provided on the CD and on the Web site is machine generated by a utility program known as javadoc. Its output is something that only a geek could love.[2] The program extracts every detail about a Java class from the source code and outputs these details using html. This section presents an anatomy of this output for the nonprogrammer.

Javadoc produces an html page for each Java object that is processed. In our case, these Java objects are Physlets. The program organizes these output files using html frames. The left frame displays a table of contents (TOC) listing the Physlet or package for which documentation has been generated. Each TOC entry is hyperlinked to the appropriate file, and clicking on this link loads the page into the right frame.

---

[2]*Geek* is a technical term used by programmers to refer to members of their community who are held in high regard. The term may be pejorative when it is used by nonprogrammers to refer to programmers since geeks were originally carnival performers who publicly ate or swallowed live animals.

```
Class Animator
java.lang.Object
  |
  +--java.awt.Component
        |
        +--java.awt.Container
              |
              +--java.awt.Panel
                    |
                    +--java.applet.Applet
                          |
                          +--edu.davidson.tools.SApplet
                                |
                                +--animator4.Animator
```

FIGURE 54: Class hierarchy for *Animator* generated by javadoc.

11.2.1 Class Hierarchy

A Physlet documentation page begins by showing the class hierarchy. For example, Figure 54 shows that the *Animator* Physlet is a subclass of SApplet, which itself is a subclass of Applet. This parent → child relationship can be followed to the root of all Java objects called, in typical engineering fashion, Object. Since a child can perform any of the methods defined in a parent, *Animator* can perform any method defined in any of its six superclasses. Since the physics that we wish to script was added to the Applet class by the SApplet and the *Animator* subclasses, we usually only need to concern ourselves with the methods listed in these last two levels.

11.2.2 Method Summary and Detail

Below the class hierarchy is a table summarizing every available public (i.e., scriptable) method. The first column indicates the data type that is returned when the method is executed. For example, the *addObject* method in Table 3 returns an integer while the *forward* method does not return any value. The keyword void indicates that no value is returned. The second column is more interesting since it lists the object name and its

TABLE 3: Two entries from the javadoc generated method summary.

Method Summary	
int	**addObject** (java.lang.String name, java.lang.String parList) Create an object and add it to the Physlet.
void	**forward**() Start the animation.

signature. The signature consists of a comma-delimited list of data types and variable names. Following the signature is a short description of the method.

Names in the method summary are actually hyperlinks to entries further on down the documentation page in the detail table. In addition to a more extensive description, entries in the method detail table usually provide descriptions of parameters and return types, as shown in Figure 55.

Javadoc generates numerous other tables, including laundry lists of methods that are defined in the various superclasses. This fire hose of information is more intimidating than useful to JavaScript programmers.[3] The exception is the SApplet superclass since it contains the animation clock and information about data connections. You should, however, click on the hyperlink to the SApplet documentation page rather than trying to decipher SApplet by studying the cryptic listing on any of its subclass documentation pages.

```
public int addObject(java.lang.String name,
                     java.lang.String parList)
```

Create an object and add it to the Physlet. The first argument is the name of the object to be added and the second is a comma-delimited list of parameters. For example, a circle can be added a follows:

```
addObject ("circle", "x = 0, y = -1.0, r = 10");
```

See the supplemental documentation for a list of object names and parameters.

Parameters:

`name` - the type of object to be created.

`parList` - a list of parameters to be set

Returns: integer id that identifies the object.

FIGURE 55: Method detail table entry for the addObject method in *Animator*.

[3]In fact, some publicly available methods, such as start and stop, should never be called by JavaScript programmers. The start method is called by the browser whenever a user enters a page and stop is called whenever a user leaves a page. They appear in the javadoc output since they have the same access permission as the methods that we have designed to make our applets scriptable.

CHAPTER 12

INHERITED METHODS

A typical Physlet will have dozens of public methods designed to control its behavior and appearance using JavaScript. Many of these methods are unique to each applet, but certain methods are of sufficient utility that they reoccur. Many if these reoccurring methods have been moved to the common Physlet superclass, SApplet. SApplet is, therefore, the logical place to start documenting Physlets.

12.1 CLOCK METHODS

SApplet contains an animation clock that is available even if the applet does not appear to use animation. The behavior of this clock can be scripted through the methods shown in Table 4. See the *DataGraph* tutorial for an example of how this clock can be used to generate data for on-screen display.

TABLE 4: Clock methods contained in the SApplet superclass.

Method	Description
cyclingClock( )	This method is called whenever the clock reaches its maximum value in cycle mode. Override this method and reset the applet into its initial state.
forward( )	Start the clock with a positive time step.
pause()	Pause the clock at the current time.
reset()	Reset the applet clock to its initial state. The clock is set to zero or to the minimum time if oneShot or cycle mode is enabled. *See also*: setDefault under miscellaneous methods.
reverse()	Start the clock with a negative time step, *dt*.
setDt(dt) **double dt** animation time step	Set the time step for each clock tick. Usually not scripted since the time step is set with the <PARAM> tag when the applet is embedded. The default behavior calls clock.setDt. Example: setDt(0.1); Note: Animation time corresponds to actual time if fps*dt=1.
setFPS(fps) **double fps** frames per second	The number of clock ticks per second. Usually not scripted since fps is set with the <PARAM> tag when the applet is embedded. Each tick will often result in a new animation frame, hence the use of FPS (Frames Per Second) in the name. Example: setFPS(10); *Note*: Animation time corresponds to actual time if fps*Dt=1.
setTimeContinuous()	Animation time will increase without limit when the clock is running.

setTimeCycle(double) **double max** the maximum time	Animation time will cycle between 0 and a maximum value. Example: setTimeCycle (10); // time will cycle at 10
setTimeOneShot(max,msg) **double max** the maximum time **String msg** a display message for the user.	Animation will run from $t = 0$ to a maximum time. When the clock reaches its maximum time the clock will stop. Example: setTimeOneShot (10, "End of animation."); Note: Some Physlets do not automatically display a message.
stepTimeBack()	Step the clock back by one time step, *dt*. If the clock is running the pause method is called.
stepTimeForward()	Step the clock forward by one time step, *dt*. If the clock is running the pause method is called.
stoppingClock()	This method is called whenever the clock reaches its maximum time in one-shot mode. The applet author should override this method to display the oneShotMsg on the screen.

12.2 DATA CONNECTIONS

Interapplet communication could not occur without the common SApplet superclass. The connection that is established is actually an object and, like any object, this connection has properties that can be set. Setting a data connection's property requires that the integer returned by the makeDataConnection method be passed as the object's identifier in the set method.

TABLE 5: Data connection methods contained in the SApplet superclass.

Method	Description
clearAllData()	Sends the clear series command to all data listeners. This usually deletes all the data in the data listener.
clearData(id) **int id** the object identifier	Clears a single data connection with the given object identifier, id.
deleteDataConnection(id) **int id** the object identifier	Breaks the data connection to the data listener. The data listener is not cleared.
deleteDataConnections()	Breaks the data connections to the all data listeners. Data listeners are not cleared.
makeDataConnection(sid, lid, series, fx, fy)	Creates a data connection between a data source and a data listener. The two string parameters, *fx* and *fy*, are mathematical functions of the data source variables. These functions are evaluated by the data connection to produce either a data point (*fx*, *fy*) or an entire array of data (*fx*[], *fy*[]) that is then passed on to the data listener.

(*continued*)

TABLE 5: (Continued)

Method	Description
int sid data source id **int lid** data listener id **idint series** data listener series **String fx** function to be evaluated for x **String fy** fuction to be evaluated for y **return int** object identifier of the connection	Returns the object identifier for the connection.
setConnectionBlock(id, block) **int id** the object identifier **boolean block** block if true.	Blocks the connection from accepting data. Unblocking can occur at any time.
setConnectionSmoothing(id,num) **int id** object identifier of the data connection **int num** number of points to average	Smooths data as they pass through the connection. If the data source sends an array of points, nearest neighbors will be averaged.
setConnectionStride(id,num) **int id** the object identifier	Sets the data connection to skip every num-1 points.
updateDataConnection(id) **int id** the object identifier	Sends data from the data source.
updateDataConnections()	Sends data from all data sources.

12.3 MISCELLANEOUS METHODS

TABLE 6: Miscellaneous methods contained in the SApplet superclass.

Method	Description
setAutoRefresh(auto) **boolean auto** true if applet should redraw itself after changes are made.	Determines if the applet will recalculate and repaint every time a parameter changes. Set this parameter to false at the beginning of a long script in order to eliminate screen flash and reduce unnecessary processing. AutoRefresh should be set to true at the end of a script.
setDefault()	Resets the applet to a predetermined initial state. The default method sets the animation time to zero and removes all existing data connections. Programmers should override this method to set other default conditions. See also reset() in the table of clock methods.

CHAPTER 13

NAMING CONVENTIONS

13.1 COMMON METHODS

Inherited methods that were listed in Chapter 12 are guaranteed to be available in every Physlet that inherits from SApplet. Common methods, on the other hand, are applet specific. These methods may or may not be implemented. Use Table 7 as a guide, but consult the online documentation for an applet to determine if the method is available.

TABLE 7: Common method names.

Method	Description
addObject(name, list) **String name** the object to be added **String list** a list of settable properties **return int** object identifier	Creates an object and adds it to the simulation. This method should be implemented in any applet that is designed to instantiate data sources using JavaScript. Instantiated objects are often, but not always, drawn on the screen. Returns an object identifier, usually the hashCode(). Example: addObject("circle","x = 2,y = 3,r = 8") *Note*: Methods in the edu.davidson.tools.SUtil class, such as parameterExist and getParam, can be used to parse the properties list.
getVX(id) **int id** the object identifier **return double** vx	Reads the *x*-velocity of an object.
getVY(id) **int id** the object identifier **return double** vy	Reads the *y*-velocity of an object.
getX(id) **int id** the object identifier **return double** x	Reads the *x*-coordinate of an object.

Method	Description
getY(id) **int id** the object identifier **return double** y	Reads the *y*-coordinate of an object.
set(id, name, list) **int id** the object identifier **String name** the object to be added **String list** a list of settable properties **return int** object identifier	Sets an object's properties. Similar to the addObject method except that the object must already exist. An id of 0 can be used to signal the default object. This is a new method that we have not yet widely implemented. We believe, however, that this method would be easier to use and less error prone for setting complex data structures. For example, `set(0,"scale","xmax = 20, autoScaleY");` can now be used to set the scale in the *DataGraph* Physlet. The *x*-axis maximum value is given a new value and the *y*-axis is set to autoScale. All other scale parameters are left unchanged. Expect to see this method adopted in new applets.
setDragable(id, drag) **int id** the object identifier **boolean drag** true if object is dragable	Makes the object with the specified id dragable. Returns true if the object exists. Example: setDragable(id,true)
setFormat(id, str) **int id** the object identifier **String str** the format string	Sets the numeric format for the display of numbers using UNIX printf conventions. Example: `str = "%-+6.2f"` or `str = "%3i"` See *DataTable* reference for further documentation.
setRGB(id, r, g, b) **int id** the object identifier **int r** red **int g** green **int b** blue **return boolean** object exists	Sets the color of the object. Returns true if the object exists. Example: `setRGB(id,255,0,0);` // sets the object's color to red
setX(id, x) **int id** the object identifier **double x** the x value	Sets the *x*-coordinate of an object.
setXY(id, x,y) **int id** the object identifier **double x** the x value **double y** the y value	Sets the *x*- and *y*-coordinates of an object simultaneously.

(continued)

TABLE 7: (Continued)

Method	Description
setY(id, y) **int id** the object identifier **double y** the y value	Sets the *y*-coordinate of an object.
setVisibility(id,vis) **int id** the object identifier **boolean vis** true if the object is visible on the screen	Sets the visibility of any onscreen object. Special case: Time display is affected if the id is that of the clock. Example: id = getClockID(); setVisibility(id, false); // disable the time display in the UI

13.2 addObject METHOD

Almost all Physlets support one or more *add methods* designed to create an object and add it to the applet. These objects usually, but not always, have an on-screen representation. For example, both *Animator* and *EField* still support the old—and deprecated—addCircle methods. These methods are invoked using the following JavaScript statements:

```
document.animator.addCircle(int r, String x,String y);
```

and

```
document.efield.addCircle(double x, double y, int r);,
```

respectively.

Notice that the signatures, that is, the parameters, of these two methods are different. *Animator* version 1 was designed to move simple geometric shapes along predefined trajectories that were passed to the object's constructor. *EField* version 1, on the other hand, was designed to plot electric fields for fixed charges. Its add method required fixed positions. Hence, *Animator* was passed two functions of time and *EField* was passed two numbers. Both applets were designed to run on 90-MHz computers with relatively slow Java virtual machines. Since each applet only had half a dozen methods, it was not difficult to remember the correct signature. However, advances in computer hardware and software have made it possible to add a great deal of functionality to these two applets. *Animator* now supports particle interactions and *EField* supports dynamic test charges. Although their capabilities have converged, their method names have not.

In order to bring some order to this method madness, version 4 of *Animator* and *EField*, along with newly written Physlets, has adopted a uniform convention for adding objects. Both applets now support an addObject method with the following signature:

```
addObject(String name, String attributes)
```

The first argument is the name of the object to be added, and the second is a comma-delimited list of parameters. A circle can now be added to either applet using the following JavaScript statement:

```
document.physletname.addObject("circle", "x = 0,y = -1.0,r = 10");.
```

Furthermore, the new method is more forgiving since not all parameters need to be specified. Default values are overridden only if the parameter appears in the list. Incorrect and unsupported parameters do not affect the applet and are ignored.[1]

Table 8 lists the name and the associated parameters for the addObject method in version 4 of the *Animator* and *EField* Physlets. An **X** in the **A** or **E** column indicates that the named object can be added. Consult the online documentation for information about other Physlets. *DataGraph*, for example, supports the addObject method for geometric shapes such as circles, boxes, and shells. *OpticsBench*, on the other hand, supports the addObject method for lenses, mirrors, and light sources.

TABLE 8: AddObject parameter names and properties.

Name	Attributes	A	E
arrow Arrows are often animation slaves of other objects. They can represent almost any vector since the *h* and *v* components can be functions of the variables.	**x-double** *x* position of the base in world units **y-double** *y* position of the base in world units **h-string** horizontal component as a function of *t*, *x*, *y*, *vx*, *vy*, *ax*, and *ay* **v-string** vertical components as a function of *t*, *x*, *y*, *vx*, *vy*, *ax*, and *ay*	x	x
box A box is a hollow rectangle.	**x-double** *x* position of the center in world units **y-double** *y* position of the center in world units **h-int** height in pixels **w-int** width in pixels **s-int** thickness of the box **m-double** the mass	x	x
circle	**x-double** *x* position of the center in world units **y-double** *y* position of the center in world units **r-int** radius in pixels **m-double** the mass	x	x

(continued)

[1] Users of old scripts should note that although the original add methods are still included in version 4 for backward compatibility, they have been deprecated. In addition, there have been structural changes in the tools package. The original STools.jar and SGraphics.jar have been combined into a single archive called STools4.jar. Consequently, the archive attribute for the Animator Physlet has changed from

archive = "Animator3_.jar,SGraphics.jar,STools,jar" to archive = "Animator4_.jar,STools4.jar."

TABLE 8: (Continued)

Name	Attributes	A	E
caption A caption is text that is centered near the top of the screen.	**text-string** text of the caption **calc-string** A function of *t* to be evaluated at every time step. The value of the function is displayed to the right of the text.	x	x
charge	**x-double** *x* position of the center in world units **y-double** *y* position of the center in world units **q-double** object's charge **r-int** radius in pixels **m-double** the mass		x
connectorline A straight line connecting two objects.	**id1-int** first object identifier. **id2-int** second object identifier. *See also*: addConnectorSpring(int,int) in the online documentation. This method is easier to use if the ids are stored as integers.	x	x
connectorspring A spring connecting two objects.	**id1-int** first object identifier. **id2-int** second object identifier. *See also*: addConnectorLine(int,int) in the online documentation. This method is easier to use if the ids are stored as integers.	x	x
cursor A circle with cross hairs.	**x-double** *x* position of the center in world units **y-double** *y* position of the center in world units **r-int** radius in pixels	x	x
exshell Extended shell is similar to shell except that the radius can be a function of time. Useful for shock waves and space-time diagrams.	**x-double** *x* position of the center in world units **y-double** *y* position of the center in world units **r-string** The radius in world units as a function of time **s-int** thickness of the shell in pixels **m-double** the mass	x	
image Add a gif image. The coordinates are the coordinates of the upper left-hand corner.	**x-double** *x* position of the left side in world units **y-double** *y* position of the top in world units **file-string** name of the gif file. The image should be located in the same directory as the jar file for the applet. **m-double** the mass	x	x
line	**x-double** *x* position of the base in world units **y-double** *y* position of the base in world units **h-string** horizontal component as a function of t, x, y, vx, vy, ax, and ay **v-string** vertical component as a function of t, x, y, vx, vy, ax, and ay.	x	x

Name	Attributes	A	E
polyshape polyshape draws an arbitrary shape by connect-the-dots. The shape is specified by passing the pixel positions of the dots starting at *x* and *y*.	**x-double** *x* position of the center in world units **y-double** *y* position of the center in world units **n-int** number of vertices in the polygon. **h-string** A slash (/) separated list of the *x* positions of the vertices in pixel units. **v-string** A slash (/) separated list of the *y* positions of the vertices in pixel units. **m-double** the mass	x	
rectangle	**x-double** *x* position of the center in world units **y-double** *y* position of the center in world units **h-int** height in pixels **w-int** width in pixels **m-double** the mass	x	x
relshape relshape draws an arbitrary shape by connect-the-dots. The shape is specified by passing the relative pixel positions of the dots starting at *x* and *y*.	**x-double** *x* position of the center in world units **y-double** *y* position of the center in world units **n-int** number of vertices in the polygon **hStr-string** A comma separated list of the *x* positions of the relative vertices in pixel units **vStr-string** A comma separated list of the *y* positions of the relative vertices in pixel units **m-double** the mass	x	
shell A circle with a hollow center	**x-double** *x* position of the center in world units **y-double** *y* position of the center in world units **r-int** radius in pixels **s-int** thickness of the shell **m-double** the mass	x	x
testcharge A charge that is free to move under the action of fixed charges and an external potential.	**x-double** *x* position of the center in world units **y-double** *y* position of the center in world units **vx-double** *x* velocity in world units **vy-double** *y* velocity in world units **q-double** object's charge **r-int** radius in pixels **m-double** the mass	x	
text A fixed text string followed by an optional calculation.	**x-double** *x* position of the left side of the text in world units **y-double** *y* position of the top of the text in world units **text-string** static text. **calc-string** A function of *t*, *x*, *y*, *vx*, *vy*, *ax*, and *ay* that is evaluated at every time step. The value of the function is displayed to the right of the static text. *Note:* Use the setFormat method described in Section 17.3.1 to change the decimal format used to output the calculation.	x	x

CHAPTER 14

ANIMATOR

Animator is designed to animate shapes and images on the screen. After an object is created, it can be scripted to move along a predefined trajectory or to move in response to forces. The default behavior is for an object to remain fixed at the position where it was created.

```
id = document.animator.addObject("circle","x = 1,y = 0");
```

In order to move an object according to the analytic functions of time xStr and yStr, the object identifier must be known and the setTrajectory method must be called.

```
document.animator.setTrajectory(id,xStr,yStr);
```

If an object is to become dynamic, that is, it interacts via forces and obeys Newton's laws, the setForce method must be called.

```
fxStr = "0";
fyStr = "-9.8";
document.animator.setForce(id,fxStr,fyStr,x,y,vx,vy);
```

Notice that the second and third parameters are string variables.

Since Newton's laws are second order differential equations, the setForce method requires initial values for both position and velocity. Initial values are fixed; they are passed as numbers, not strings. The solution to these equations of motion is calculated using a Runge-Kutta, RK4/5, algorithm with adaptive step size. This algorithm is not foolproof, but it is a general-purpose algorithm that solves most dynamics problems.

The design goal for *Animator* was flexibility, not speed. Most computers can easily handle four objects interacting via $1/r^2$ forces, but most cannot handle 20. The reason is twofold. First, the number of calculations scale as n squared. Therefore, increasing the number of interacting particles by a factor of 5 increases the calculation time by 25. Second, *Animator* makes extensive use of parsers, not predefined force laws. Interactions between objects are specified using strings that are passed to the applet using JavaScript. These strings must be evaluated repeatedly to determine the object's position and rate of movement. Although this allows any force to be entered, it is considerably slower than knowing the force whereby particles interact, coding this force function in Java, and then compiling the applet for optimum performance. Other

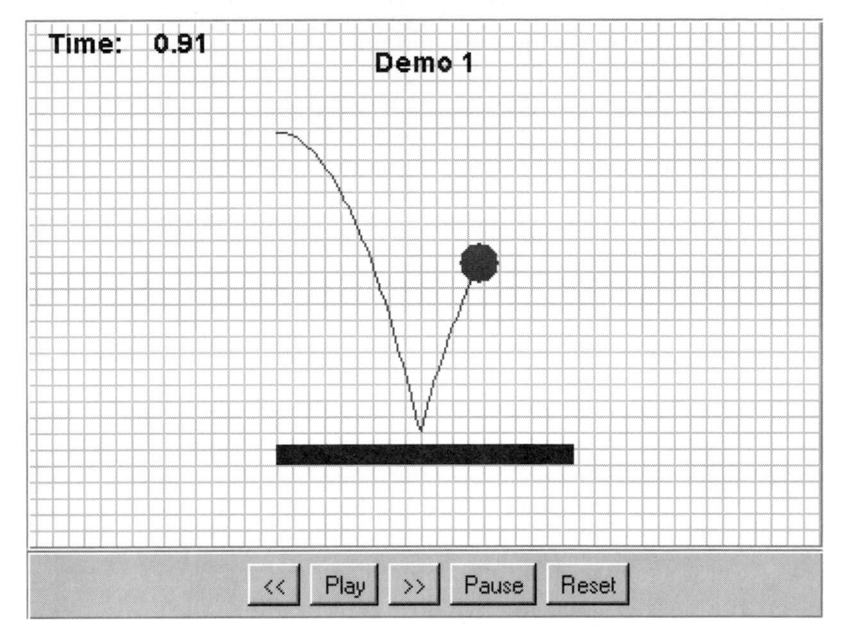

FIGURE 56: *Animator* scripted to show a rectangle and a ball.

Physlets, such as those in the molecular dynamics package discussed in Section 18.7, were designed for higher computational speed with large numbers of particles and therefore have predetermined force laws. These Physlets can be downloaded directly from the Davidson WebPhysics server.

14.1 EMBEDDING

Animator version 4 is embedded into an html page as follows.

```
<applet codebase = [location] code = "animator4.Animator.class"
archive = "Animator4_.jar,STools4.jar" name = [name] id = [name]>
```

Notice the pairing of version 4 of this applet with version 4 of the Science Tools library.

Important attributes are codebase, code, and archive as described in section 4.2.2. As with other Physlets, name and id must be set in order to use JavaScript. Height and width attributes can also be set when the applet is embedded, and usually are. Parameter tags, such as those shown in Table 9 for *Animator*, are also usually included. Most embedding parameters have a one-to-one correspondence to methods, so this gives the author a choice in where to set these values.

The <applet> tag must, of course, be closed by an </applet> tag after the embedding parameter tags have been coded.

TABLE 9: *Animator* embedding parameters.

Parameter	Value	Description
FPS	10	Frames per second.
Dt	0.1	Animation time step per frame.
ShowControls	true	Show VCR buttons at bottom of applet.
GridUnit	1.0	The grid spacing. A value of 0 will suppress the grid.
PixPerUnit	10	Conversion factor from pixel units to world units.

14.2 DATA SOURCES

Various objects in *Animator* have the ability to generate data. Table 10 lists objects that have this ability, how each object's identifier is obtained, and the state variables that these objects provide to a data connection.

There is an important distinction between $m*a_x$ and fx when an object is being dragged. Dragging is most commonly associated with holding an object, and we have found that users expect the acceleration and the velocity of a particle to be set to zero when an object is being dragged. Strictly speaking, the holding force should also be included in the total force, but this external force is difficult to calculate since it depends on computer-specific mouse settings. Since we cannot anticipate how a user will actually move the mouse, we have chosen to store only the dynamic forces acting on an object within *Animator* when it is being dragged. These values are set equal to the data source variables fx and fy.

14.3 METHODS

Since *Animator* was the prototype for our naming conventions, it supports most of the recommended methods listed in Table 7. The setX, setVX, and setMass methods all work as expected. The addObject method is clearly the most complex of these

TABLE 10: *Animator* data sources and data source variables.

Source Object	Identifier	Variables
Geometric shapes such as circles, rectangles, arrows, etc.	The id is returned when an object is created using an add method. id = addObject("circle","r = 10")	$t, x, y, vx, vy, ax, ay, m, fx, fy$
clock	id = getClockID()	t
ensemble The collection of all shapes.	id = getEnsembleID()	$t, xcm, ycm, px, py, m, ke$

methods, since the first parameter can take on 15 different values and these can be combined with a flexible attribute list as the second parameter. These combinations are described in Table 8. This section discusses those methods that need further explanation.

14.3.1 Loading Images

Animator is able to load gif images from either of two directories: (1) the codebase where the java files are stored, or (2) the document base where the html file is located using the methods

```
document.animator.addImageFromCodeBase(String file, String xStr, String yStr);
```

and

```
document.animator.addImageFromDocumentBase(String file, String xStr, String yStr);
```

respectively. The two string parameters, xStr and yStr, specify the trajectory of the image. The document base is the directory containing the html file and the codebase is the directory containing the html file. Either of these methods will work if the necessary resources (image, Java archives, and html file) are located on the same Web server. However, there is a bug in Internet Explorer—at least we think it is a bug—that prevents images from loading from the document base if the html file is loading from a local hard drive or CD. This operation generates a security exception error. Luckily, loading an image from the codebase directory works in all instances, and it is even possible to specify a relative path into a subdirectory.

We recommend that you use the new addObject method with an "image" parameter string. This method searches for image files in both the codebase and the document base. Use the codebase for consistency. If you have many images, you may want to consider creating a special images subdirectory of the codebase. For example, the following fragment loads the pencil.gif image from the images subdirectory of the codebase.

```
document.animator.addObject("image", "file = images/pencil.gif");.
```

14.3.2 Sticky Objects

It is sometimes useful to stop an animation when two objects come into contact, because the interesting physics is at an end, or we want the students to answer a question at that point, or there is a mathematical singularity that causes the equations of motion to become unstable. Since we may not know exactly when this condition will occur, *Animator* has a simple collision detection mechanism that is invoked at the end of every time step. This mechanism depends on an object's *sticky* property, which is set true using the setSticky method.

```
document.animator.setSticky(id,true);
```

A sticky object will stop the clock upon collision with another sticky object. Both objects must be sticky, and one of the objects must be moving under the action of forces. Objects that have trajectories will not stick since we assume that the script author wanted the object to follow the specified path even if it passes under or over other objects. Sticky objects must not be too thin in order for the sticky mechanism to function. Since *Animator* tests for a sticky collision at the end of every time step, *dt*, it is possible that an object can be on one side of a sticky object at the beginning of the time step and on the other side at the end of the time step.

Dragging a sticky object into another sticky object will also stop a simulation, but the script author should be careful since dragging with the mouse is not a continuous process. In order to conserve computational resources, the operating system does not track the mouse. Rather, it sends occasional mouse-move events to the applet whenever it has time. If the mouse is moving slowly, every new pixel position generates a new event and the applet responds accordingly. However, if the object is dragged quickly, position data may be generated only once for every 40 or 50 pixels of mouse motion and the sticky object may not be encountered.[1]

14.3.3 *Z*-Order

The order in which objects are drawn on the screen is known as the *z*-order. Normally, the *z*-order is determined by the order in which objects are created. Objects that are created later in a script will therefore appear to pass over the top of objects that were created earlier. This convention is easy to remember and works well except for connector objects. Connector lines and connector springs require two object identifiers when they are created using the following methods:

```
id3 = document.animator.addConnectorLine(id1,id2);
id4 = document.animator.addConnectorSpring(id1,id2);
```

Connectors will be drawn on top of the objects being connected since they were created last. But it is easy to change the drawing sequence by invoking the swapZOrder method.

```
document.animator.swapZOrder(id1,id3);
```

---

[1]A possible solution is to create a second sticky object that is much larger than the first and set its visibility property to false.

`public boolean` **`setAnimationSlave`**`(int masterID, int slaveID)`

Force an object to follow another object on the screen.

Parameters:

`masterID` - The id of the master object.

`slaveID` - The id of the slave object.

Returns:

true if successful.

`public boolean` **`setDisplayOffset`**`(int id, int xOff, int yOff)`

Offset the object's position on the screen from its default drawing position.

Parameters:

`id` - the id of the object.

`xoff` - the x offset.

`yoff` - the y offset.

Returns:

true if successful.

FIGURE 57: Animation slave method and display offset method details.

If the object with id1 was created before the object with id2, the connector line will now be drawn before either object.

14.3.4 Building Compound Objects with Animation Slaves

Animator, *EField*, and other Physlets provide the ability to link objects together through the setAnimationSlave method. This method is usually used in conjunction with the setDisplayOffset method to adjust the on-screen position. (See the javadoc ouput in Figure 57.)

For example, a simple cart with two black wheels can be constructed by slaving two circles to a rectangle.

```
id1 = document.animator.addObject("rectangle", "h = 30,w = 110");
document.animator.setRGB(id1,255,0,0);
id2 = document.animator.addObject("circle","r = 7");
document.animator.setDisplayOffset(id2,-50,-15);
document.animator.setAnimationSlave(id1,id2);
id3 = document.animator.addObject("circle","r = 7");
document.animator.setDisplayOffset(id3,+50,-15);
document.animator.setAnimationSlave(id1,id3);
```

This cart can now be animated by scripting only the rectangle using trajectories, forces, and interactions. An important point to note is that the inertial mass of the combined object is the mass of the rectangle only. Animation slaves do not contribute any physical properties to the master object and are purely a drawing convenience.

Both test objects and arrows take on their master's properties when they become slaved. This is convenient for showing the values of dynamic properties as objects move. Figure 58 shows how the previous script can be extended with a slaved text object that displays the kinetic energy and a slaved arrow that displays the net force using the following JavaScript statements.

```
id4 = document.animator.addObject("arrow", "h = m*ax, v = m*ay");
document.animator.setDisplayOffset(id4,+50,-20);
document.animator.setAnimationSlave(id1, id4);
id5 = document.animator.addObject("text", "text = KE, calc = m*vx*vx/2");
document.animator.setDisplayOffset(id5,0,+25);
document.animator.setAnimationSlave(id1,id5);
```

Notice that the height and the width of the arrow, the *h* and *w* attributes in the first addObject method, are function strings. Any of the dynamic variables of the master object can be used when defining a slaved arrow's components.

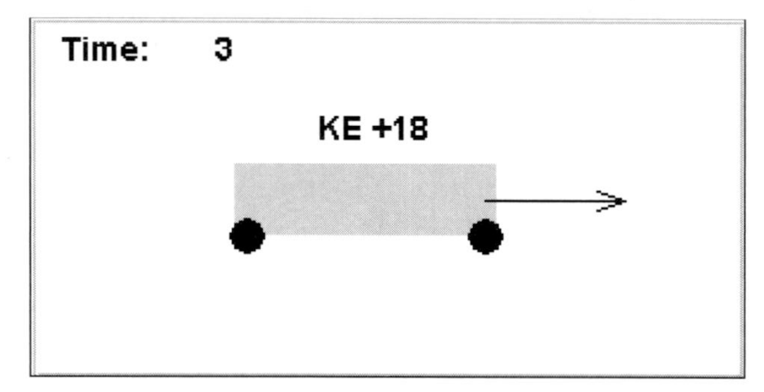

FIGURE 58: A cart created with animation slaves.

CHAPTER 15

EFIELD

Although *Animator* and *EField* have many superficial similarities, their design philosophies are different. *EField* is based almost entirely on potential energy, whereas *Animator* is based on forces. *EField* version 1 was conceived to show the relationship between a scalar field and its gradient. In electrostatics, this is the relationship between a potential and the associated electric field. The potential that is shown on the screen can be specified by both an external function and embedded point charges. Test charges, that is, charges that move under the influence of the field but are too small to affect the field, were added in version 2 of *EField*.

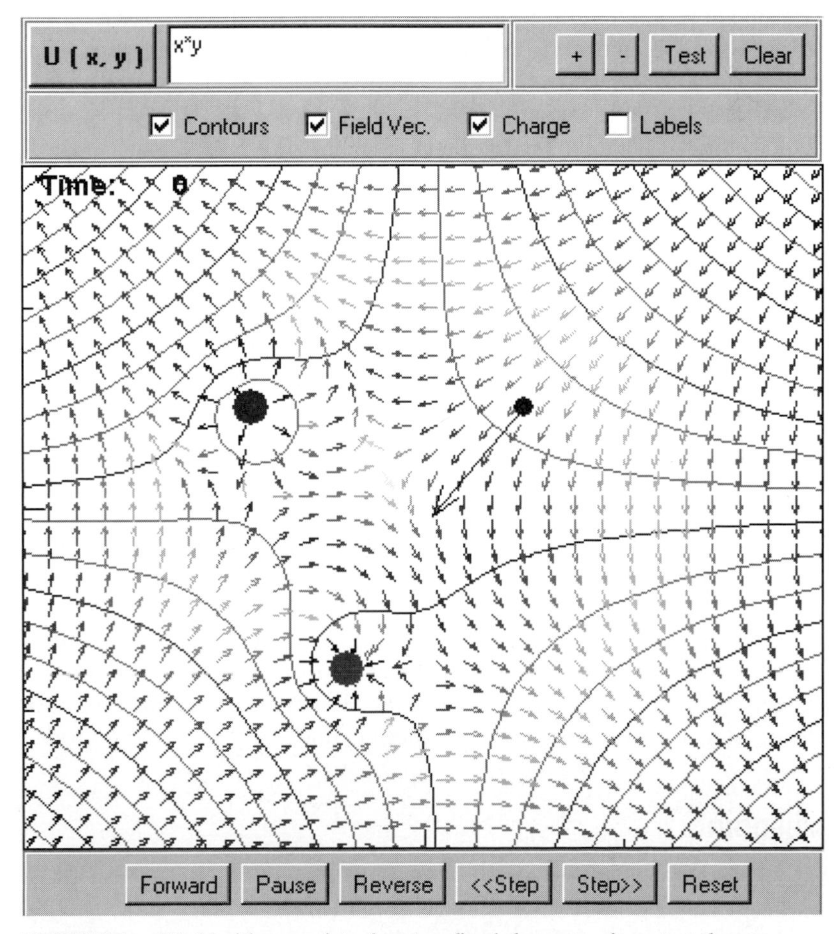

FIGURE 59: *EField* with a user interface, two fixed charges, and one test charge.

Since much of the computational machinery necessary to move charges and draw objects was in place in the basic applet, it was straightforward to extend *EField* to include other topics in a typical electricity and magnetism course. Cylindrical as well as spherical charges, electric flux calculations, and magnetic fields (albeit only in the z-direction) are now supported. Nevertheless, at its core, this Physlet does its calculations using the potential energy and its gradient. Interested readers may wish to investigate the *Poisson* Physlet (Section 18.9) for another approach to presenting electric field and potential problems to students.

A potential function can be defined in *EField* in one of two ways. The most direct way is to call the setPotential method.

```
potStr = "1/(x*x+y*y)";
document.efield.setPotential(potStr,xmin,xmax, ymin,ymax);
```

The potential string can be any analytic function of x and y. The horizontal range will be set to xmin and xmax, but the vertical range will vary depending on the embedded size since the applet enforces a 1:1 aspect ratio for the two axes.

Potentials may also be defined by using the addObject method to add charge.

```
param = "q = 1";
document.efield.addObject("charge", param);
```

Both of the preceding examples will produce the same potential. The first uses a parser to evaluate the string, while the second uses a preprogrammed Coulomb formula. Either method will then cause potential to be evaluated on a grid in the (x, y) plane. This grid is then used to generate the contour and field plots.

Notice the use of so-called dimensionless units for charge. The Coulomb constant has, in effect, been set equal to one. Because of the strength of the electrostatic interaction—two unit charges interacting with a 1-meter separation does, after all, produce quite a large force in MKS units—it is often desirable to scale problems so that results are of the order of unity. When authoring problems in MKS units, we often phrase the text as if distance were in μm and the charge in mC.

Also note that the field from a charge can be changed from a $1/r^2$ point charge dependence to a $1/r$ cylindrical charge dependence if the PointChargeMode embedding parameter is set to false or if the setPointChargeMode method is called. Cylindrical mode is, in fact, a very natural interpretation for a simulation since a circle on a two-dimensional computer screen naturally represents the cross section of an object that extends in the z-direction. However, since most textbook problems are stated in terms of point charges and not cylinders, point charge mode is the default.

15.1 EMBEDDING

EField is embedded into an html page as follows. Notice the pairing of version 4 of this applet with version 4 of the Science Tools library.

```
<applet codebase = [location] code = "eField4.EField.class"
archive = "EField4_.jar,STools4.jar">
```

The tags shown in Table 11 may be included when *EField* is embedded, and usually are. Embedding tags are described in Section 4.2.2. Since *EField* was originally designed as a stand-alone applet without JavaScript support, it has more embedding parameters than most Physlets. These parameters may be set using parameter tags. Many of these parameters can be ignored if the applet is scripted, since they are overridden by the setDefault method, which usually appears at the beginning of a script.

The <applet> tag must, of course, be closed by an </applet> tag after the embedding parameter tags have been coded.

15.2 DATA SOURCES

Various objects in *EField* have the ability to generate data. Table 12 lists the objects that have this ability, how the object's identifier is obtained, and the state variables that these objects provide via a data connection.

TABLE 11: *EField* embedding parameters.

Parameter	Value	Data Type	Description
FPS	10	double	Frames per second during animation.
Dt	0.1	double	Animation time step per frame.
ShowControls	true	boolean	Show VCR buttons at bottom of applet.
ShowContours	true	boolean	Show equipotential contours.
ShowPoles	true	boolean	Show fixed charges.
ShowLabels	true	boolean	Add labels to contour lines.
ShowFieldLines	false	boolean	Draw field lines. Computationally EXPENSIVE.
ShowFieldVectors	true	boolean	Draw direction arrows to represent field.
PointChargeMode	true	boolean	Select between point charge and line charge mode. Use $1/(r*r)$ dependence for fixed charge if true. Use $1/r$ dependence if false.
Potential	0	string	The potential function, $U(x, y)$.
Range	"-1,1, -1,1"	string	Approximate x- and y-coordinate range. X-range takes precedence to ensure $1:1$ aspect ratio. MUST be a string: "-1,1,-1,1"
GridSize	64	int	Sets the size of the n-by-n grid for the evaluation of the potential function.

15.3 METHODS

EField and *Animator* define similar methods including most of the recommended methods and attributes listed in Tables 7 and 8. Since the main difference between these two applets is *EField*'s use of a grid and potential functions, this section focuses on these aspects.

15.3.1 Grid Size

The GridSize embedding parameter determines the size of the n-by-n array, (i.e., the grid) that is used to generate contour and field plots. The contour algorithm uses the potential at four neighboring grid points to see if any contour lines pass through this rectangular cell. If so, a linear extrapolation is performed to draw straight-line segments between the edges where the contours enter and leave the cell. Too coarse a grid will produce a very jagged contour and an unreliable plot. The default grid size of 64 usually produces smooth and accurate contours, although values as low as 32 can be used to increase computational speed.

The grid size also determines the number of vectors that will be drawn. Field vectors are calculated by taking a second-order numerical derivative of the potential at the grid points. However, since drawing a vector at every grid point often produces too many vectors, we have defined a parameter to lower the drawing density.

```
resolution = 2;
document.efield.setFieldResolution(resolution);
document.efield.showFieldVectors(true);
```

The default value for the resolution is two. It works well with 64 grid points and an applet embedding size of 300 × 300.

Field vectors are of uniform length and are approximately equal to the spacing between grid points. They are color-coded from blue to red to black in order to provide a visual clue as to the field's magnitude. This technique allows us to represent four orders of magnitude in a compact format. Numeric values of the electric field are most easily obtained using the yellow message box that appears during a mouse click-drag after the setShowEOnDrag method has been invoked.

```
document.efield.showEOnDrag(true);
```

TABLE 12: *EField* data sources and data source variables.

Source Object	Identifier	Variables
charge	The id is returned when an object is created using an add method.	$t, x, y, vx, vy, ax, ay, fx, fy, p, m, q$ *Note*: p is the potential and q is the magnitude of the charge.
shapes: circle, rectangle, box, etc.	The id is returned when an object is created using an add method. addObject("circle","r = 5");	$t, x, y, vx, vy, ax, ay, f$ *Note*: f is the flux passing out of a cylindrical surface.
clock	id = getClockID()	t

15.3.2 Set Method

The number of display options in *EField* is quite large, and so, rather than defining dozens of set methods, we have implemented many scaling options using a *set* method that always has two parameter strings.

```
document.efield.set(int, string, string);
```

This scriptable method is similar to the addObject method. The first parameter specifies the object, the second defines the name of the property to be set, and the last defines a list of attributes. For example, the x-scale can be set using the following statement:

```
document.efield.set(0,"scale","xmin = -5, xmax = 5");
```

The y-scale and other attributes are set similarly as shown in Table 13.

In order to make scripting easier, applets that implement the set method interpret an id of zero to be a default object. This default object is usually the on-screen graphic element in which the animation takes place. In the case of *EField*, it is the contour plot. In the *DataGraph* Physlet it is the graph, and in the *DataTable* it is the table.

Contour attributes in *EField* are also controlled by the set parameter. By default, contour levels autoscale their range to the maximum and minimum values on the grid. Special assumptions are, however, made for point charges since the range is infinite. You can override these settings by using the zmin, zmax, and zlevels attributes. For example, the following script sets the contour plot to have 11 levels, starting at -1 and ending at $+1$.

```
document.efield.set(0,"scale","zmin = -1,zmax = 1,zlevels = 11");
```

TABLE 13: Parameters and attributes for the *EField* set method.

Name	Attributes
potential	**v-string** the potential energy function, $v(x, y)$.
scale Set the scale of the x- and y-axis and the contour levels.	**xmin-double** minimum value on the x-axis. **xmax-double** maximum value on the x-axis. **ymin-double** minimum value on the y-axis. **ymax-double** maximum value on the y-axis. **zmin-double** minimum contour level. **zmax-double** maximum contour level. **zlevels-int** the number of contour levels. **delta-double** the spacing between levels if the maximum z is not set.
style	**labels-boolean** show labels on selected contours. **vectorSize-int** the size of the field vectors.

It is also possible, and often easier, to specify the first contour level and the contour spacing.

```
document.efield.set(0,"scale","zmin = -1,delta = 1,zlevels = 11");
```

If point charges are used in the simulation, the script author should check the resulting plot to ensure that contours are being drawn close to the on-screen representation of the charge. Otherwise, the absence of contour lines near a charge may suggest that the potential is uniform when, in fact, it is rapidly approaching infinity.

15.3.3 Field Lines

Generating textbooklike field-line plots on a two-dimensional page is difficult and may be impossible.[1] Textbook figures are usually produced by artists, not programs. Nevertheless, *EField* attempts to produce such a figure if the setShowFieldLines method is invoked.

```
document.efield.setShowFieldLines(true);
```

Caveat emptor! The field-line algorithm is slow and may fail. The algorithm first determines the dominant charge type inside the applet. Assume, for the sake of argument, that the dominant type is positive. The algorithm attempts to calculate six field lines per unit charge beginning at each positive charge. These field lines should either extend to infinity or end on a negative charge. But some field lines may wander into a region where the electric field is zero—where the calculation will fail—while others may be too long to make calculation practical. Field lines often bunch if more than two charges are present, thereby giving misleading information if one assumes that the field-line density is proportional to the field strength.

The calculation is even more difficult if an arbitrary potential is defined using the setPotential method. Any potential that has nonzero divergence is difficult to calculate, since there is no good way to determine on which charge the field line originates or ends. Even nondiverging fields are problematic since ad hoc decisions about where to start a field-line calculation are necessary.

For both numerical and pedagogic reasons, it is therefore desirable to let students start their own field-line calculation using the setShowFieldLineOnDoubleClick method.

```
document.efield.setShowFieldLineOnDoubleClick(true);
```

Students can now double-click anywhere within the applet to start a field-line calculation. This will reinforce the idea that field-line plots are an artificial construction and that field lines do not, in fact, exist. Field lines can be drawn an infinite number of ways. A conventional drawing that shows a straight field line connecting two charges in a dipole, for example, is just one of may possible drawings.

---

[1]See for example the article "Electric field line diagrams don't work" by Alan Wolf, *Am. J. Phys.* Vol. 64, pp. 714–724 (1996).

CHAPTER 16

DATAGRAPH

DataGraph plots both data sets and functions. A data set is part of a Java object known as a data *series* that contains the coordinates, (x, y), together with drawing attributes such as color and style. Each series is identified by a series number, usually starting at one for the first series. Whenever data are added, the appropriate series number must be specified. Note that a series number is not the object identifier that is used in data connections. Object identifiers are assigned by the operating system whereas series numbers are assigned by the script author. The object identifier for a series can be obtained by calling the method

```
id = document.datagraph.getSeriesID(int);
```

if it is required.

16.1 EMBEDDING

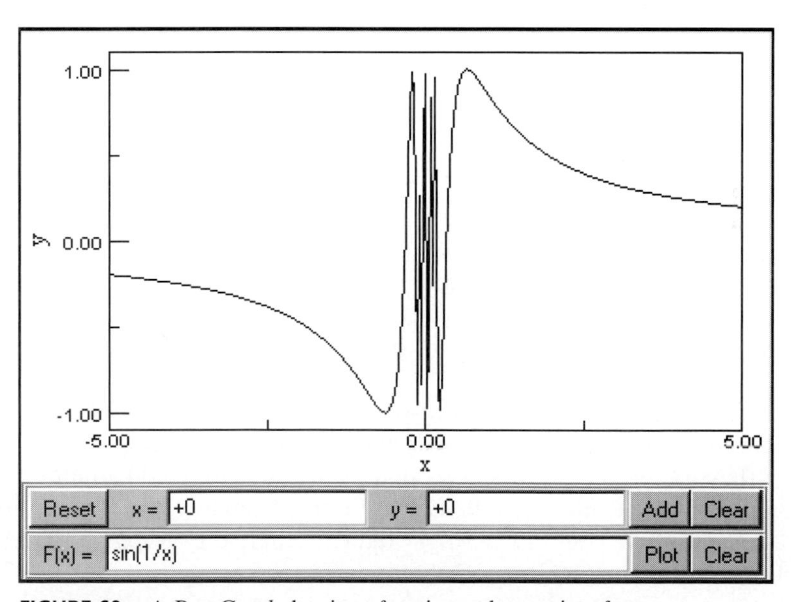

FIGURE 60: A *DataGraph* showing a function and a user interface.

TABLE 14: *DataGraph* embedding parameters.

Parameter	Value	Type	Description
Function	null	string	An analytic function, $f(x)$.
XMin	−1.0	double	Minimum value along x-axis.
XMax	1.0	double	Maximum value along x-axis.
YMin	−1.0	double	Minimum value along y axis.
YMax	1.0	double	Maximum value along y-axis.
AutoScaleX	true	boolean	Autoscale the x-axis.
AutoScaleY	true	boolean	Autoscale the y-axis.
DataFile	null	string	Data file to read when the applet is initialized.
ShowControls	true	boolean	Show the user interface.

DataGraph is embedded into an html page using the standard html tags.

```
<applet codebase = [location] code = "dataGraph.DataGraph.class"
archive = "DataGraph4_.jar, STools4.jar " name = [name] id = [name]>
```

Other tags may be included when *DataGraph* is embedded, and usually are. Embedding tags are described in Section 4.2.2. The parameter tags listed in Table 14 set the initial scale of the graph and specify an initial plotting function as shown in Figure 60.

The <applet> tag must, of course, be closed by an </applet> tag after the embedding parameter tags have been coded.

16.2 DATA SOURCES AND LISTENERS

Although *DataGraph* is used most often as a data listener, it also contains data sources as shown in Table 15. For example, the clock data source is available since this applet inherits from the SApplet superclass. The clock is usually used to generate data at a uniform rate that is then added into a *DataGraph* series, as demonstrated in the *DataGraph* section of the tutorial, Section 6.4.

Functions are data sources that send an array of data as well as the data's numeric derivatives to its data listeners. The length of this array is roughly equal to the applet's embedding width so that the function generates one datum per screen pixel.

Each *DataGraph* series is also a data source. This allows other data listeners to obtain the entire data set whenever data is added to a series.

Finally, the geometric shapes that were designed to add annotation or emphasis to a graph are data sources that deliver their on-screen location, x and y, to data listeners.

DataGraph contains a single data listener, the graph itself. The listener id is obtained using the getGraphID method.

```
id = document.datagraph.getGraphID();
```

TABLE 15: *DataGraph* data sources and data source variables.

SourceObject	Identifier	Variables
clock	id = getClockID()	t

function	The id is returned when a function is created using an add method.
	
	id = addObject("function", "var = x f = sin(x), xmin = 0, xmax = 20");
	
	id = addFunction("x", "sin(x)");
		x, y, v, a
		
		y is the value of the function at x.
		
		v and a are the first and second derivates of the function at x.

regression	id = getRegressionID(sid, start, end);	m, dm, b, db m is the slope and dm is the uncertainty. b is the y intercept and db is the uncertainty.
series	id = getSeriesID(sid);	x, y, dx, dy dx and dy are the differences between neighboring data points, $\Delta x, \Delta y$
shapes: circle, rectangle, box, etc.	The id is returned when an object is created using an add method. id = addObject("circle", "r = 5");	x, y

16.3 METHODS

DataGraph supports both the addObject and set methods. The addObject method creates geometric shapes using the same syntax as *EField* and *Animator*. The following shapes are supported: arrow, box, circle, caption, cursor, function, rectangle, shell, and text.

```
id = document.datagraph.addObject("rectangle","x = 1,w = 10,h = 30");
```

16.3.1 Set Method

The most convenient and flexible method for adjusting the appearance of a *DataGraph* is the set method, with the following signature:

```
set(int id , String name, String attributes);
```

The first parameter specifies the object identifier, the second defines the type of property to be set, and the last defines a list of attributes. For example, the x-scale can be set using the following statement:

```
document.datagraph.set(0,"scale","xmin = -5,xmax = 5");
```

The y-scale can be set similarly or both scales can be set at the same time.

```
document.datagraph.set(0,"scale","xmin = -5,xmax = 5, ymin = -10,ymax = 10");
```

As in *EField*, zero is used to signal that we wish to set the attributes of the default object, the graph itself.

In order to be consistent for all objects, it is necessary to use the object identifier rather than the series number when changing a series style.

```
id = document.datagraph.getSeriesID(1);
document.datagraph.set(id,"style","connected = true,markerstyle = 1");
```

Although the old setSeriesStyle method can still be used, it has been deprecated.

```
document.datagraph.setSeriesStyle(1,true,1);
```

The setSeriesStyle method may appear to be easier to use, but it is not consistent, and the resulting script is not as easy to read. Other style attributes, such as the marker size, would require additional methods, whereas the set method can easily incorporate them by appending new attributes to the last argument.

16.3.2 Data Files

Data files can be read into a *DataGraph* either upon initialization using the DataFile embedding parameter or by using the loadDataFile method.

```
document.datagraph.loadDataFile("test.txt");
```

TABLE 16: Parameters and attributes for the *DataGraph* set method.

Name	Attributes
scale Set the scale of the x- and y-axis.	**autoscaleX-boolean** autoscale the x-axis. **autoscaleY-boolean** autoscale the y-axis. **xmin-double** minimum value on the x-axis. **xmax-double** maximum value on the x-axis **ymin-double** minimum value on the y-axis **ymax-double** maximum value on the y-axis
style	**connected-boolean** connect the points **markerstyle-int** the marker style **markersize-int** the marker style

The path to the text file must be relative to the embedding html document. Data files are simple comma-delimited ASCII text files that can include comments and the series identifier. A sample data file with two series and one comment would contain the text shown in Script 20.

```
# Loading series data into DataGraph
series 2
1 2
2 4
3 8.2

series 3
2 3
3 5
4 9.2
```

Script 20: Data file format.

Any data existing in a series are cleared during a load operation. You can, of course, use script to set the series style before or after the series is loaded. *DataTable* supports a similar load operation although only a single numeric value is specified on each line.

16.3.3 Linear Regression

Curve fitting is one of the most common data analysis techniques. Currently, *DataGraph* supports only linear regression, but we intend to add other analysis options in the future. The two regression coefficients, slope and intercept, and their uncertainties are obtained using the getRegressionID method. This method returns an object identifier for a data source.

```
start = 1;
end = 10;
sid = document.datagraph.getRegressionID(1,start,end);
```

The first parameter is the series number, while the second and third integer parameters define the subset of data to be analyzed. If the end parameter is less than or equal to the start parameter, the regression analysis will use all data in the data set. The source id can be used to display the regression coefficients by establishing a data connection to a table. The regression parameters can also be read directly as a string as described in Section 17.3.4.

Finally, it is possible to plot a regression line without bothering to obtain the regression coefficients using the plotRegression method.

```
start = 1;
end = 10;
ans = document.datagraph.plotRegression(1,start,end);
```

Figure 61 shows how a *DataGraph* can be combined with a *DataTable* to provide a simple numerical methods application. The complete html page, regression.html, can be found in the Chapter 16 directory on the CD.

16.3.4 Special Situations

DataGraph begins to process a curve the moment a data source sends new values. There are, however, situations where it may be desirable to suppress this processing. For example, the animation clock will usually send a datum at every clock tick even if the data source has not changed its state. This repeated data can produce very large data sets and may skew a regression analysis. *DataGraph* can be set to ignore repeated data using the setAddRepeatedDatum method.

```
document.datagraph.setAddRepeatedDatum(1,false);
```

The default value for this property is true.

Some data sources, such as the transverse wave simulation, *Pipes*, send an array of data at every time step. Since *Pipes* is sending an entirely new complete data set, it

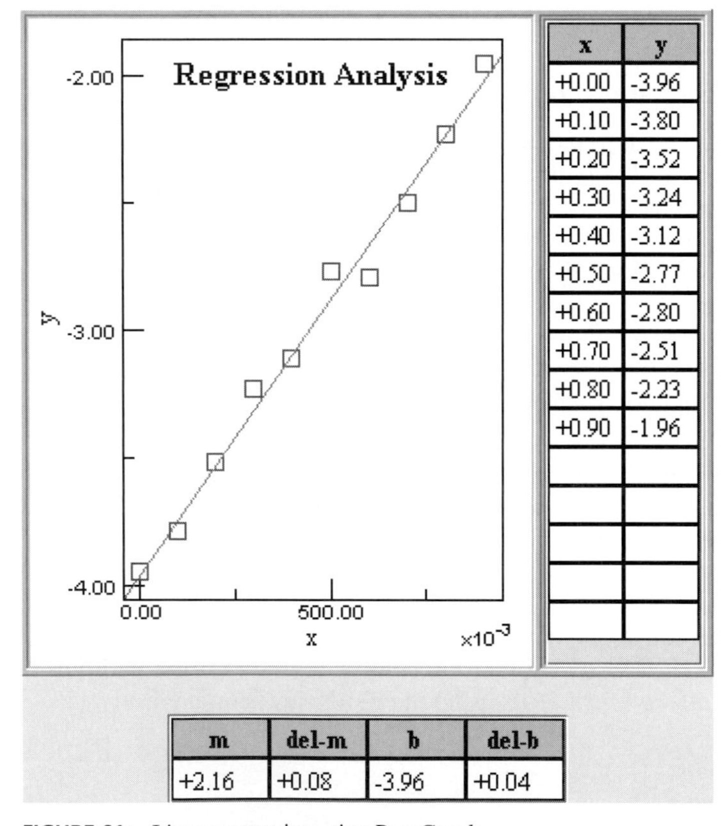

x	y
+0.00	-3.96
+0.10	-3.80
+0.20	-3.52
+0.30	-3.24
+0.40	-3.12
+0.50	-2.77
+0.60	-2.80
+0.70	-2.51
+0.80	-2.23
+0.90	-1.96

m	del-m	b	del-b
+2.16	+0.08	-3.96	+0.04

FIGURE 61: Linear regression using *DataGraph*.

clears the old data set first. However, clearing data forces *DataGraph* to do a screen redraw even though new data will shortly become available. This unnecessary drawing causes the screen momentarily to display an empty curve. This screen flash can be disabled by calling the setAutoReplaceData method.

```
document.datagraph.setAutoReplaceData(1,true);
```

This method disables the clearSeries method and forces a single screen redraw when a data connection sends array data.

C H A P T E R 1 7

DATATABLE

DataTable is designed to be used as a data listener in order to show a row-column table of numbers. Just as in *DataGraph*, each column is a series with a unique style. Whenever data are added, the appropriate series number must be specified.

17.1 EMBEDDING

DataTable, shown in Figure 62, is embedded into an html page using the following tags.

```
<applet codebase = [location] code = "dataTable.DataTable.class"
archive = "DataTable4_.jar, STools4.jar" name = [name] id = [name]>
```

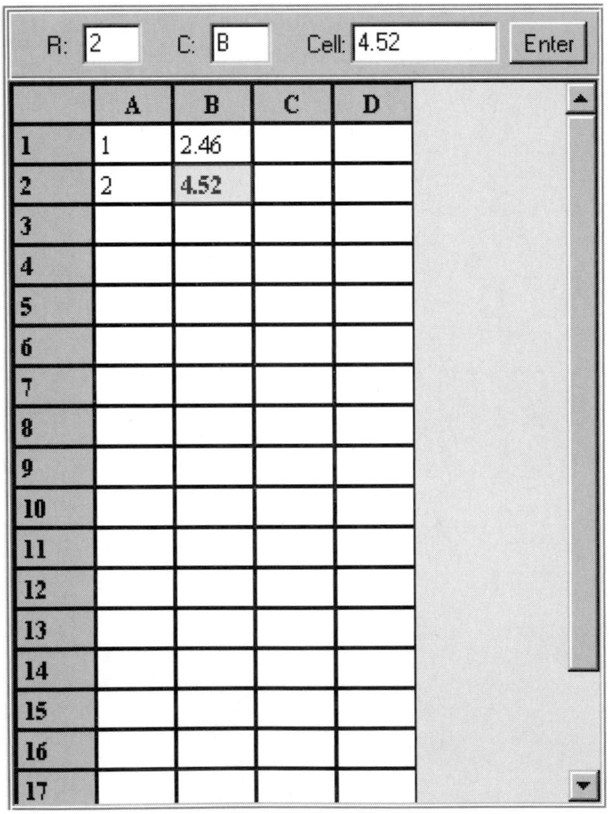

R: 2 C: B Cell: 4.52 Enter

	A	B	C	D
1	1	2.46		
2	2	4.52		
3				
4				
5				
6				
7				
8				
9				
10				
11				
12				
13				
14				
15				
16				
17				

FIGURE 62: A *DataTable* with the user interface enabled.

TABLE 17: *DataTable* embedding parameters.

Parameter	Value	Type	Description
NumRows	10	int	The number of rows.
NumCols	2	int	The number of columns.
CellWidth	40	int	The size of a cell in pixels.
DataFile	null	string	Data file to read when the applet is initialized.
ShowRowHeader	true	booelan	Show the top header row.
ShowRowHeader	true	boolean	Show the left most header column.
LastOnTop	false	boolean	Add data to the top of the table rather than the bottom.
ShowScrollBars	true	boolean	Create a scrollable table. Useful for large data sets.
SizeToFit	false	boolean	Make the data table fit the embedding size.
ShowControls	true	boolean	Show the user interface.

The embedding tags shown in Table 17 may be included when *DataTable* is embedded, and usually are. Embedding tags are described in Section 4.2.2.

The ShowScrollBars parameter allows the display of large data sets. It is rarely used since good Physlet problems focus on interaction with the animation and not on the presentation of large quantities of numeric data. Our advice is to keep the number of rows and columns small. Specify appropriate height and width attributes and set the SizeToFit attribute to true. *DataTable* will then adjust the cell size to fill the embedding area. The following html markup is typical.

```
<applet codebase = [location] code = "dataTable.DataTable.class"
archive = "DataTable4_.jar, STools4.jar"
name = "datatable" id = "datatable">
  <param name = "ShowScrollBars" value = "true">
  <param name = "LastOnTop" value = "false">
  <param name = "SizeToFit" value = "true">
  <param name = "CellWidth" value = "40">
  <param name = "NumRows" value = "1">
  <param name = "NumCols" value = "3">
  <param name = "ShowRowHeader" value = "true">
  <param name = "ShowColHeader" value = "true">
  <param name = "ShowControls" value = "false">
</applet>
```

Headers are not included in the row or column count since their visibility is controlled by separate parameters.

17.2 DATA SOURCES AND LISTENERS

Although *DataTable* is used most often as a data listener, Table 17 shows that it contains data sources similar to those in *DataGraph*.

DataTable contains a single data listener, the table itself. The listener id is obtained using the getTableID method.

```
id = document.datatable.getTableID();
```

TABLE 18: *DataTable* data sources and data source variables.

Source Object	Identifier	Variables
clock	id = getClockID()	t
series	id = getSeriesID(1)	x, dx *Note*: dx is the difference between x-values.

17.3 METHODS

17.3.1 Display Format

The display format for a column of numbers can be changed using either the setNumericFormat or the setFormat methods. The setFormat method is the recommended method in most Physlets, but the setNumericFormat is usually the easier method to use since it takes a series number as the first argument rather than the object identifier. For example, the statement

```
document.datatable.setNumericFormat(1, "%-+6.2f");
```

sets the numeric output for the first column to use a floating point decimal representation with six characters, two of which are to the right of the decimal point.

The format string follows UNIX printf conventions. That is, the string has a prefix, a format code, and a suffix with the following structure:

- % (required)
- **modifiers** (optional)
  - **+** forces display of + for positive numbers
  - **0** show leading zeroes
  - **–** align left in the field
  - **space** prepend a space in front of positive numbers
- **an integer** denoting field width (optional)
- **a period** followed by an integer denoting precision (optional)
- **a format descriptor** (required)
  - **f** floating point number in fixed format
  - **e or E** floating point number in exponential notation (scientific format). The E format results in an uppercase E for the exponent (1.14130E+003), the e format in a lowercase e.
  - **g or G** floating point number in general format (fixed format for small numbers, exponential format for large numbers). Trailing zeroes are suppressed. The G format results in an uppercase E for the exponent (if any), the g format in a lowercase e.
- **d or i** integer in decimal

17.3.2 Adding Data

Each column in *DataTable* is a series with its own display attributes. Data can be added to a series from JavaScript using the addDatum method with the following signature.

```
addDatum(int id, double x)
```

The first parameter is the series number, while the second is the datum. The data will be displayed on either the top or the bottom of the column depending on the value of the LastOnTop embedding parameter. If the LastOnTop parameter is set to true, the most recent value will always be displayed, since new values appear at the top and push old values off the bottom of the table. However, if LastOnTop is set to false, values that are added after the table is full will be lost.

Although only a single value is needed to specify a *DataTable* datum, the makeDataConnection method still requires string arguments for both an ordinate and an abscissa. This is because methods that are used to handle data connections are defined in the superclass, and these methods need to remain consistent across all Physlets.[1] Since the second string function is not used, it should be set to zero when connecting to a *DataTable*.

17.3.3 Active Cells

A single cell, a row, or a column can be highlighted in order to draw attention to its datum. Highlighting can be done by the user with a mouse click if the setShowActiveCell method has been called.

```
document.datatable.setShowActiveCell(true);
```

Alternately, the setActiveCell method can be called from JavaScript with row and column parameters.

```
row = 2;
col = 3;
document.datatable.setActiveCell(row,col);
```

Whenever a single cell is highlighted either by a mouse click or by script, it can be accessed using the setActiveCellValue and getActiveCellValue methods without knowing the exact row or column position.

---

[1]It is possible to define methods with the same name but different signatures in Java. However, many browsers are unable to choose the appropriate method when invoked from JavaScript. This bug may be due to the ambiguous nature of JavaScript data types since it is not possible to know if a variable in JavaScript is a number or a string.

17.3.4 Reading Any Data Source

The advanced JavaScript programmer can design custom data analysis routines by reading an array of data from a data source and processing these data with script. Physlets that inherit from the SApplet superclass implement the getSourceVariables and the getSourceData methods.

```
variables = document.datagraph.getSourceVariables(int id);
values = document.datagraph.getSourceData(int id, string var);
```

The first argument in either method is the source identifier. The second argument in getSourceData is a data source variable, such as x, y, fx, fy, or t.

Since it is not possible to return an array of numbers in a browser-independent manner, the getSourceData method returns a comma-delimited string of numbers. The result must be processed using the JavaScript *split* command. This JavaScript command breaks a string into an array of substrings using the argument to separate the values.

```
data = values.split(",");
```

The return value, data in this case, is an array of strings. The string at the ith position can now be accessed using JavaScript array notation, data[i].

The following JavaScript fragment shows how to read a column from a *DataTable*, break the result string into substrings, and calculate the average value.

```
values = document.datagraph.getSourceData(sid, "x");
data = values.split(",");
sum = 0;
num = data.length;
for(i = 0;i<num;i++)
   sum = sum +eval(data[i]);
avg = sum/num;
```

Notice the use of the JavaScript *eval* function to convert the substrings in the data array into numbers. The data array is an array of strings, and these strings would be concatenated without the conversion.

CHAPTER 18

VERSION 4 PHYSLETS

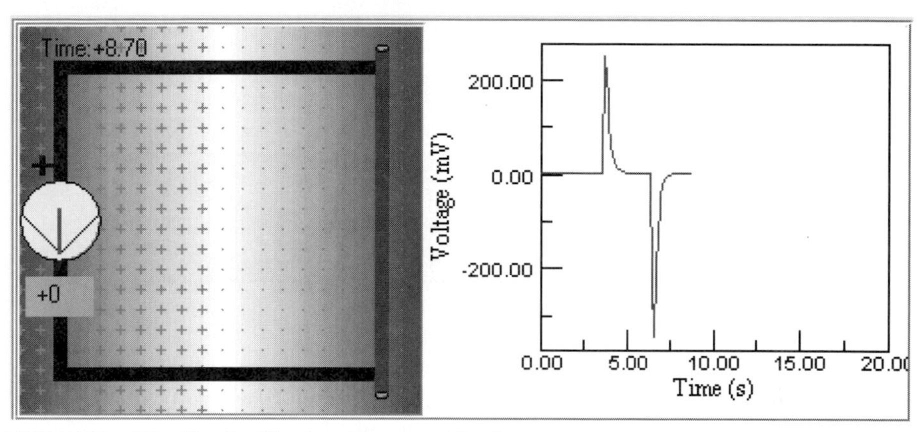

FIGURE 63: The *Faraday* Physlet with a dragable wire.

This chapter provides a brief overview of other Physlets that implement inter-applet communication and portions of the addObject syntax that was introduced in version 4 of the STools archive. Unlike Part Two of this book, these examples are not examples of pedagogy; they are proof-of-concept scripts that we have used to check the basic functionality of each Physlet. Copy these examples to your local hard drive and view their source. Then modify them to suit your local curricular requirements.

Script authors should be aware that not all methods listed in Table 7 are available and that there are minor inconsistencies among the method names and parameter signatures. These Physlets are, after all, upgrades of older versions. Consult the online documentation to determine if the method that you are using is supported by the Physlet.

As an example of what to watch out for, the *Faraday* Physlet, shown in Figure 63, allows the user to specify the magnetic field, $B(x,t)$, as a function of position and time and the wire position, $x(t)$, as a function of time.[1] Consequently, only variations in the x-coordinate are significant, and the setTrajectory method for the slideable wire in *Faraday* has a parameter signature that takes the object identifier and a string that specifies the wire's x-coordinate:

```
document.faraday.setTrajectory(int id,String xStr);
```

[1]It is, of course, not possible to let the user drag an object and to have the object follow an analytic trajectory. The dragable property should be set to false before a trajectory is set.

Animator and *EField*, on the other hand, require a signature that specifies both position coordinates as functions of time.

```
document.animator.setTrajectory(int id, String xStr, String yStr);
```

Finally, another deprecated method, setPosFunction(*String xStr*), that has the same purpose as setTrajectory is still available for backward compatibility.

18.1 *BAR*

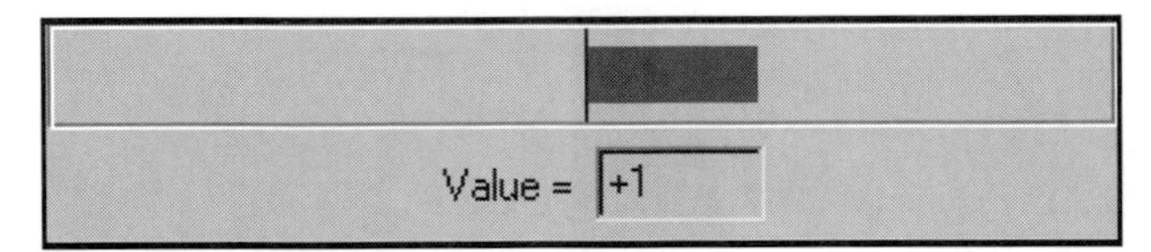

FIGURE 64: *Bar* Physlet.

Bar implements a simple two-color bar graph with an optional numeric output field located below the bar as shown in Figure 64.. The orientation can be either vertical or horizontal. This Physlet is usually used as a data listener in conjunction with other Physlets. It ignores the series id parameter in a data connection since it supports only a single bar.

Bar is embedded into an html page using the following tags[2]

```
<applet codebase = [location] code = "dataGraph.Bar.class"
archive = "Bar4_.jar, STools4.jar" name = [name]>
```

The embedding parameters shown in Table 19 can be specified as part of this applet tag.

TABLE 19: *Bar* embedding parameters.

Parameter	Value	Type	Description
Value	10	double	The initial value of the bar's height.
Min	0	double	The value at the minimum height.
Max	100	double	The value at the maximum height.
BarWidth	10	int	The width of the bar in pixels.
Label	null	string	A label to the left of the output field.
Vertical	true	boolean	Sets the orientation to vertical or horizontal.
AutoScale	false	boolean	Autoscale the max and min values.
ShowControls	true	boolean	Show the numeric output field.

[2]Note that although the *Bar* Physlet is in the *dataGraph* package, it is not in the *dataGraph* jar file. Both the DataGraph4_. jar file and the Bar4_.jar file should be listed in the archive tag if these Physlets are to be used on the same html page.

The <applet> tag must, of course, be closed by an </applet> tag after the embedding parameter tags have been coded.

18.2 *BFIELD*

BField allows users to simulate the magnetic field produced by long straight wires or by circular loops, as shown in Figure 65. In addition, an external field may be included by specifying B_x and B_y.

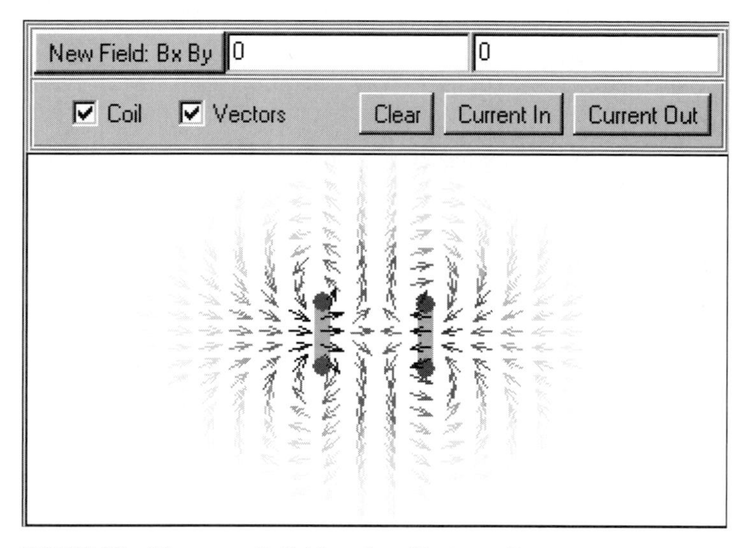

FIGURE 65: The magnetic field produced by two coils.

BField is embedded into an html page using the following tags:

```
<applet codebase = [location] code = "bfield.BField.class"
archive = "BField4_.jar, STools4.jar" name = [name]>
```

The embedding parameters shown in Table 20 can be specified as part of this applet tag.

TABLE 20: *BField* embedding parameters.

Parameter	Value	Type	Description
BxFunction	null	string	The x-component of the external field.
ByFunction	null	string	The y-component of the external field.
GridSize	24	int	The number of cells in the vector field plot.
Range	1,−1,1,1	string	The approximate range for the x- and y-axes.
ShowControls	true	boolean	Show the user interface.

The <applet> tag must, of course, be closed by an </applet> tag after the embedding parameter tags have been coded.

18.2.1 Data Sources

As can be seen in Table 21, data sources in *BField* are similar to *EField*. All on-screen objects supply their position coordinates, x and y. Wires, that is, long straight wires as well as coils, supply current, i, and field components, bx and by, in addition to their position.

TABLE 21: *BField* data sources and data source variables.

Source Object	Identifier	Variables
wire	The id is returned when an object is created using an add method.	t, x, y, bx, by, i *Note*: i is the current in the wire.
shapes: circle, rectangle, box, etc.	The id is returned when an object is created using an add method. `addObject("circle","r = 5");`	t, x, y
clock	id = getClockID()	t

18.2.2 addObject Method

BField allows the creation of geometric shapes, coils, and straight wires listed in Table 22 using the addObject method described in Section 13.2.

TABLE 22: *BField* addObject parameter names and properties.

Name	Attributes
arrow Arrows are often animation slaves of other objects. They can represent almost any vector since the *h* and *v* components can be functions of the variables.	**x-double** x position of the base in world units **y-double** y position of the base in world units **h-string** horizontal component as a function of t, x, y, vx, vy, ax, and ay **v-string** vertical components as a function of t, x, y, vx, vy, ax, and ay
box A box is a hollow rectangle.	**x-double** x position of the center in world units **y-double** y position of the center in world units **h-int** height in pixels **w-int** width in pixels **s-int** thickness of the box
caption A caption is text that is centered near the top of the screen.	**x-double** x position of the center in world units **y-double** y position of the center in world units **text-string** text of the caption

Name	Attributes
circle	**x-double** x position of the center in world units **y-double** y position of the center in world units **r-int** radius in pixels
coil	**x-double** x position of the center in world units **y-double** y position of the center in world units **i-double** the current in the wire **r-int** radius in pixels *See also* wire.
compass	**x-double** x position of the center in world units **y-double** y position of the center in world units **r-int** radius in pixels
cursor A circle with cross hairs.	**x-double** x position of the center in world units **y-double** y position of the center in world units **r-int** radius in pixels
field The vector-field plot.	**No parameters.** Include an empty string as a placeholder in the addObject method. Example: addObject("field","");
rectangle	**x-double** x position of the center in world units **y-double** y position of the center in world units **h-int** height in pixels **w-int** width in pixels
shell	**x-double** x position of the center in world units **y-double** y position of the center in world units **r-int** radius in pixels
text A fixed text string followed by an optional calculation.	**x-double** x position of the left side of the text in world units **y-double** y position of the top of the text in world units **text-string** static text **calc-string** An analytic function to be evaluated. The calculation is displayed to the right of the static text Text objects are often slaved to other objects. If a text object is slaved to an object, it takes on the properties of that object. In this Physlet, a slaved text object can evaluate a function of t, x, y, bx, by, i, curl, and path. *Note*: The curl of the field is evaluated at the position, (x, y). The path is the line integral evaluated along the perimeter of a shell or a box object. *Note*: Use the setFormat method described in Section 17.3.1 to change the decimal format of the displayed calculation.
wire A long straight wire.	**x-double** x position of the center in world units **y-double** y position of the center in world units **i-double** the current in the wire. Positive currents flow out of the screen. *See also* coil.

18.3 *CIRCUITS*

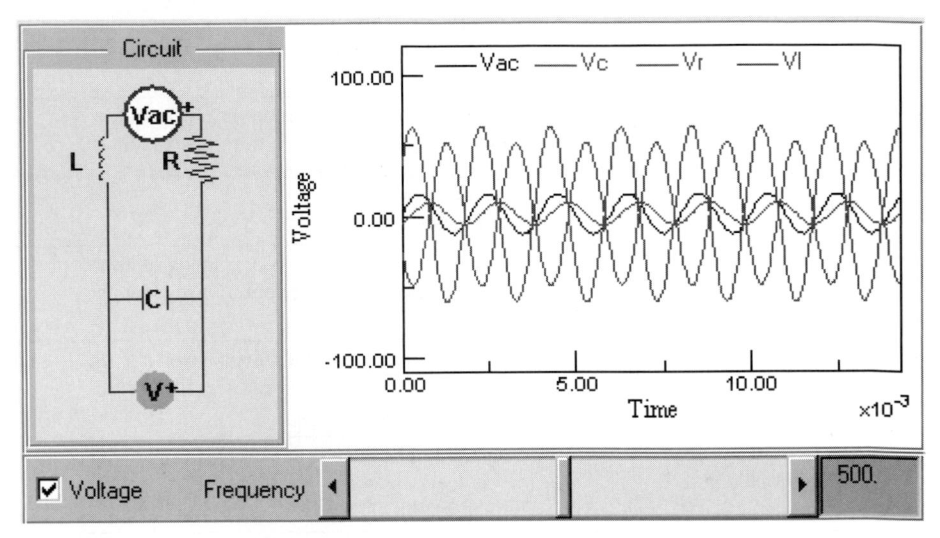

FIGURE 66: LRC circuit response close to resonance.

The *Circuits* package contains five Physlets designed to solve the common ac and dc circuit problems. In addition to the usual LRC simulation shown in Figure 66, there is an applet to plot nonlinear *I* versus *V* response and an applet to plot frequency response.

Although most circuit applets have a graph that is internal to the applet, as shown in Figure 67, it is also possible to use a data connection to present a custom graph.

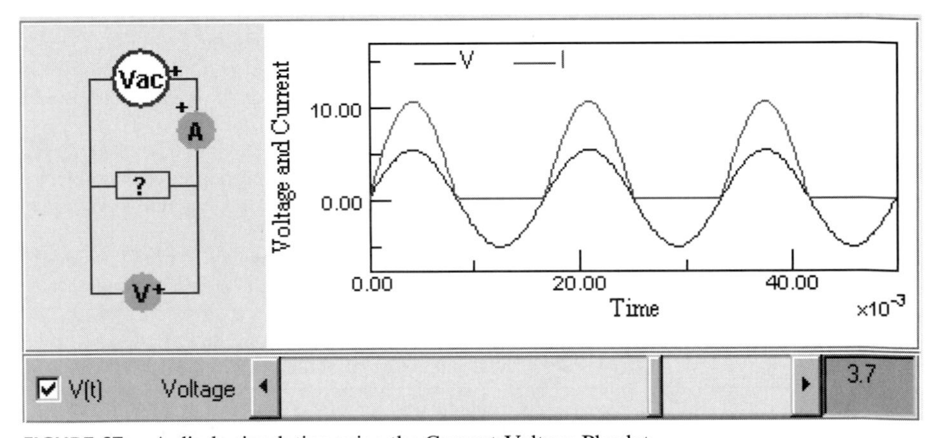

FIGURE 67: A diode simulation using the Current-Voltage Physlet.

18.3.1 Current-Voltage (*IVPhyslet*)

The *IVPhyslet* shown in Figure 67 is designed to simulate current-voltage relations and show how these relationships affect an ac signal. *IVPhyslet* is embedded into an html page using the following tags:

```
<applet codebase = [location] code = "circuit.IVApplet.class"
archive = "Circuit4_.jar, STools4.jar" name = [name]>
```

The embedding parameters shown in Table 23 can be specified as part of this applet tag. The <applet> tag must, of course, be closed by an </applet> tag after the embedding parameter tags have been coded.

TABLE 23: *IVPhyslet* embedding parameters.

Parameter	Value	Type	Description
Current	2*V	string	Current as a function of voltage, $I(V)$, of the load.
Vmin	10	double	The voltage slider minimum.
VMax	1000	double	The voltage slider maximum.
FPS	10	int	Frames per second when the load is used as a data source and the clock is running.
Dt	0.1	double	Time step per frame.
PixPerCell	60	int	The size of a schematic component in pixels.
IVGraphType	false	boolean	Plot the rms current as a function of rms voltage.
DefaultCircuit	true	boolean	Draw the default circuit shown in Figure 67.
ShowControls	true	boolean	Show the user interface.
ShowCheckBox	true	boolean	Show the check box in the user interface.
ShowGraph	true	boolean	Show the graph.

18.3.2 *IVPhyslet* Data Sources

The unknown component, that is, the load, is a data source as shown in Table 24.

TABLE 24: *IVPhyslet* data sources and data source variables.

Source Object	Identifier	Variables
load	id = getUnknownID()	t, v, i, f
clock	id = getClockID()	t

18.3.3 Current-Impedance (*IZPhyslet*)

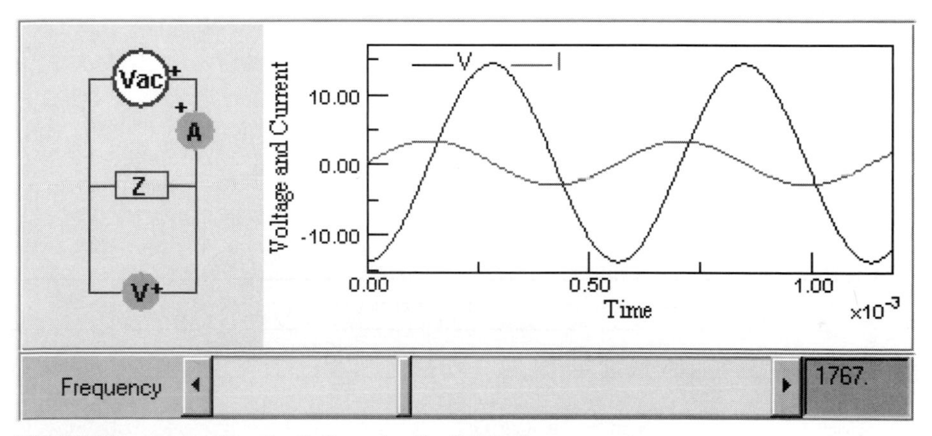

FIGURE 68: A capacitor simulation using the *IZPhyslet*.

IZPhyslet is designed to be scripted with two analytic functions: (1) current as a function of frequency, $I(f)$; and (2) phase as a function of frequency, Phase(f). This Physlet can be used to simulate an arbitrary ac circuit as a "black box." The script author must, of course, calculate the desired impedance function. *IZPhyslet* is embedded into an html page using the following tags:

```
<applet codebase = [location] code = "circuit.IZApplet.class"
archive = "Circuit4_.jar, STools4.jar" name = [name]>
```

The embedding parameters shown in Table 25 can be specified as part of this applet tag.

TABLE 25: *IZPhyslet* embedding parameters.

Parameter	Value	Type	Description
Current	2*V	string	Current as a function of frequency, $I(f)$.
Phase	0	string	Phase as a function of voltage, Phase(f).
Fmin	10	double	The frequency slider minimum.
FMax	500	double	The frequency slider maximum.
FPS	10	int	Frames per second when the load is used as a data source and the clock is running.
Dt	0.1	double	Time step per frame.
PixPerCell	60	int	The size of a schematic component in pixels.
ImpedanceGraphType	false	boolean	Plot the impedance as a function of frequency.
DefaultCircuit	true	boolean	Draw the default circuit shown in Figure 68.
ShowControls	true	boolean	Show the user interface.
ShowCheckBox	true	boolean	Show the user interface.
ShowGraph	true	boolean	Show the user interface.

The <applet> tag must be closed by an </applet> tag after the embedding parameter tags have been coded.

18.3.4 *IZPhyslet* Data Sources

The unknown component is a data source.

TABLE 26: *IZPhyslet* data sources and data source variables.

Source Object	Identifier	Variables
load	id = getUnknownID()	t, v, i, f
clock	id = getClockID()	t

18.3.5 Internal Resistance

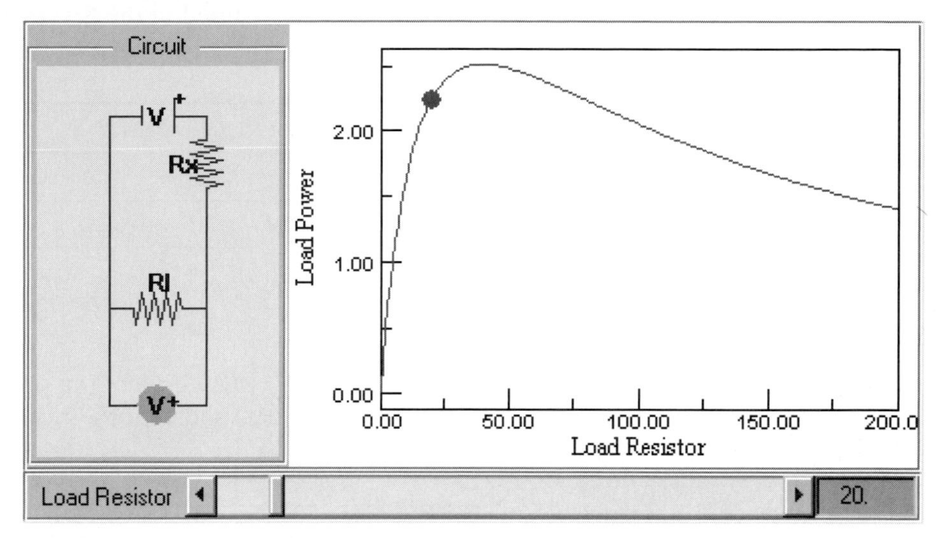

FIGURE 69: Power as a function of load resistance.

The *LoadPhyslet* shown in Figure 69 is designed to demonstrate how the output voltage of a battery depends on the resistive load. Although a graph of either output voltage or power can be shown, this applet is often embedded with the show graph parameter set to false. The user can rest the mouse on a circuit element in order to read its parameters, thereby making the necessary measurements to determine the unknown internal resistance.

LoadPhyslet is embedded into an html page using the following tags:

```
<applet codebase = [location] code = "circuit.LoadApplet.class"
archive = "Circuit4_.jar, STools4.jar" name = [name]>
```

The embedding parameters shown in Table 27 can be specified as part of this applet tag.

TABLE 27: *LoadPhyslet* embedding parameters.

Parameter	Value	Type	Description
Unknown	20	double	The unknown internal resistance of the battery.
Battery	12	double	The battery voltage.
RMin	0	double	The resistance slider minimum.
RMax	100	double	The resistance slider maximum.
PixPerCell	60	int	The size of a schematic component in pixels.
PowerGraphType	false	boolean	Plot the output power as a function of load resistance.
DefaultCircuit	true	boolean	Draw the default circuit shown in Figure 69.
ShowControls	true	boolean	Show the user interface.
ShowGraph	true	boolean	Show the user interface.

18.3.6 *LoadPhyslet* Data Sources

LoadPhyslet only supports the clock data source contained in the SApplet superclass as shown in Table 28.

TABLE 28: *LoadPhyslet* data sources and data source variables.

Source Object	Identifier	Variables
clock	id = getClockID()	*t*

18.3.7 Capacitive Reactance (*RCPhyslet*)

RCPhyslet models the current and voltage in an RC circuit driven by a sinusoidal voltage source. Figure 70 shows this Physlet acting as a data source for a *DataGraph*.

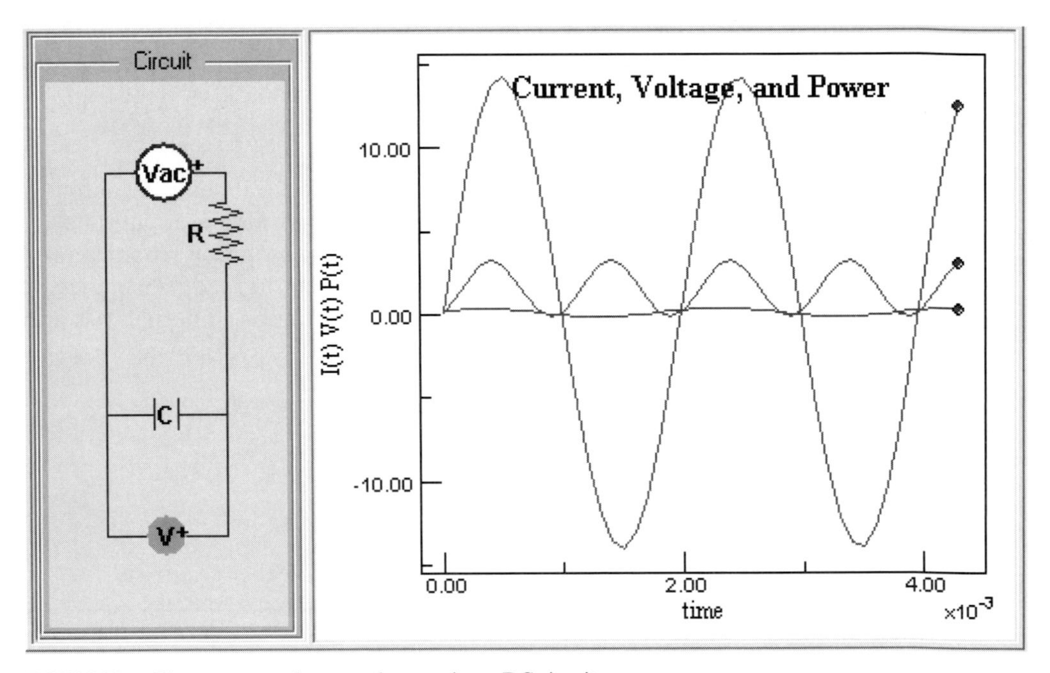

FIGURE 70: The current, voltage, and power in an RC circuit.

RCPhyslet is embedded into an html page using the following tags:

```
<applet codebase = [location] code = "circuit.RCApplet.class"
archive = "Circuit4_.jar, STools4.jar" name = [name]>
```

The embedding parameters shown in Table 29 can be specified as part of this applet tag.

TABLE 29: *RCPhyslet* embedding parameters.

Parameter	Value	Type	Description
Resistance	50	double	The resistance value in Ohms.
Capacitor	10	double	The capacitance value in microfarads.
Fmin	10	double	The frequency slider minimum.
FMax	1000	double	The frequency slider maximum.
FPS	10	int	Frames per second when the load is used as a data source and the clock is running.
Dt	0.1	double	Time step per frame.
PixPerCell	60	int	The size of a schematic component in pixels.
ImpedanceGraphType	false	boolean	Plot the impedance as a function of frequency.
DefaultCircuit	true	boolean	Draw the default circuit shown in Figure 70.
ShowControls	true	boolean	Show the user interface.
ShowCheckBox	true	boolean	Show the user interface.
ShowGraph	true	boolean	Show the user interface.

The <applet> tag must be closed by an </applet> tag after the embedding parameter tags have been coded.

18.3.8 *RCPhyslet* Data Sources

All circuit components are data sources.

TABLE 30: *RCPhyslet* data sources and data source variables.

Source Object	Identifier	Variables
capacitor	id = getCapacitorID()	t, v, i, f, c
resistor	id = getResistorID()	t, v, i, f, r
source	id = getSourceID()	t, v, i, f
clock	id = getClockID()	t

18.3.9 Resonance (*LRCPhyslet*)

LRCPhyslet models the current and voltage in an LRC circuit driven by a sinusoidal voltage source. Figure 71 shows this Physlet acting as a data source for a *DataGraph*.

LRCPhyslet is embedded into an html page using the following tags:

```
<applet codebase = [location] code = "circuit.LRCApplet.class"
archive = "Circuit4_.jar, STools4.jar" name = [name]>
```

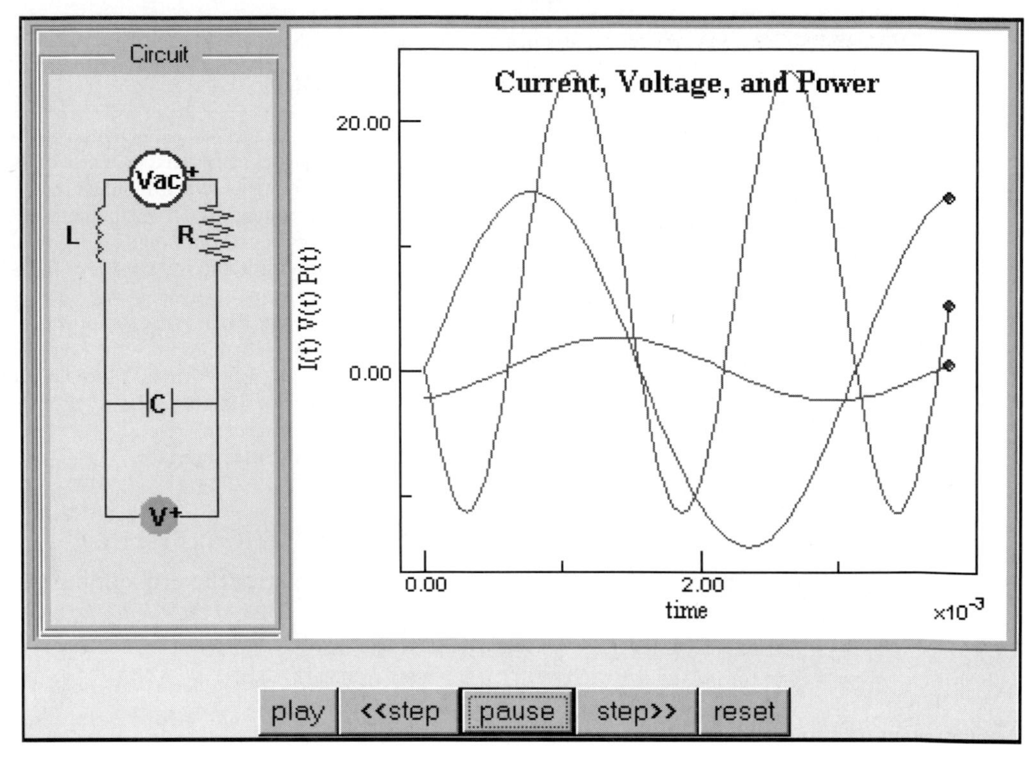

FIGURE 71: The current, voltage, and power in an LRC circuit.

The embedding parameters shown in Table 31 can be specified as part of this applet tag.

TABLE 31: *RCPhyslet* embedding parameters.

Parameter	Value	Type	Description
Resistance	25	double	The resistance value in Ohms.
Capacitor	10	double	The capacitance value in microfarads.
Inductance	98	double	The resistance value in millihenrys.
Fmin	10	double	The frequency slider minimum.
FMax	500	double	The frequency slider maximum.
FPS	10	int	Frames per second when the load is used as a data source and the clock is running.
Dt	0.1	double	Time step per frame.
PixPerCell	60	int	The size of a schematic component in pixels.
ImpedanceGraphType	false	boolean	Plot the impedance as a function of frequency.
DefaultCircuit	true	boolean	Draw the default circuit shown in Figure 71.
ShowControls	true	boolean	Show the user interface.
ShowCheckBox	true	boolean	Show the user interface.
ShowGraph	true	boolean	Show the user interface.

The <applet> tag must be closed by an </applet> tag after the embedding parameter tags have been coded.

18.3.10 *LRCPhyslet* Data Sources

All circuit components data sources.

TABLE 32: *RCPhyslet* data sources and data source variables.

Source Object	Identifier	Variables
capacitor	id = getCapacitorID()	t, v, i, f, c
inductor	id = getInductorID()	t, v, i, f, l
resistor	id = getResistorID()	t, v, i, f, r
source	id = getSourceID()	t, v, i, f
clock	id = getClockID()	t

18.4 EIGENVALUES AND *QM* WAVEFUNCTIONS

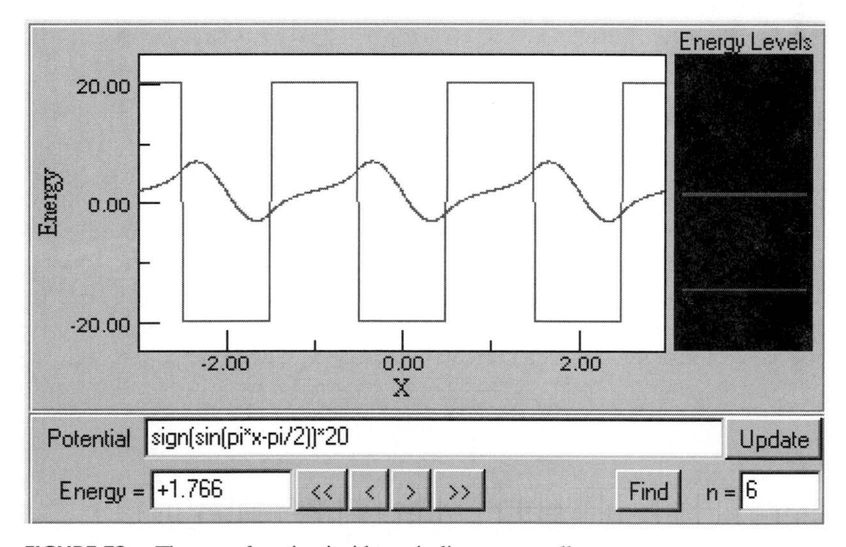

FIGURE 72: The wavefunction inside periodic square wells.

EnergyEigenvalue uses the shooting method to solve the time-independent Schrödinger equation in order to find the quantum wavefunction and its energy eigenvalue.

EnergyEigenvalue is embedded into an html page using the following tags:

```
<applet codebase = [location]
code = " energyEigenvalue.EnergyEigenvalue.class"
archive = " EnergyEigenvalue4_.jar, STools4.jar" name = [name]>
```

The embedding parameters shown in Table 33 can be specified as part of this applet tag.

TABLE 33: *EnergyEigenvalue* embedding parameters.

Parameter	Value	Type	Description
XMin	−5	double	Minimum value of x
XMax	5	double	Maximum value of x
YMin	0	double	Minimum value of y
YMax	25	double	Maximum value of y
Potential	x*x	string	Potential function
HBarTwoM	1	double	Energy scale, $hbar^2/(2m)$
Lowest	1	int	Lowest quantum number to calculate
Highest	1	int	Highest quantum number to calculate
NumPts	200	int	Number of points to calculate in the wavefunction.
BreakValue	1.0e12	double	Wavefunction divergences if this value is reached
MaxIterations	40	int	Maximum number of calculational attempts
Tolerance	1.0e-8	double	Error below which convergence is assumed
ScaleToArea	false	boolean	Normalize the probability if true. Otherwise, scale the function so that all functions have the same maximum amplitude.
AutoScaleY	true	boolean	Scale the energy axis Wavefunctions are offset by their energy eigenvalue in the function plot.
ShowSpectrum	true	boolean	Show the user interface.
ShowFunctions	true	boolean	Show the user interface.
ShowPotential	true	boolean	Show the user interface.
ShowControls	true	boolean	Show the user interface.

The large number of parameters in *EnergyEigenvalue* allows this Physlet to be used without script. The potential energy as well as the number of wavefunctions to pre-calculate are specified. Additional wavefunctions can be calculated by entering either the quantum number or energy eigenvalue into the appropriate field in the user interface shown in Figure 72. Click-dragging the mouse inside the energy level diagram will display the appropriate wavefunction when the mouse crosses the eigenvalue. Right-click dragging will display the shooting-method calculation as it is being performed, thereby showing how only certain energy values satisfy the appropriate boundary conditions.

18.4.1 Data Sources

Wavefunctions are data sources as shown in Table 34.

TABLE 34: *EnergyEigenvalue* data sources and data source variables.

Source Object	Identifier	Variables
wavefunction	id = getActiveWavefunctionID() id = getWavefunctionID(int)	x, p, psi, energy
	The active wavefunction is the wavefunction that has been selected by a mouse action in the energy level display.	*Note*: The wavefunction, *psi*, and the potential function, p, are evaluated at every point in the x array.
clock	id = getClockID()	t

18.4.2 addObject Method

EnergyEigenvalue does not support the addObject method. Use a data connection to a *DataGraph* if you wish to add geometric objects or other annotation to a wavefunction plot. Examples are available on the CD.

18.5 *FARADAY*

Faraday models a wire placed in a magnetic field. Both the position of the wire and strength of the magnetic field may be time dependent. (It is usually not a good idea to have both types of changes occurring simultaneously.) Figure 73 shows the default configuration consisting of a U-shaped wire with a slider.[3] Script can be used to create other wires, such as circular loops and rectangles. Setting the dragable property of the active wire to true will disable the position function, $x(t)$, and allow the user to experiment and observe the relationship among the slider velocity, field strength, and voltage.

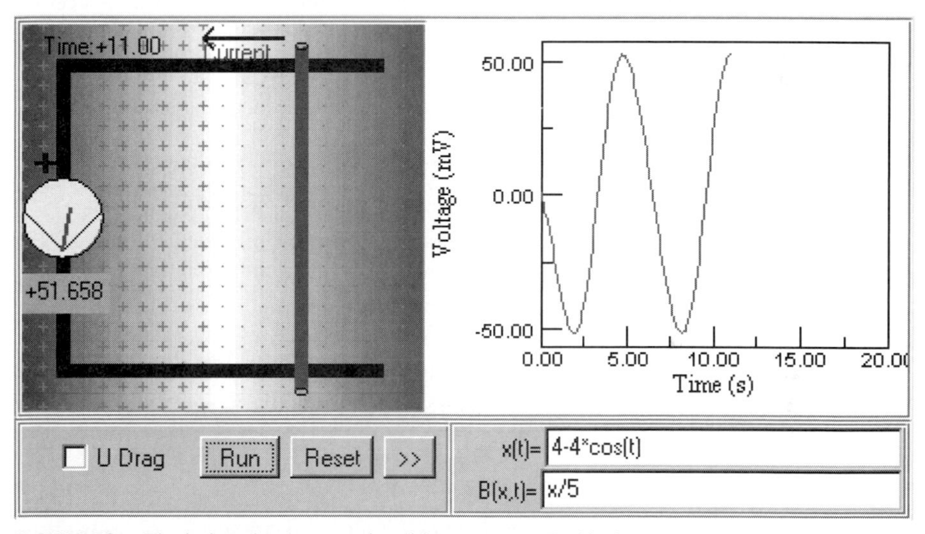

FIGURE 73: The induced voltage as the slider moves on the U-wire.

Faraday is embedded into an html page using the following tags:

```
<applet codebase = [location] code = "faraday.Faraday.class"
archive = "Faraday4_.jar, STools4.jar" name = [name]>
```

The embedding parameters shown in Table 35 can be specified as part of this applet tag.

[3]This Physlet uses the convention that $B > 0$ points out of the screen and emf > 0 is counterclockwise.

TABLE 35: *Faraday* embedding parameters.

Parameter	Value	Type	Description
FieldFunction	10* sin (pi*x/5)	string	The magnetic field into or out of the screen as a function of x, $B_z(x)$.
PositionFunction	1.0 + 3.0*t	string	The x-position of the active wire. Only one wire can be active at a time.
MaxTime	10	double	The maximum time. The clock will reset to zero and clear the internal graph when this time is reached.
PixelsPerUnit	10	int	The scale for the schematic area.
FPS	10	int	The frames per second for the animation.
Dt	0.1	double	The time step per frame. FPS*dt should equal 1.0 for real-time animation.
SchematicWidth	200	int	The width of the schematic area where the wires will be drawn.
ShowSchematic	true	boolean	Show the schematic area.
ShowColor	true	boolean	Show the field using color. Red represents a field that is pointing out of the screen.
ShowGrid	true	boolean	Show the direction of the field, using x and · to represent into and out of the screen, respectively.
ShowCurrentArrow	true	boolean	Show the direction of the induced current using an arrow.
ShowMeter	true	boolean	Show the galvanometer. Only available on the U-wire.
ShowGraph	true	boolean	Record the voltage induced in the active wire on the internal graph. The clock must be running for the voltage to be recorded.
DragMode	false	boolean	Allow the user to drag the active wire.
ShowControls	true	boolean	Show the user interface.

Notice the large number of embedding parameters needed to control this applet. This is a good indicator that it may be more efficient to use script.

18.5.1 Data Sources

Faraday data connections are similar to *EField*. All on-screen objects supply their position. Wires (that is boxes, shells, and the U-wire shown previously) supply their flux and their voltage. However, the voltage can only be calculated when the clock is running since it depends on the rate of change of flux and this rate is calculated in real time.

TABLE 36: Faraday data sources and data source variables.

Source Object	Identifier	Variables
wire	The id is returned when an object is created using the addObject method.	t, x, y, flux, v *Note*: v is the induced voltage.
clock	id = getClockID()	t

18.5.2 addObject Method

Faraday allows the creation of geometric shapes and wires listed in Table 37 using the addObject method described in Section 13.2. There are three types of wire objects: uwire, box, and shell.

TABLE 37: *Faraday* addObject parameter names and properties.

Name	Attributes
arrow	**x-double** *x* position of the base in world units **y-double** *y* position of the base in world units **h-double** constant horizontal component **v-double** constant vertical component
box A hollow rectangle.	**x-double** *x* position of the center in world units **y-double** *y* position of the center in world units **h-int** height in pixels **w-int** width in pixels **s-int** thickness of the box
caption A caption is text that is centered near the top of the screen.	**x-double** *x* position of the center in world units **y-double** *y* position of the center in world units **text-string** text of the caption
circle	**x-double** *x* position of the center in world units **y-double** *y* position of the center in world units **r-int** radius in pixels
constraint	**xmin-double** minimum *x* value **xmax-double** maximum *x* value **ymin-double** minimum *y* value **ymax-double** maximum *y* value A constraint, by itself, will have no effect. It must be attached to a dragable object using the setConstraint method.
rectangle	**x-double** *x* position of the center in world units **y-double** *y* position of the center in world units **h-int** height in pixels **w-int** width in pixels
shell	**x-double** *x* position of the center in world units **y-double** *y* position of the center in world units **r-int** radius in pixels
text A fixed text string followed by an optional calculation.	**x-double** *x* position of the left side of the text in world units **y-double** *y* position of the top of the text in world units **text-string** static text **calc-string** An analytic function to be evaluated. The calculation is displayed to the right of the static text. Text objects are often slaved to other objects. If a text object is slaved to a wire, it takes on the properties of that wire and can evaluate a function of *t*, *x*, *y*, flux, and *v*. *Note*: Use the setFormat method described in Section 17.3.1 to change the decimal format of the displayed calculation.
uwire A U-shaped wire with a slider on the right.	**x- double** *x* position of the slider in world units

18.6 *HYDROGENIC*

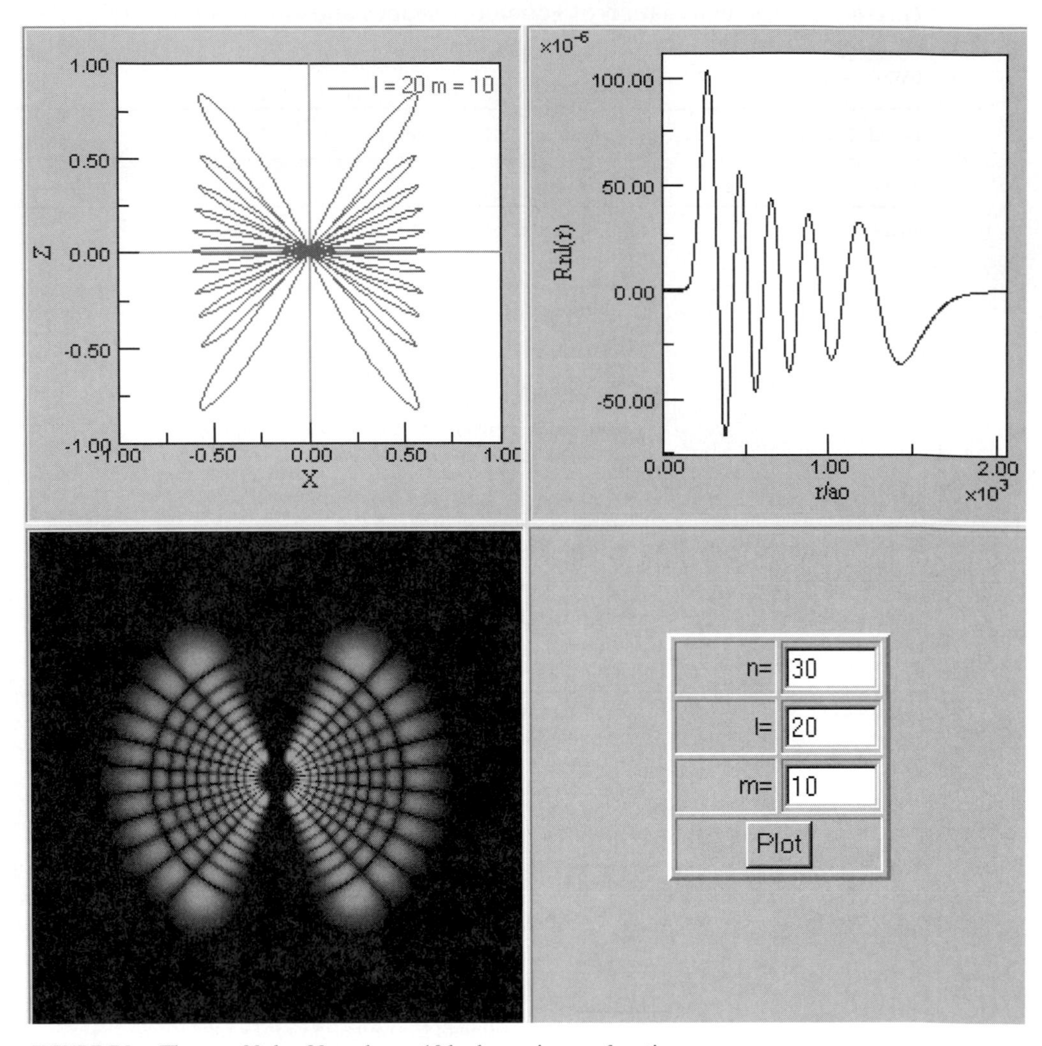

FIGURE 74: The $n = 30, l = 20$, and $m = 10$ hydrogenic wavefunction.

The *Hydrogenic* package contains three Physlets, shown in Figure 74, that plot the radial wavefunction, angular wavefunction, and the wavefunction density in the x-z plane. Principal quantum numbers from $n = 1$ to $n = 50$ are supported.

18.6.1 Radial Wavefunction

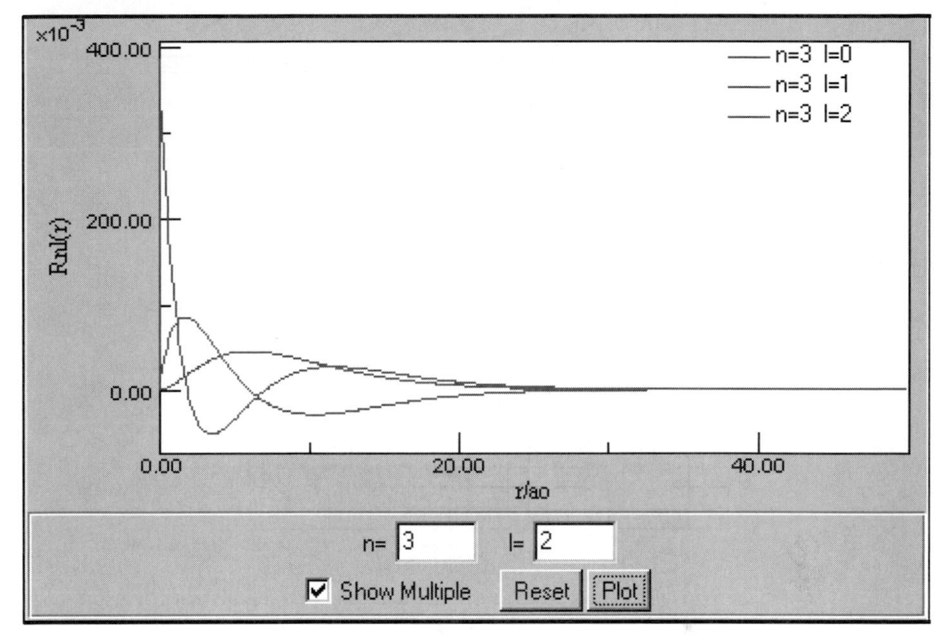

FIGURE 75: The $n = 3$ and $l = 2$ radial wavefunction.

Radial plots the radial wavefunction (see Figure 75) or the radial probability distribution. *Radial* is embedded into an html page using the following tags:

```
<applet codebase = [location] code = " hydrogenic.Radial.class"
archive = "Hydrogenic4_.jar, STools4.jar" name = [name]>
```

The embedding parameters shown in Table 38 can be specified as part of this applet tag.

TABLE 38: *RadialApplet* embedding parameters.

Parameter	Value	Type	Description
n	1	int	The principal quantum number.
l	0	int	The angular quantum number.
MultiPlot	false	boolean	Enable more than one wavefunction to be shown.
ShowAmplitude	true	boolean	If true, plot the wavefunction amplitude. If false, plot the probability density.
ShowControls	true	boolean	Show the user interface.

18.6.2 Angular Wavefunction

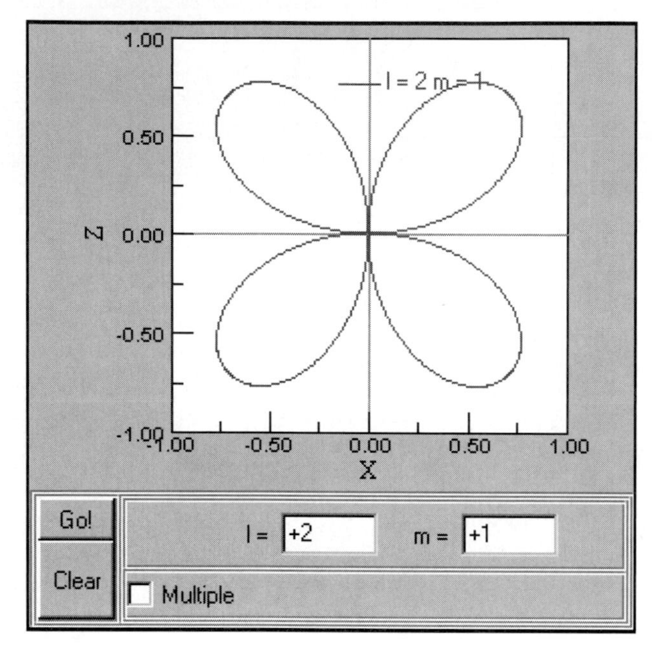

FIGURE 76: The $l = 2$ and $m = 1$ angular wavefunction.

Angular plots the angular wavefunction as shown in Figure 76. *Angular* is embedded into an html page using the following tags:

```
<applet codebase = [location] code = " hydrogenic.Angular.class"
archive = "Hydrogenic4_.jar, STools4.jar" name = [name]>
```

The embedding parameters shown in Table 39 can be specified as part of this applet tag.

TABLE 39: *Angular* embedding parameters.

Parameter	Value	Type	Description
l	2	int	The angular quantum number.
m	1	int	The azimuthal quantum number.
MultiPlot	false	boolean	Enable more than one wavefunction to be shown.
Normalize	true	boolean	Normalize the wavefunction.
ShowControls	true	boolean	Show the user interface.

18.6.3 Probability Density

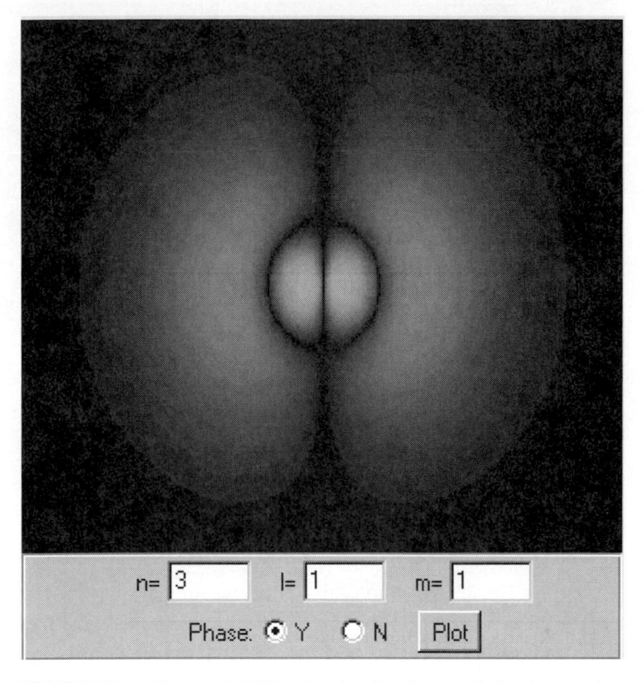

FIGURE 77: The probability density for the $n = 3, l = 1, m = 1$ hydrogenic wavefunction.

Density represents the wavefunction as a color-coded density plot. The intensity is proportional to the amplitude of the wavefunction as shown in Figure 77. The color represents the phase. *Density* is embedded into an html page using the following tags:

```
<applet codebase = [location] code = " hydrogenic.Density.class"
archive = "Hydrogenic4_.jar, STools4.jar" name = [name]>
```

The embedding parameters shown in Table 40 can be specified as part of this applet tag.

TABLE 40: *Density* embedding parameters.

Parameter	Value	Type	Description
n	1	int	The principal quantum number.
l	0	int	The angular quantum number.
m	0	int	The azimuthal quantum number.
ShowPhase	true	boolean	Show the phase as a color.
ShowControls	true	boolean	Show the user interface.

18.7 *MOLECULAR*

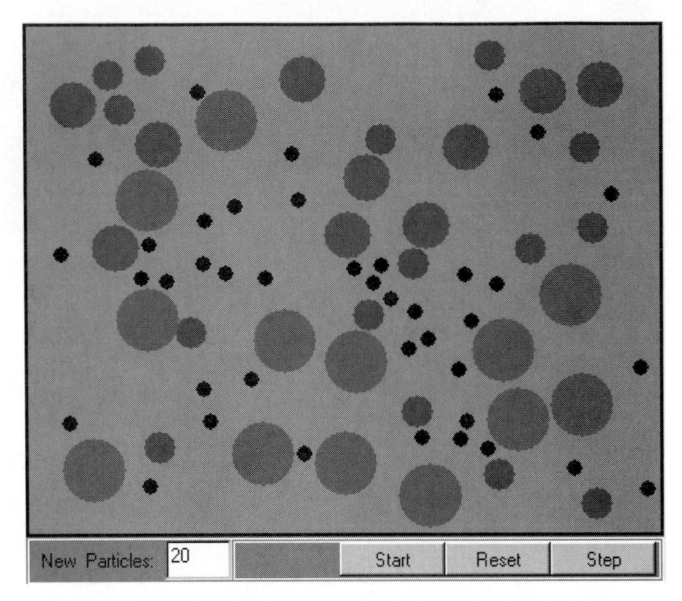

FIGURE 78: Molecular dynamics simulation using hard disks.

The *Molecular* package contains a number of Physlets that are based on a simple hard-disk molecular model. As shown in Figure 78 and in the examples on the CD, each disk can be set to have a unique size, mass, and color. In addition, each disk can act as a data source that delivers position and velocity information. Finally, the entire ensemble can act as a data source that delivers ensemble averages and histogram data to data listeners.

The *Molecular* package, along with most other Physlets, solves dimensionless equations. It is possible to set k, Boltzmann's constant, to something other than 1 using the setBoltzmann method. It is the responsibility of the script author to decide on the appropriate units and to convey these units to the student in the text. Physlets make minimal assumptions about the scale of the system being modeled.

18.7.1 MolecularApplet

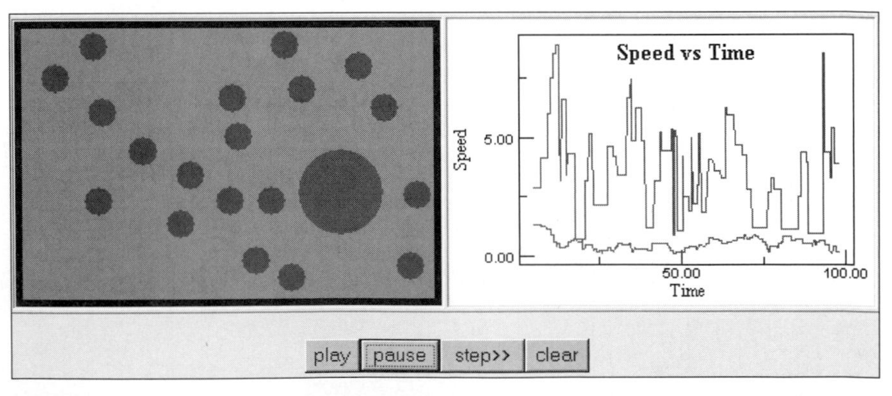

FIGURE 79: Speed as a function of time for a light and a massive particle.

MolecularApplet is the simplest applet in the molecular package. It consists of a single ensemble with either periodic or reflecting boundary conditions. Figure 79 shows this Physlet being used with a DataGraph to demonstrate Brownian motion.

MolecularApplet is embedded into an html page using the following tags:

```
<applet codebase = [location] code = " molecular.MolecularApplet.class"
archive = "Molecular4_.jar, STools4.jar" name = [name]>
```

The embedding parameters shown in Table 41 can be specified as part of this applet tag.

TABLE 41: *MolecularApplet* embedding parameters.

Parameter	Value	Type	Description
InitialP	20	int	The initial number of particles.
MaxParticles	80	int	The maximum number of particles allowed.
FPS	10	int	Frames per second when the load is used as a data source and the clock is running.
Dt	0.1	double	Time step per frame.
PixPerUnit	10	int	The scale factor from pixel to world units.
Border	5	int	The size of the border surrounding the ensemble.
PeriodicH	false	boolean	Enable horizontal periodic boundary conditions.
PeriodicV	false	boolean	Enable vertical periodic boundary conditions.
ShowControls	true	boolean	Show the user interface.

18.7.2 *MolecularApplet* Data Sources

TABLE 42: *MolecularApplet* data sources and data source variables.

Source Object	Identifier	Variables
ensemble	id = getEnsembleID()	time, $p, v, t, n, qt, qr, qb, ql, pt, pr, pb, pl, dv, dt$ State variables: p, v, and t are the pressure, volume, and temperature. Heat flow: qr, ql, qt, and qb are the heat flow into the ensemble at the right, left, top, and bottom walls when the wall temperature is set. Pressure: pr, pl, pt, and pb are the momentum change per unit time at the right, left, top, and bottom walls, respectively.
histogram	id = getHistogramID (bins, vmin, vmax) bins: The number of bins vmin: The velocity of the first bin vmax: The velocity of the last bin	v, n The speed and number of particles at that speed.
particle	id = getparticleID(i)	t, x, y, vx, vy, m
clock	id = getClockID()	t

18.7.3 *MolecularPiston*

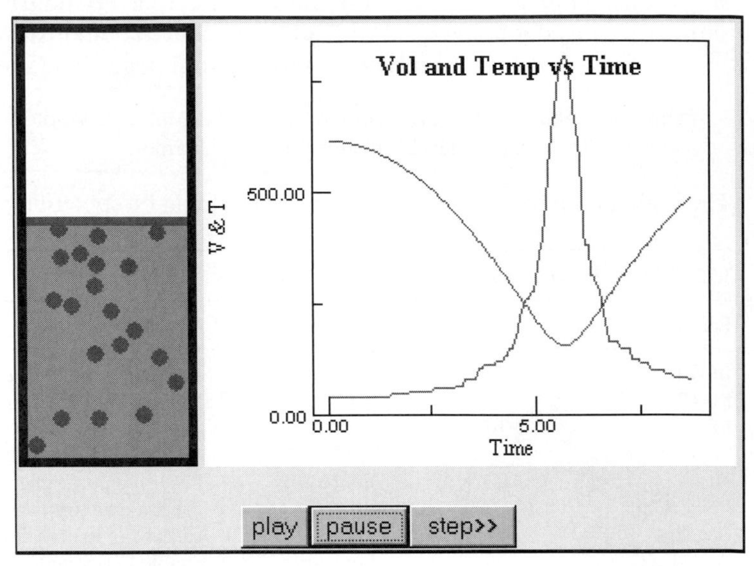

FIGURE 80: The change in state of a gas being compressed by a falling piston.

MolecularPiston adds a movable wall, that is, a piston, to the basic ensemble class. The piston can "fall" under the action of a gravitational force as shown in Figure 80, it can follow an analytical trajectory, or it can be made dragable.

MolecularPiston is embedded into an html page using the following tags:

```
<applet codebase = [location] code = " molecular.MolecularPiston.class"
archive = "Molecular4_.jar, STools4.jar" name = [name]>
```

The embedding parameters shown in Table 43 can be specified as part of this applet tag.

TABLE 43: *MolecularPiston* embedding parameters.

Parameter	Value	Type	Description
InitialP	20	int	The initial number of particles.
MaxParticles	100	int	The maximum number of particles allowed.
FPS	10	int	Frames per second when the load is used as a data source and the clock is running.
Dt	0.1	double	Time step per frame.
PixPerUnit	10	int	The scale factor from pixel to world units.
InitialPistonPos	30	double	The initial piston position in world units.
PistonWidth	5	int	The width of the piston.
PistonMass	20	double	The mass of the piston.
Gravity	−1	double	The acceleration of gravity.
ParseString	5*sin(t)	string	The analytic function for the piston's position.
ParseMode	false	boolean	Set to true if the piston should follow an analytic function.
Orientation	vertical	string	Sets the piston to be vertical or horizontal.
Dragable	false	boolean	Let the user drag the piston.
ShowControls	true	boolean	Show the user interface.

The default mode models a piston falling under the action of a constant force. As the piston falls the gas heats and the pressure increases until the piston rebounds, as shown in Figure 80.

18.7.4 *MolecularPiston* Data Sources

TABLE 44: *MolecularPiston* data sources and data source variables.

Source Object	Identifier	Variables
ensemble	id = getEnsembleID()	time, $p, v, t, n, qt, qr, qb, ql, pt, pr, pb, pl, dv, dt$ State variables: p, v, and t are the pressure, volume, and temperature. Heat flow: qr, ql, qb, and qt are the heat flow into the ensemble at the right, left, top, and bottom walls when the wall temperature is set. Pressure: pr, pl, pb, and pt are the momentum change per unit time at the right, left, top, and bottom walls.
histogram	id = getHistogramID (bins, vmin, vmax) bins: The number of bins vmin: The velocity of the first bin vmax: The velocity of the last bin	v, n The speed and number of particles at that speed.
particle	id = getparticleID(i)	t, x, y, vx, vy, m
clock	id = getClockID()	t

18.7.5 *TwoEnsembles*

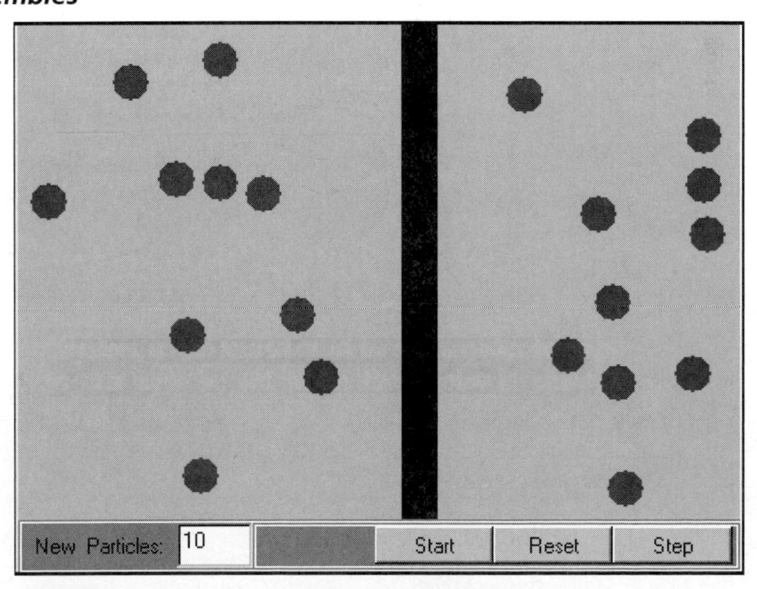

FIGURE 81: A movable piston acted on by two ensembles.

TwoEnsembles adds a second ensemble to the *MolecularPiston* Physlet. The piston will respond to a collision from either ensemble as shown in Figure 81.

TwoEnsembles is embedded into an html page using the following tags:

```
<applet codebase = [location] code = " molecular.TwoEnsembles.class"
archive = "Molecular4_.jar, STools4.jar" name = [name]>
```

The embedding parameters shown in Table 45 can be specified as part of this applet tag.

TABLE 45: *TwoEnsembles* embedding parameters.

Parameter	Value	Type	Description
InitialP	20	int	The initial number of particles.
MaxParticles	80	int	The maximum number of particles allowed.
FPS	10	int	Frames per second when the load is used as a data source and the clock is running.
Dt	0.1	double	Time step per frame.
PixPerUnit	10	int	The scale factor from pixel to world units.
PistonPosition	0.5	double	The initial piston position in world units.
PistonWidth	5	int	The width of the piston.
PistonMass	20	double	The mass of the piston.
ShowControls	true	boolean	Show the user interface.

18.7.6 *TwoEnsembles* Data Sources

Data can be collected from either ensemble by specifying the appropriate left or right method name.

TABLE 46: *TwoEnsembles* data sources and data source variables.

Source Object	Identifier	Variables
ensemble	id = getEnsembleLeftID() id = getEnsembleRightID()	time, $p, v, t, n, qt, qr, qb, ql, pt, pr, pb, pl, dv, dt$ State variables: p, v, and t are the pressure, volume, and temperature. Heat flow: qr, ql, qb, and qt are the heat flow into the ensemble at the right, left, top, and bottom walls when the wall temperature is set. Pressure: pr, pl, pb, and pt are the momentum change per unit time at the right, left, top, and bottom walls.
histogram	id = getHistogramLeftID (bins, vmin, vmax) id = getHistogramRightID (bins, vmin, vmax) bins: The number of bins vmin: The velocity of the first bin vmax: The velocity of the last bin	v, n *Notes*: The speed, v, and number of particles at that speed, v, are arrays. The size of these arrays is the bins parameter.
particle	id = getparticleLeftID(i) id = getparticleRightID(i)	t, x, y, vx, vy, m
clock	id = getClockID()	t

18.8 *OPTICS*

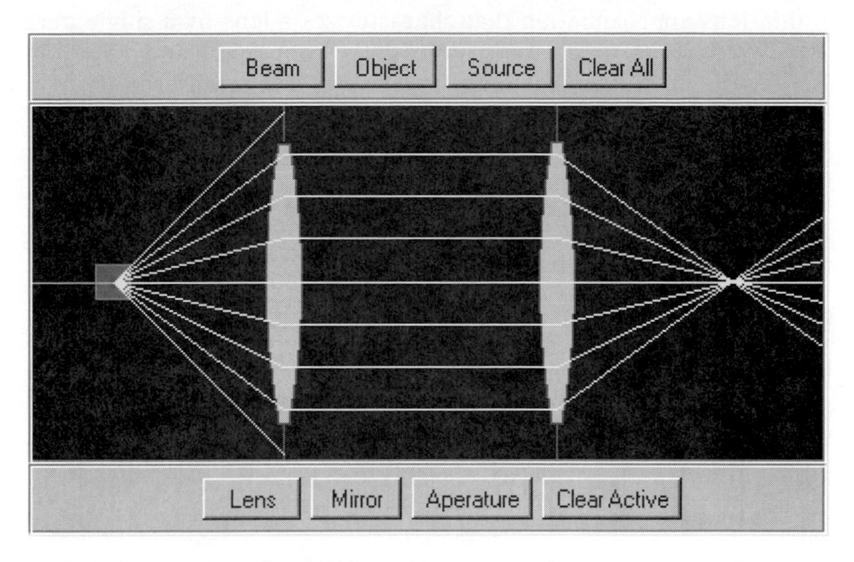

FIGURE 82: An *Optics* bench with two lenses and a point source.

Optics allows users to simulate standard optic elements (lenses, mirrors, dielectrics, sources, apertures) and observe the ways that light rays propagate through these elements. This Physlet is also scriptable, but some physics teachers have provided instructions for the user interface shown in Figure 82 in order to incorporate it into tutorials. Element properties, such as position and focal length, can be adjusted using a click-and-drag metaphor.

Curriculum authors should point out that this Physlet uses a thin-lens model. This model does not include aberrations and bends light rays at the center of the lens rather than at the surface. Aberrations can be demonstrated by constructing a lens using two changes in index of refraction. (See Section 9.9 on the CD.)

Optics is embedded into an html page using the following tags:

```
<applet codebase = [location] code = "optics.OpticsApplet.class"
archive = "Optics4_.jar,STools4.jar " name = [name]>
```

The single embedding parameter shown in Table 47 can be specified as part of this applet tag.

TABLE 47: *Optics* embedding parameter.

Parameter	Value	Type	Description
ShowControls	true	boolean	Show the user interface.

18.8.1 Data Sources

All objects listed in Table 48 supply their position. Optic elements, such as lenses and mirrors, supply other parameters, such as focal length, f, and index of refraction, n, to data listeners. Script authors should note that the index of refraction is only set for dielectrics. Although a real lens obviously has an index of refraction, our model uses a thin-lens approximation that characterizes a lens by a single parameter, the focal length. Although we do try to indicate the converging and diverging nature of a lens by drawing it with convex and concave surfaces of differing radii, the drawing radii are only intended to be visual clues. The index of refraction and the radius of curvature of the glass are not adjustable parameters in our lens object.

TABLE 48: *Optics* data sources and data source variables.

Source Object	Identifier	Variables
geometric objects rectangles, circles, and text	The id is returned when an object is created using the addObject method.	x, y
dielectrics	The id is returned when an object is created using the addObject method.	x, y, n
lenses and mirrors	The id is returned when an object is created using the addObject method.	x, y, f
clock	id = getClockID()	t

18.8.2 addObject Method

Optics allows the creation of geometric shapes as well as the optic elements listed in Table 49 using the addObject method described in Section 13.2.

TABLE 49: *Optics* addObject parameter names and properties.

Name	Attributes
aperture Restricts the propagation of light rays.	**x-double** x position of the center in world units **opening-double** the opening size as a fraction from 0 to 1.0
box A hollow rectangle.	**x-double** x position of the center in world units **y-double** y position of the center in world units **h-int** height in pixels **w-int** width in pixels **s-int** thickness of the box
circle	**x-double** x position of the center in world units **y-double** y position of the center in world units **r-int** radius in pixels
constraint	**xmin-double** The minimum x value. **xmax-double** The maximum x value. **ymin-double** The minimum y value. **ymax-double** The maximum y value. A constraint, by itself, will have no effect. It must be attached to a dragable object using the setConstraint method.

Name	Attributes
dielectric	**x-double** x position of the center in world units **dn-double** the change in index of refraction **r-double** radius or curvature of the index change Models a change in index of refraction using Snell's law. Aberrations are observed. *See also* refraction.
lens	**x-double** x position of the center in world units **f-double** the focal length Example: addObject("lens","x = 1, f = -2");
mirror	**x-double** x position of the center in world units. **f-double** the focal length **spherical** include spherical aberrations Example: addObject("mirror","x = 1, f = -2, spherical");
rectangle	**x-double** x position of the center in world units **y-double** y position of the center in world units **h-int** height in pixels **w-int** width in pixels
refraction	**x-double** x position of the center in world units. **dn-double** the change in index of refraction. **r-double** radius or curvature of the index change Models a change in index of refraction using the small angle approximation. *See also* dielectric.
screen	**x-double** x position of the center in world units
shell	**x-double** x position of the center in world units **y-double** y position of the center in world units **r-int** radius in pixels

source	**x-double** x position of the center in world units
	
	Choose either infinite or point selectors. The three principal rays will be drawn if neither selector is specified.
	
	infinite a beam of diverging light rays
	s-int the size of the beam in pixels
	slope-double the slope of the rays. −1<angle<1
	
	point a beam of diverging light rays
	inc-double the slope increment between rays
	slope-double the slope of the rays. −1<angle<1.
	
	Examples:
	addObject("source","x = 1, point, inc = 0.1, slope = 1.0");
	addObject("source","x = 1, infinite, s = 30, slope = 0.5");

(*continued*)

TABLE 49: (continued)

Name	Attributes
text A fixed text string followed by an optional calculation.	**x-double** *x* position of the left side of the text in world units **y-double** *y* position of the top of the text in world units **text-string** static text **calc-string** An analytic function to be evaluated. The calculation is displayed to the right of the static text. Text objects are often slaved to other objects. For example, if a text object is slaved to a wire, it takes on the properties of that wire and can evaluate a function of *t*, *x*, *y*, flux, and *v*. *Note*: Use the setFormat method described in Section 17.3.1 to change the decimal format of the displayed calculation.

18.9 *POISSON*

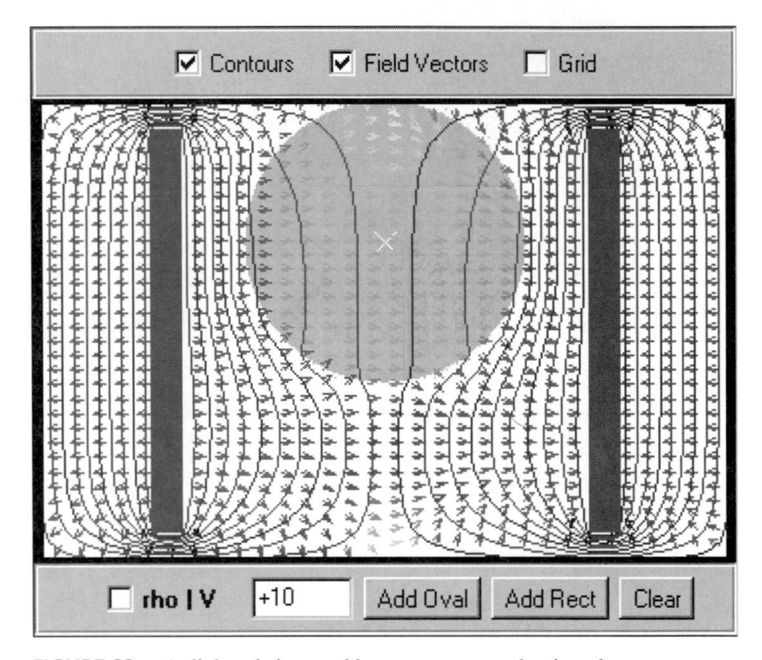

FIGURE 83: A dielectric inserted between two conducting plates.

Poisson calculates electrostatic potentials by numerically solving Poisson's equation in the presence of spherical and rectangular conductors, charge densities, and dielectrics as shown in Figure 83. The gradient of this potential is then calculated in order to display the electric field and the divergence is calculated to display the induced charge.[4]

[4]Calculating the induced charge greatly increases the computation time, particularly if dielectrics are present. The default computational tolerance of 10^{-3} for the potential values on the grid produces an error of 1 to 5% in the induced charge calculation. The setTolerance method can be used to increase this precision.

Although *Poisson* may be used in interactive mode using a click-and-drag metaphor, it is designed to be scripted. Many options, such as dielectrics, are only available through script. Figure 83, for example, shows the field and potential as a dielectric cylinder is dragged between two conducting plates.

Poisson is embedded into an html page using the following tags:

```
<applet codebase = [location] code = "poisson.Poisson.class"
archive = "Poisson4_.jar, STools4.jar" name = [name]>
```

The embedding parameters shown in Table 50 can be specified as part of this applet tag.

TABLE 50: *Poisson* embedding parameters.

Parameter	Value	Type	Description
Gutter	0	int	The number of extra grid points outside the viewing area.
GridSize	32	int	The number of grid points in the horizontal or vertical direction. The voltage is evaluated at these grid points.
Range	−1,1,−1,1	string	The approximate range for the x- and y-dimensions.
ShowFieldVectors	true	boolean	Show the electric field.
ShowContours	true	boolean	Show the equipotential contours.
ShowControls	true	boolean	Show the user interface.

18.9.1 Data Sources

Data sources in *Poisson* are similar to *EField*. All on-screen objects supply their position coordinates, x and y. Electrostatic objects, that is, conductors as well as charge densities and dielectrics, supply additional data as shown in Table 51.

TABLE 51: *Poisson* data sources and data source variables.

Source Object	Identifier	Variables
conductors charge-density dielectric	The id is returned when an object is created using the addObject method.	x, y, q, v, and p

	Examples:
	addObject("conductor","circle, r = 5, v = 2");
	addObject("charge","w = 5, h = 2, q = -2");
	Example:
	addObject("dielectric","circle, chi = 5, r = 70");
		Notes:
		q is the total charge on the object.
		p is the potential energy.
		v is the voltage if the object is a conductor.

(continued)

TABLE 51: (continued)

Source Object	Identifier	Variables
induced charge array	id = getChargeArray();	None. It is assumed that all data being passed are a charge value at a grid point. Array data source should be passed to a two-dimensional plotting program such as SPlotter.
shapes: circle, rectangle, box, etc.	The id is returned when an object is created using an add method. Example: addObject("circle","r = 5");	x, y
test charge	The id is returned when an object is created using an add method. Example: addObject("testcharge","r = 5");	x, y, ax, ay, v *Notes*: v is the voltage measured at the test charge. The acceleration components assume a unit charge.
voltage array	id = getPotentialArray();	None. It is assumed that all data being passed are a voltage value at a grid point. Array data source should be passed to a two-dimensional plotting program such as SPlotter.
clock	id = getClockID()	t

18.9.2 addObject Method

Poisson allows the creation of geometric shapes as well as conductors and dielectrics listed in Table 52 using the addObject method described in Section 13.2. In addition, it supports circular and rectangular conductors, dielectrics, and charge distributions.

TABLE 52: *Poisson* addObject parameter names and properties.

Name	Attributes
arrow Arrows are often animation slaves of other objects. They can represent almost any vector since the h and v components can be functions of the variables.	**x-double** x position of the base in world units **y-double** y position of the base in world units **h-string** horizontal component as a function of t, x, y, vx, vy, ax, and ay **v-string** vertical components as a function of t, x, y, vx, vy, ax, and ay

Name	Attributes
box A box is a hollow rectangle.	**x-double** x position of the center in world units **y-double** y position of the center in world units **h-int** height in pixels **w-int** width in pixels **s-int** thickness of the box
caption A caption is text that is centered near the top of the screen.	**x-double** x position of the center in world units **y-double** y position of the center in world units **text-string** text of the caption
charge A charge density.	**x-double** x position of the center in world units **y-double** y position of the center in world units **q-double** the charge density **w-int** the width of the object in pixels **h-int** the height of the object in pixels Choose either circle, ring, or box selectors. A rectangle will be drawn if no selector is specified. **circle** a circular conductor **r-int** the radius of the circle **box** a hollow rectangular conductor **s-int** the thickness of the box walls in pixels **ring** a hollow circular conductor **r-int** the radius of the circle **s-int** the thickness of the box walls in pixels Examples: addObject("conductor","box, x = 1, y = 2, w = 20, h = 50, s = 4");
circle	**x-double** x position of the center in world units **y-double** y position of the center in world units **r-int** radius in pixels
conductor	**x-double** x position of the center in world units **y-double** y position of the center in world units **v-double** the voltage **w-int** the width of the object in pixels **h-int** the height of the object in pixels Choose either circle, ring, or box selectors. A rectangle will be drawn if no selector is specified. **circle** a circular conductor **r-int** the radius of the circle **box** follow rectangular conductor **s-int** the thickness of the box walls in pixels **ring** a circular conductor **r-int** the radius of the circle **s-int** the thickness of the box walls in pixels Examples: addObject("conductor","box, x = 1, y = 2, w = 20, h = 50, s = 4");

(*continued*)

TABLE 52: (continued)

Name	Attributes
contours The contour lines.	**No parameters.**
cursor A circle with cross hairs.	**x-double** *x* position of the center in world units **y-double** *y* position of the center in world units **r-int** radius in pixels
dielectric	**x-double** *x* position of the center in world units **y-double** *y* position of the center in world units **chi-double** the dielectric permittivity **w-int** the width of the object in pixels **h-int** the height of the object in pixels Choose either circle, ring, or box selectors. A rectangle will be drawn if no selector is specified. **circle** a circular conductor **r-int** the radius of the circle **box** hollow rectangular conductor **s-int** the thickness of the box walls in pixels **ring** a circular conductor **r-int** the radius of the circle **s-int** the thickness of the box walls in pixels Examples: addObject("conductor","box, x = 1, y = 2, w = 20, h = 50, s = 4");
field The vector-field arrows.	**No parameters.**
function	**f-string** the analytic function, $f(x)$ **var-string** the independent variable, x **n-int** the number of points used during the evaluation **xmin-double** the first value of x to evaluate **xmax-double** the last value of x to evaluate
rectangle	**x-double** *x* position of the center in world units **y-double** *y* position of the center in world units **h-int** height in pixels **w-int** width in pixels
shell	**x-double** *x* position of the center in world units **y-double** *y* position of the center in world units **r-int** radius in pixels
testcharge Used to show the *electric field* vector or as a data source.	**x-double** *x* position of the center in world units **y-double** *y* position of the center in world units **r-int** radius in pixels

Name	Attributes
text A fixed text string followed by an optional calculation.	**x-double** x position of the left side of the text in world units **y-double** y position of the top of the text in world units **text-string** static text **calc-string** An analytic function to be evaluated. The calculation is displayed to the right of the static text. Text objects are often slaved to other objects. If a text object is slaved to an object, it takes on the properties of that object. In this Physlet, a slaved text object can evaluate a function of x, y, q, v, and p as defined in Table 51 for electrostatics objects. *Note*: Use the setFormat method described in Section 17.3.1 to change the decimal format of the displayed calculation.

18.10 SURFACE PLOTTER

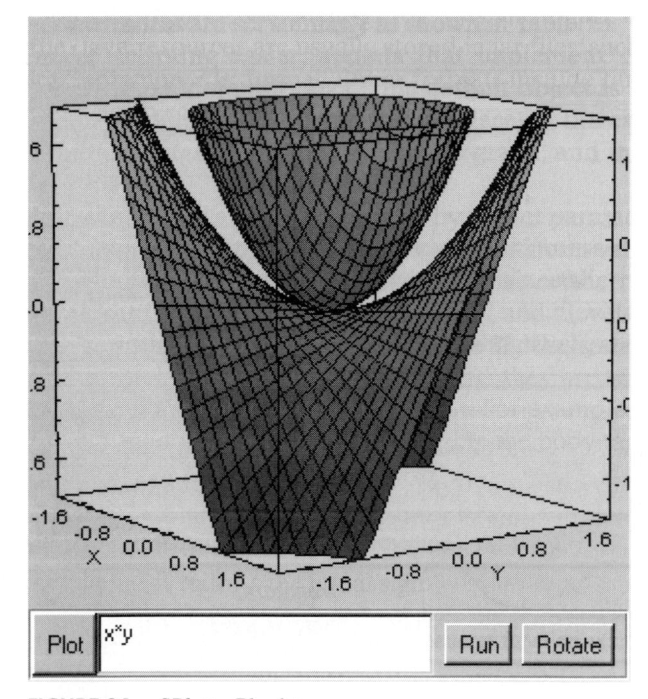

FIGURE 84: *SPlotter* Physlet

As shown in Figure 84, *SPlotter* can display analytic functions of x, y and t as 3-D, contour, and density plots. In addition, it can also be a data listener for array data. The *SPlotter* Physlet is based on the Surface Plotter applet written by Yanto Suryono.[5] It was modified to work with Physlets and interapplet communication by Wolfgang Christian.

[5]Contact Yanto Suryono for information about *SPlotter*. E-mail: d0771@cranesv.egg.kushiro-ct.ac.jp

SPlotter is embedded into an html page using the following tags:

```
<applet codebase = [location] code = "splotter.SPlotter.class"
archive = "SPlotter4_.jar, STools4.jar" name = [name]>
```

The embedding parameters shown in Table 53 can be specified as part of this applet tag.

TABLE 53: *SPlotter* embedding parameters.

Parameter	Value	Type	Description
Function	null	string	The initial function string, $f(x, y, t)$.
Function2	null	string	The initial function string, $g(x, y)$.
FPS	10	int	Frames per second when the clock is running.
Dt	0.1	double	Time step per frame.
GridPts	32	int	The number of grid points along a side for a surface plot.
Levels	10	int	The number of levels for a contour plot.
ScaleFactor	10	double	The size of the bounding box.
Type	threed	string	The type of plot: contour, density, or threed.
ShowControls	true	boolean	Show the user interface.

18.10.1 Set Method

Many of *SPlotter*'s drawing attributes can be changed using the set method. This method is similar to the addObject method. It takes an object identifier and two string parameters. The first string specifies the property to be set and the second specifies the attribute. For example, the x-scale can be set as follows:

```
document.splotter.set(id, "scale", "xmin = 0, xmax = 2, autoscalex = false");.
```

For convenience, the object identifier, id, can be zero when scripting. This signals the Physlet that the default object, that is, the surface plot, is the object to be set.

Not all attributes need be specified. Current values are overridden only if they appear in the parameter list. Incorrect or unsupported attributes have no effect and are ignored.

TABLE 54: *SPlotter* set parameters and attributes.

Name	Attributes
font	**family-string** the font family **style-double** the font style **size-double** the font size Example: set(id, "font","family = Helvetica, style = bold, size = 12");
mode drawing options	**dualshade** **grayscale** **hidden** **mesh** **spectrum** Select one from the above list. Example: set(id,"mode","grayscale");
scale	**xmin-double** the x-axis minimum **xmax-double** the x-axis maximum **ymin-double** the y-axis minimum **ymax-double** the y-axis maximum **zmin-double** the z-axis minimum **zmax-double** the z-axis maximum
style plot style options	**autoscalez-boolean** autoscale the z-axis **box-boolean** draw a 3-D bounding box **facegrid-boolean** draw the 3-D face grids **gutter-int** the gutter around contour plots in pixels. **mesh-boolean** draw a mesh on the 3-D surface **scaledbox-boolean** force x-, y-, and z-scales to be equal **xyticks-boolean** draw ticks on the x- and y-axes **zticks-boolean** draw ticks on the z-axis
type plot type	**contour** **density** **threed** Select one from the above list. Example: set(id,"type","threed");
view observer perspective	**distance-int** 3-D perspective **elevationangle-double** the elevation in degrees **rotationangle-double** the rotation in degrees **scalefactor-int** 3-D scale

BIBLIOGRAPHY

[Beichner 1997] Beichner, R., "The Impact of Video Motion Analysis on Kinematics Graph Interpretation Skills," *American Journal of Physics*, *64*, 1272–1277 (1997).

[Bonham 1999] Bonham, S. W., Risley, J. S., and Christian, W., "Using Physlets to Teach Electrostatics," *The Physics Teacher*, *37*, 276–280 (1999).

[Bransford 1999] Bransford, J., Brown, A., and Cocking, R. *How People Learn: Brain, Mind, Experience and School* (National Academy Press, Washington, DC, 1999).

[Christian 1998] Christian, W. and Titus, A., "Developing Web-Based Curricula Using Java Applets," *Computers in Physics*, *12*, 227–232 (1998).

[Dancy 2000] Dancy, M., Titus, A., and Beichner, R., "The Effect of Animation on Students' Responses to Conceptual Questions." *The Physics Teacher* (in preparation).

[Derby 1999] Derby, N. and Fuller R.,. "Reality and Theory in Collision," *The Physics Teacher* *37*, 24–27 (1999).

[Eckel 1998] Eckel, B., *Thinking in Java*, (Prentic Hall, Upper Saddle River, NJ, 1998)

[Giancoli 1998] Giancoli, D., *Physics: Principles with Applications*, (Prentic Hall, Upper Saddle River, NJ, 1998).

[Goldfarb 1997] Goldfarb, C. and Prescod, P., XML Handbook (Prentice Hall, Upper Saddle River, NJ, 1997).

[Hake 1998] Hake, R., "Interactive-engagement versus Traditional Methods: A Six-thousand-student Survey of Mechanics Test Data for Introductory Physics Courses," *American Journal of Physics*, *66*, 64–74 (1998).

[Hale 1985] Hale, M. E., Oakey, J. R., Shaw, E. L., and Burns, J. "Using Computer Animation in Science Testing," *Computers in the Schools*, *2*, 83–90 (1985).

[Heller 1992] Heller, P., Keith, R., and Anderson, S. "Teaching Problem Solving Through Cooperative Grouping. Part 1: Group versus Individual Problem Solving," *American Journal of Physics*, *60*, 627–636 (1992).

[Hestenes 1992] Hestenes, D., Wells, M., and Swackhamer, G., "Force Concept Inventory," *The Physics Teacher 30*, 141–158 (1992).

[Larkin 1980] Larkin, J. H., McDermott J., Simon, D. P., and Simon, H. A. "Expert and Novice Performance in Solving Physics Problems," *Science*, *208*, 1342–1355 (1980).

[Laws 1997] Laws, P., "Millikan Lecture 1996: Promoting Active Learning Based on Physics Education Research in Introdutory Physics Courses," *American Journal of Physics*, *65*, 13–21 (1997).

[Leonard 1996] Leonard, W. J., Dufresne, R. J., and Mestre, J. P. "Using Qualitative Problem Solving Strategies to Highlight the Role of Conceptual Knowledge in Solving Problems." *American Journal of Physics 64*, 1495–1503 (1996).

[McDermott 1998] McDermott, L. and Shaffer, P. S., *Tutorials in Introductory Physics*, (Prentice Hall, Upper Saddle River, NJ, 1998).

[Mazur 97] Mazur, E., *Peer Instruction: A User's Manual*, (Prentice Hall, Upper Sadle River, NJ 1997).

[Meyer 1997] Meyer, J. and Downing, T. *Java Virtual Machine* (O'Reilly & Associates, Sebastopol, CA, 1997).

[Novak 1999] Novak, G., Patterson, E., Gavrin, A., and Christian, W., *Just-in-Time Teaching: Blending Active Learning with Web Technology*, (Prentice Hall, Upper Saddle River, NJ, 1999).

[Rieber 1994] Rieber, L., *Computers, Graphics, and Learning*, (Brown & Benchmark, Madison, WI, 1994).

[Russo 1989] Russo, J. E., Johnson, E. J., and Stephens, D. L., "The Validity of Verbal Protocols," *Memory & Cognition*, *17*, 759–769 (1989).

[Schoenfeld 1985] Schoenfeld, A. H. Mathematical Problem Solving. (Academic Press, Orlando, FL, 1985).

[Shapiro 1994] Shapiro, M. A., "Think-Aloud and Thought-List Procedures in Investigating Mental Processes," *Measuring Psychological Responses to Media,*" Annie Lang (Ed.), pp. 1–14. (Lawrence Erlbaum Associates, Hillsdale, NJ, 1994).

[Sokoloff 1997] Sokoloff, D. R., "Using interactive lecture demonstrations," *Physics Teacher*, Vol. 35 No. 6, 340-347 (1997).

[Thacker 1994] Thacker, B., "Comparing Problem Solving Performance of Physics Students in Inquiry-based and Traditional Introductory Physics Courses," *American Journal of Physics*, *62*, 627–633 (1994)

[Titus 1998]Titus, A., *Integrating Video and Animation with Physics Problem Solving Exercises on the World Wide Web*, Ph.D. dissertation, North Carolina State University, Raleigh, NC (1998).

[Wells 1995] Wells, M., Hestenes, D., and Swackhamer, G., "A Modeling Method," *The American Journal of Physics*, *63*, 606–619 (1995).

[Zollman 1994] Zollman, D, and Fuller, R., "Teaching and Learning Physics with Interactive Video," *Physics Today*, *47*, 41–47 (1994).

APPENDIX A: GLOSSARY OF HTML AND JAVA TERMINOLOGY

APPLET

A self-contained program written in the Java language that is designed to be embedded in an html page.

ARCHIVE

A collection of resources, such as documents, programs and images, that have been stored in a single file. Java resources are usually stored in jar files, such as Doppler.jar, a variant of the more common ZIP format. Other formats include tar on UNIX computers and bin on Macintosh computers.

ATTRIBUTE

A characteristic of an object. For example, a red ball could be said to have a color attribute whose value is red and a mass attribute whose value is 0.5.

In html, attributes are extensions or modifications to a tag. Tag attributes are listed after the tag name. If multiple tag attributes are specified, they are separated by commas. Examples of attributes are color, width, and height. For example, the background color of the body of an html document can be specified in the body tag as follows:

```
<body bgcolor = "#FFFFFF">.
```

If an attribute takes a value, it follows the equal sign.

CLASS

A file containing a the compiled code for a Java objcct, such as Animatior4.class. Can also be used to refer to the definition of a Java object, as in the Button class.

CODEBASE

The directory on a disk or Web server where a browser will begin to search for the compiled Java code.

```
codebase="../classes"
```

or

```
codebase = "http://webphysics.davidson.edu/applets/classes"
```

CSS

Cascading Style Sheet. An addition to html for controlling presentation. A CSS is a text document that can control color, typography, and alignment for any html tag. For example, the CSS for Physlet problems on the CD contains the following definition for the <h1> tag:

```
h1{
     color : #800000;
     font-weight : normal;
     text-align : center;
     font-family : "Times New Roman", serif;
     font-size : 22pt;
}
```

DEPRECATED

A Java method, html tag, or attribute that is outdated and discouraged from use in favor of newer constructs.

DOCUMENT BASE

The directory on a disk or Web server from which the current html page was loaded.

ENCAPSULATION

The ability of an object to hide both its data and how the data is manipulated. Only the functionality of an object—its public methods—is accessible from outside the object.

IDENTIFIER

See *Object Identifier*.

INSTANCE

A realization of an object that can, in principle, occur multiple times. George Washington is, for example, an instance of a person. Just as each instance of a person can have different genes, each instance of an object, such as a Java applet, can change its internal data so that it appears very different. Instances can be assigned a name attribute so that they can be uniquely identified.

INHERITANCE

A programming feature that allows a programmer to pass functionality from one object to another. Physlets, for example, are able to communicate with a browsers by virtue of the fact that they are a type—or subclass—of applet. They inherit their ability to communicate.

INVOKE

The act of calling an object's method.

JAVA

A cross-platform object-oriented programming language created by Sun Microsystems in 1995. Java programs are compiled into machine-independent class files that are later executed by a machine-dependent Java VM (Virtual Machine).

JAVASCRIPT

A programming language designed by Netscape Communications to control html pages. Unlike Java, JavaScript is interpreted. That is, the actual code is displayed in the embedding document. It is not suitable for complex programming projects.

MARKUP

The process of adding tags to a document in order to define a document's logical structure. For example, surrounding the word Physlets with the tags For example, surrounding the word Physlets with the tags <title> and </title> indicates the function of this word in the document. Markup can also refer to the actual tags.

METHOD

The ability of an object to perform a certain action. A method is similar to a function or subroutine in procedural programming languages, such as Basic or FORTRAN. However, method in an object-oriented language is very tightly coupled to the object

OBJECT

An object is a structure that contains both variables and the methods to act upon and manipulate these variables.

OBJECT IDENTIFIER

A unique integer that identifies an Object. Physlets use the hash code assigned by the Java Virtual Machine as the object identifier.

OBJECT-ORIENTED

A style of programming that uses encapsulation, inheritance, and polymorphism to build independent pieces of code.

PARSE

The act of converting a sequence of characters—a string—into a number or a sequence of operations. For example, a user types the eight characters sin(0.5) on a keyboard. These characters are converted to a number with the value 0.4794255 by a parser.

PARSER

See *Parse.*

POLYMORPHISM

The ability of an object to change behaviors that are inherited from a superclass. All on-screen objects understand the paint method, for example. However, each object invokes its own unique paint method.

PROPERTY

See *Attribute.*

SCRIPT

A weakly typed or untyped interpreted language that has few provisions for complex data structures. Scripts are designed to interact either with other programs—as glue—or with a set of functions provided by the interpreter. See also JavaScript.

SIGNATURE

The number, order, and type of data that must be passed to a method when it is invoked. The sin function, for example, has a signature consisting of a single floating point number since it is invoked as *sin(double angle)*. The setDragable method that is widely used with Physlets, on the other hand, has a signature that consists of an integer followed by a boolean, setDragable(*int id, boolean drag*).

STRING

A sequence of characters. See also *Parse.*

SUBCLASS

In object-oriented programming, a class that is derived from a superclass by inheritance. The subclass contains all the features of the superclass, but may have new features added or redefine existing features.

SUPERCLASS

A class that is the base class for more specialized objects. See *Subclass.*

TAG

A label that is used to indicate functionality or structure in a document. An html tag is made up of a tag name followed by an optional list of attributes all of which appear between angle brackets. For example, the following image tag has source, width, and height attributes:

```
<img src="apparatus.gif" width="300" height="250">
```

Many html tags are containers, meaning that a tag has a matching end tag. For example, a section of code can be italicized using <I> and </I>. Stand-alone tags, such as the preceeding image tag, are usually used to placed elements on a page and have no ending tag.

VIRTUAL MACHINE, VM

A program that converts a Java class files into native machine code so that it can be executed on the local computer. Although it is possible to build a computer chip that executes a Java class file directly, these chips are not yet available.

XHTML

Extensible Hypertext Markup Language. A strict form of html that conforms to the XML standard. XHTML requires that documents be well formed by requiring the use of a Document Type Definition. Elements and attributes must be in lowercase. All attribute values must be quoted. Nonempty elements require end tags. Empty elements require either an end tag or a termination.

XML

Extensible Markup Language. A definition for creating languages for marking up documents. HTML is, more or less, an instance of an XML language. XML allows authors to create custom tags to define additional functionality and document structure.

APPENDIX B: COPYRIGHT AND CONDITIONS OF USE

Physlets, (i.e., the Java applets themselves) are copyrighted by Wolfgang Christian. A trademark has been applied for. Physlets may be used under the conditions for non-commercial use as outlined below.

Physlet Problems (i.e., the text and associated script) for problems in this book are copyrighted by Prentice Hall unless otherwise noted.

COPYRIGHT

Instructors may post Prentice Hall copyrighted Physlet problems from the Physlet CD for non-commercial use on their own class-related websites provided those problems copyrighted by Prentice Hall are clearly marked as such and the following copyright notice is given on each page containing Prentice Hall problems:

"Problems x, xx, xxx Copyright Prentice Hall. All rights reserved."

Instructors may not otherwise distribute, or publish them without express written permission from the Publisher.

Prentice Hall Companion Web sites contain Physlet Problems that are copyrighted by Prentice Hall. These sites include *Principles with Applications* (5th ed.) and *Physics for Scientists and Engineers*, both by Douglas Giancoli, and *College Physics* by Jerry Wilson and Anthony Buffa. These problems are not included in the permission to post granted above.

Instructors wishing to author and post **their own Physlet problems** should consult the conditions of use below.

Commercial use of Physlet Problems from this text, or edited versions thereof, in printed or electronic form requires the written permission of the copyright holder.

CONDITIONS OF USE

Under the conditions noted below, Physlets, that is, the applets themselves, may be used to author new problems and these problems may be distributed along with the Physlet jar files for non-profit, educational purposes without requesting permission under the following conditions:

1. That the text and script of Physlets Problems, that is, problems that make use of Physlets to provide animation, visualization, or other types of educational content, be made available to others by placing them in the public domain.
2. That Davidson College should be credited as the source of the Physlet on at least one HTML page of the instructional unit where these applets are being used. (An instructional unit would be a course or a collection of problems with a

unifying theme.) A small Davidson logo or simply the name "Davidson College" with a link to the Davidson College Physlet Archive serves this purpose.

3. Credit does NOT have to be given on every problem or HTML page that contains a Physlet problem. We encourage authors to make the integration of Physlets into their curriculum material as seamless as possible.
4. Publications which result from using Physlets will reference the Physlet site http://webphysics.davidson.edu/applets/applets.html.

Please share your work. Authors who have written Physlet problems are encouraged to send Wolfgang Chrsitan or Mario Belloni short email with a link to their problems.

Commercial use:

Commercial use of Physlets, (i.e. the Java Applets) requires the written permission of Wolfgang Christian.

INDEX

Physlets: Teaching Physics with Interactive Curricular Material
CD-ROM
0-13-029341-5
Prentice-Hall, Inc.

System Requirements

PC running Windows 95/98/NT/2000 (for other systems, see README file on CD)
166 MHz processor
32 MB RAM
800 x 600 pixel screen resolution
4x CD-ROM drive
Requires a browser supporting the Java 1.1 Virtual Machine and JavaScript to Java communication:
Internet Explorer 4.01 or higher or Netscape Navigator 4.08 or higher.
(Netscape Navigator 4.08 included on CD)